MW01520611

Brownfield Sites
Assessment, Rehabilitation
and Development

FIRST INTERNATIONAL CONFERENCE ON BROWNFIELD SITES:
ASSESSMENT, REHABILITATION AND DEVELOPMENT

BROWNFIELDS

CONFERENCE CHAIRMEN

C.A. Brebbia
Wessex Institute of Technology, UK

D. Almorza
University of Cadiz, Spain

H. Klapperich
Technical University of Freiberg, Germany

INTERNATIONAL SCIENTIFIC ADVISORY COMMITTEE

R. Annaert	D. Grimski	R. Pottler
M. Bruel	E. Laginestra	J. Slavich
A. Callaba de Roa	M. Larran	C.V. Sole
J.L. de Velasco	F. Lopez-Aguayo	W. Timmermans
G. Deplano	J. Lourenço	B.J. Trilling
I. Erzi	D. Nuyens	E. Walkowiak
F.E. Falcone	R.L. Olsen	J.D. Weiss
J. Grima	J.B. Park	G. Whelan

Organised by
Wessex Institute of Technology, UK
University of Cadiz, Spain

Brownfield Sites
Assessment, Rehabilitation
and Development

Edited by

C.A. Brebbia
Wessex Institute of Technology, UK

D. Almorza
University of Cadiz, Spain

H. Klapperich
Technical University of Freiberg, Germany

WITPRESS Southampton, Boston

C.A. Brebbia
Wessex Institute of Technology, UK

D. Almorza
University of Cadiz, Spain

H. Klapperich
Technical University of Freiberg, Germany

Published by

WIT Press
Ashurst Lodge, Ashurst, Southampton, SO40 7AA, UK
Tel: 44 (0) 238 029 3223; Fax: 44 (0) 238 029 2853
E-Mail: witpress@witpress.com
http://www.witpress.com

For USA, Canada and Mexico

Computational Mechanics Inc
25 Bridge Street, Billerica, MA 01821, USA
Tel: 978 667 5841; Fax: 978 667 7582
E-Mail: info@compmech.com
US site: http://www.compmech.com

British Library Cataloguing-in-Publication Data

A Catalogue record for this book is available
from the British Library

ISBN: 1-85312-918-6

Library of Congress Catalog Card Number: 2002102680

*The texts of the papers in this volume were set
individually by the authors or under their supervision.
Only minor corrections to the text may have been carried
out by the publisher.*

Preface

The development of brownfield sites involves an interdisciplinary approach from several parties. It not only requires a direct response to environmental demands but also provides an opportunity for lateral thinking to give creative solutions to difficult problems.

However, brownfield regeneration embraces wider issues than cleanup and economic redevelopment. It covers such matters as legislation and the prominent roles of ownership and investment. On an international scale, liability and legal structures vary considerably, as does public and private financing. This in turn leads to varying strategies for business and finance and different attitudes to land use.

The development of brownfield sites will also involve strong community participation. The ecological advantages may be obvious to all, but the benefits of reintegration of land into the economic cycle of the community will need to be promoted strongly to take effect, as will its economic and social dimensions for the reinvigoration of cities.

The interdisciplinary nature of the work is essential for a successful regeneration of such sites.

Brownfields 2002, the 1st International Conference on Assessment, Rehabilitation and Development of Brownfield sites has brought together excellent contributions on a broad range of engineering, science and land issues on the subject.

The Editors would like to thank the contributors for their papers and the International Scientific Advisory Committee for their able assistance in helping to review abstracts and papers.

The Editors
2002
Cadiz

Contents

Interdisciplinary approach for brownfields

H. Klapperich
Competence centre for Interdisciplinary Brownfield - CiF e. V.,
Freiberg/Berlin
Geotechnical Institute, Freiberg University of Mining and Technology,
Germany

Abstract

The topic of "Remediation of previously used sites" occupies professional engineers in the field of Geo-Technology, Civil Engineering, Process Engineering and Mining, Engineering Geologists, Biologists and Micro-Biologists, Geologists specialized in Hydrology and Geo-Chemistry, Physicists and Chemists, Town Planners, Landscape Architects and, beside those disciplines of technology and design, the property sector, banks, insurance sector and politics including planning authorities and environmental Legislation. The central position, of course, is taken up by the land owner, respectively the investor.

The complexity of the subject requires a successful approach with the aim of "re-use", taking into account and mastering risks and many obstacles. This calls for ideas which demonstrate solutions and innovative concepts. An interdisciplinary Competence Centre for remediation, that concentrates various activities, is being established at the Freiberg University of Mining and Technology by the author and three colleagues covering Geotechnique, Economics and Business Administration, especially Construction Business Management, and Public law.

The relevance of the theme is underlined by the loss of open land and a large number of urban brownfields, that in many German regions represent in equal parts a structural, planning and environmental handicap. A systematic approach to the management of land with local and regional involvement is a prime prerequisite for the management of property development in open areas as well as for the mobilization of urban brownfields. Activities throughout the European Unity, under the banner "City of Tomorrow", will set standards for indicators of sustainability - including the strongly propagated concept of "land consciousness".

Regional land management including used sites is in need for an interdisciplinary remediation. The successful adoption of land management has to consider the property sector and thus associated financial models, their risk potential and avoidance. The insurance sector offers solutions to mitigate these stumbling blocks to investment.

The paper will cover above topics and will focus on the rehabilitation of former mining sites - open pit mines.

1 Introduction

"Land recycling is the use-related re-incorporation of plots of land in the economic and natural cycle which have lost their former function or use ... by means of planning, environmental or economic policy measures."

The interdisciplinary recycling of land acts as a useful instrument to limit the development of greenfield land in order to shape the necessary structural changes, which have resulted from the demise of traditional industrial sectors, together with political decision makers.

Interdisciplinary land recycling serves the re-integration of previously used real estate into the economic cycle. Redevelopment under such circumstances requires the practical applications of risk rating and management as well as financial modelling. Our Competence Centre for Interdisciplinary Land Recycling is a partner for institutions, public sector agencies, trade organisations and companies, which are concerned with such topics as technology, construction, economics, ecology, risk rating, environmental law and project management as well as with the management of public sector bodies. Research and development focus the named objectives in co-operation with private and public sector partners from a federal, state and local level as well as with the mining and construction sectors, real estate, financial and insurance sectors. In addition to the technical solution of a practical redevelopment the criteria of property owners and investors in respect of project realisation of inner city real estate, rehabilitation of open cast mining areas and "conversion areas", are important for the success of an all embracing approach to land recycling in order to prevent the continuing development of greenfield sites (Azzam, Heinrich, Klapperich, 2001).

2 Sustainable Development

Urban planning in Central Europe means dealing with limitations but also taking chances for developing different types of "land".

Environmental impact assessment and also regarding environmental legislation and policy in progress the sustainable development is meanwhile an excepted goal and so far a prominent overall issue.

Prof. Thoenes in his contribution "Bodenreflexionen in unserer Gesellschaft und der Beitrag des Flächenrecyclings" (Reflections about land in our society and the contribution that land recycling makes) describes the values of land.

Quote:

"The strategies for the mobilisation of land consciousness with the help of land recycling include the analysis of proposals for the further development of economic regulations with a real estate component. This concerns the proposal of a combined real estate value and land tax as an incentive and control instrument for the development of land. Expert discussions express time and again the idea that there will not be any old burdens in twenty years time, but more contaminated soil and land and in twenty years it will not be soil limiting values that govern the situation, but land-use values that have to be achieved. It should be straight away that long term strategies are developed in respect of these future developments." (Thoenes (1998))

Following the regulation for the groundwater-protection, now soil has gained an equivalent status due to the Soil Protection Act, which passed German Parliament in 1998.
The management of brownfields in Germany is mainly influenced by the legal regulations;

- the Federal Soil Protection Act (with its sub legal regulation) and
- the Federal Building Code.

The Federal Soil Protection Act and the accompanying Federal Soil Protection Ordinance do not claim explicitly the reuse of brownfields instead of greenfields. They contain however the requirements which have to be met if contaminated sites shall be reused. Warding off dangers is required.
The Federal Soil Protection Act so far contributes substantially to an ecologically sound use-related remediation, but it does not guide to a minimization of land consumption.
The Federal Building Code however lays down that land cover has to be limited to the necessary extent and that the utilization of land has to be carried out in a careful and sparing way. Furthermore the possibility is given to remove buildings in order to unseal land.
Waste-land recycling is a challenge between formal-legal request and informal administration practice.
Sustainable developments in remediation of contaminated sites is to be achieved by the interaction of all involved parties including public-private partnership.

3   Real properties - plant and land - mine remediation

The ecological advantages that a remediation brings are obvious to all, whereas the benefits of reintegration of land into the economic cycle and their reuse needs to be marketed more strongly, also from an economic and social perspective for the re-invigoration of cities. Tailor-made development scenarios and model solutions are activities undertaken at our Competence Centre.

The relation between real properties, plant and land as well as the interaction between use - notional use - chance of use and e. g. sale is illustrated in Fig. 1. For mine remediation a special situation is given in Germany with respect to the abandonment under BbergG (federal mining law) - Fig. 2.

4   Economic assessment

Economic assessment of ecological burdens on the criterion of reduction of useful value plays an important role for the decision how to use the land and properties - especially for companies or owners with large real estate.
Reasons are: creation of legal security, establishment/check of securing/refurbishment requirement, creation of basis for financial future planning by assessing the assets for balance sheets and other lists of assets, Determination/check of possible provisions for contaminated sites, Creation of planning security for investment projects, Determination of important information as the basis for successful portfolio management.

Some definitions:

- Ecological burdens in the context of real properties assessment are all environmentally relevant matters which reduce the useful value of a property.
- Contaminated sites are abandoned waste disposal facilities and plots of land where waste is treated, stored or dumped (former dumps) and plots of land for abandoned plants and other plots of land on which environmentally hazardous substances have been handled (former sites).

Fig. 3 shows the interaction of contaminated sites and ecological burdens considering mandatory clearance.

Real properties, plant and land
inspecting, advising, planning

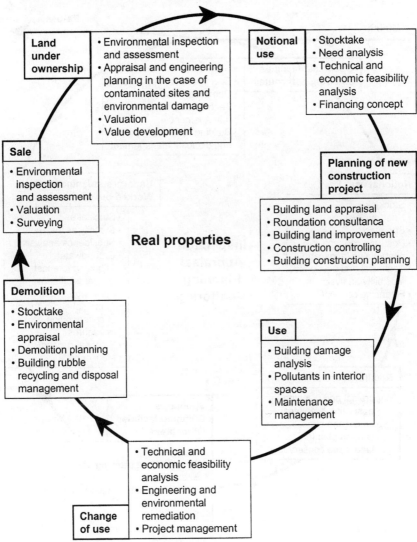

Land under ownership
- Environmental inspection and assessment
- Appraisal and engineering planning in the case of contaminated sites and environmental damage
- Valuation
- Value development

Notional use
- Stocktake
- Need analysis
- Technical and economic feasibility analysis
- Financing concept

Sale
- Environmental inspection and assessment
- Valuation
- Surveying

Real properties

Planning of new construction project
- Building land appraisal
- Roundation consultanca
- Building land improvement
- Construction controlling
- Building construction planning

Demolition
- Stocktake
- Environmental appraisal
- Demolition planning
- Building rubble recycling and disposal management

Use
- Building damage analysis
- Pollutants in interior spaces
- Maintenance management

Change of use
- Technical and economic feasibility analysis
- Engineering and environmental remediation
- Project management

Fig. 1

Mine remediation

**Abandonment under BBergG (mining law):
preventing hazards, ensuring restoration for use**

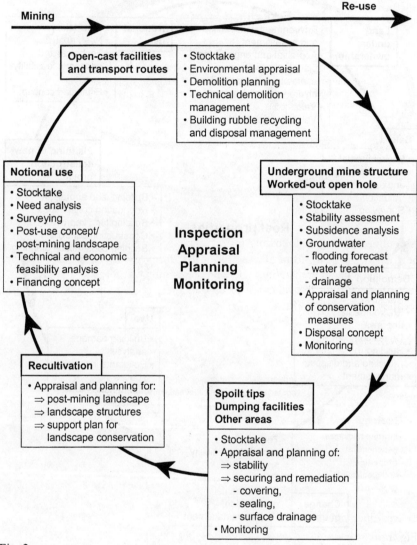

Fig. 2

Assessment of ecological burdens
Definitions

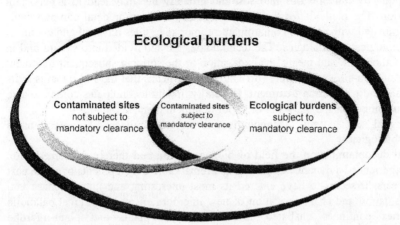

Fig. 3

As outlined in Fig. 1 following aspects have to be considered hazard control, new use (own need), form of use ≤ prior use, form of use > prior use, new use (sale), sale without preparation, sale with preparation - fit for use -.

As an example for the valuation of expenditure - related site factors, the topic location is outlined in Fig. 4.

Location

	Influence costs	Uncertainties/ need for investigation
• Geology		
- stratum thickness	+	o
- geochemical properties	+	o
- soil-mechanical properties	o	+
• Hydrogeology		
- groundwater level	++	o
- horizons	+	o
- gradients	+	o
• Exploration density		
- preliminary investigations		–
- surrounding area		–

Fig. 4

5 Remediation - in situ, ex situ securing

The input of contaminants into soil and groundwater may lead to a persistant pollution. The methods for the remediation of the environmental compartments contaminated with organic contaminants comprise besides physical and chemical also biological technologies. The technologies are subdivided into ex situ and in situ methods. Ex situ means the excavation of the soil and subsequent treatment at the site (on-site) or elsewere (off-site). In situ means that the soil remains in its natural condition during treatment. Generally, in situ technologies comprise also ex situ components, e. g., water or vapor treatment plants.

The goal of in situ technologies is to mineralize the contaminants microbiologically.

In soil decontamination the field of biodegradation and thereby bioremediation has experienced a dynamic evolution and remarkable developments over the past few years. It seems to have entered its most interesting and intense phase yet. The isolation and characterization of new microorganisms with novel catabolic activities continues unabated, and the use of plants and plant-microbe associations in bioremediation is expanding strongly. The continuously growing knowledge on catabolic pathways and critical enzymes provides the basis for the rational genetic design of new and improved enzymes and pathways for the development of more performant processes.

There exist different commercially methods of in situ and ex situ bioremediation, including advanced strategies proposing to use the abilities of plant-microbe associations called phytoremediation, and the enhancement of natural attenuation processes, especially biostimulation and bioaugmentation, and, last but not least, the potential of genetic modifications of the microorganisms applied (J. Klein (2000)).

One has to consider that each site has to be investigated not only for its geological, hydrogeological, and contaminant situation, but also for the site-specific degradability.

One thing that all procedures have in common is that the success of the remediation work can be directly controlled to a greater or lesser extent and that direct adjustment to the needs and follow-up use is possible. But this has to be set against the securing measures, where the decision regarding effectiveness and safety - discussed by way of preparation in the context of a feasibility study - is taken by the owner of the land or the investor with due consideration of the costs involved.

Regardless of the follow-up use, hazard prevention alone gives rise to constant cost. But the proceeds from the sale of the land depend to a crucial extent on the quality of the practicable form of use. Follow-up use for residential housing will certainly create the highest market value for the land. This may mean that the balance sheet for a piece of land may also be negative, if for example provisions of the development plant stipulate only industrial or purely commercial forms of use for the whole area. In terms of its perspective, planning law must become more flexible here so that investors will be willing to conduct more than a mere hazard prevention on their own site.

Surface confinement technique

A combined surface confinement technique using soil/geosynthetics is a perspective drawing in order to seal the contaminated subsoil in a cost effective way and creating a foundation level for housing or new industrial areas. Fig. 5 shows the DMT-GE0safe-system with the sandwich construction including the different geosynthetic-elements for the purposes of sealing and reinforcement of the soil. The different drainage systems are aiming for the safety of the structure above as well as for the overall goal which means the protection of the groundwater (Genske et al (1993), Klapperich (1999), Klapperich, Azzam (2000 & 2001)).

Perspective drawing

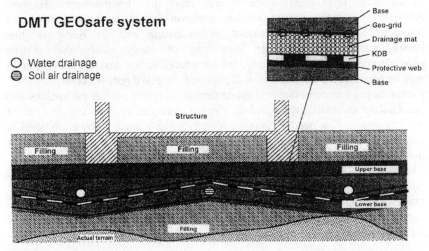

Fig. 5

6 Commercial real property and environmental issues

There only exist a few environmental issues which have captured the imagination of so many business, government and community interests as BROWNFIELDS.

The overall picture shows - above and beyond the immediacy of cleanup of contamination, that brownfields are closely related with economic redevelopment, job creation, community restoration and re-use of developed areas to counter urban sprawl.

The role of financing of real property transactions needs a wide understanding of so many concerns related to environmentally distressed property. Risk assessment is defently a key issue.

6.1   Financial and business strategies

From the perspective of a real estate developer there are following topics to consider:

- Motivation of a developer in respect of brownfields with an understanding economic decision making processes
- Overall Pro Forma financial analysis with control loop of land acquisition, return on investment - financial viability and resaleability of completed projects.

The banks as well as property developers have been sting in the past and quite often avoid contaminated land like the plague.

In the U. S. there exists since several years the "Environmental Bankers Association" which aims to identify common factors on redevelopment of brownfields through a consistent understanding and to recognize that environmental constraints are only one of many issues that restrict redevelopment including security, infrastructure, access and transportation as well as administrative and political factors and - to make positive contributions.

Restructuring and corporate real estate means analysing market evaluations and potential for increasing value and developing strategies of revitalisation - one of our key issues. This needs "planning security" which needs to be established in case of old buildings and contamination so that one expects a yield when investing in real estate.

The "Identify and Use Potential" leads to categories of properties with and without development potential regarding location, use, size, contamination and miscellaneous like legal frame work conditions.

Fig. 6 and 7 show objectives for the one or the other under consideration of influencing factors.

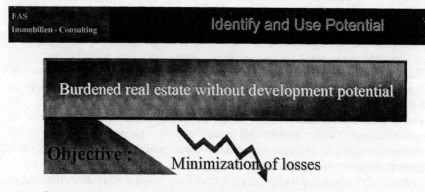

Fig. 6

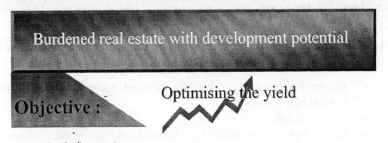

FAS
Immobilien - Consulting

Identify and Use Potential

Burdened real estate with development potential

Objective :

Optimising the yield

- ⩒   priorities according to economic risk
- ⩒   development of utilisation scenarios using best case / worst case considerations
- ⩒   cost estimate (for ecological contamination) for the respective utility scenarios

Fig. 7

In an international comparison, there are very different tools and instruments to promote the redevelopment.

One of the basic reasons are that the laws for Brownfields cleanup and remediation are quite different among countries and states.

When we look at the financial aspect of Brownfield projects in the USA there are six basic tools being used to increase the potential possibility for Brownfield redevelopment. These are: bank, insurance, real estate, tax, loan funds and grants. Among them bank, insurance and real estate are more market-based tools, instead, tools such as tax, loan funds, grants are more policy-orientated. The latter ones are targeted especially for initiation of the recycling projects, while the market-based tools are targeted mainly for reducing the projects risks, meanwhile, setting up standard financial operation systems for the long-run.

In Germany there aren't so many types of financial tools used for Brownfield remediation projects. Funds are the mostly common financial resource for Brownfield projects in Germany. In general remediation projects could be financed by loan funds through the programs of Federal State, German Federal Republic and European Union. In addition, there are also some kinds of environmental insurances used for remediation projects in Germany. However other types of financial instruments are still need to be defined and developed.

Public programs can help to leverage private brownfield financing by reduce lenders risk, reduce borrower's costs, improve the borrower's financial situation and provide resources directly.

To enhance the climate for investment there are non-cash financing tools that also cities can promote. A prominent role plays the "Environmental Insurance" - a tool which is under wide use in the US and Great Britain and now developing on the European continent (D. Jacob (2000), E. Henry (2000), H. H. Katte (2000)).

6.2   Inter-related commercial lending risks - risk evaluation of brownfields & safety concepts

There are three main risks which ca be quantifed, diminished or eliminated - focussing on a greater ability to manage the collective risk:

- Business or economic risk implications are embodied in the concept of brownfields investment
- Technical risk
- Legal risk implications are inherent in the overall context but particularly embodied where perceived contamination carries as much concern as real contamination.

Further use of contaminated land and brownfield areas is an urgent task to reduce the consumption of virgin land. When redeveloping such areas as assessment of unacceptable risks has to be carried out and, if necessary, remedial action has to be taken. The aim of risk assessment normally is focused on criteria of human health.

As it is known from technological risks, also remedial measures cannot absolutely avoid future risks. The reasons are the scientific, technological and economic limits of knowledge, according to the state-of-the-art of identification, assessment, remediation and containment of contaminants in soil under real conditions.

In addition to risks which are quantifiable or can be evaluated by quantifiable parameters also non-quantifiable risks have to be considered. They may be important for the acceptance of a certain further usage.

In addition to risks which are quantifiable or can be evaluated by quantifiable parameters also non-quantifiable risks have to be considered. They may be important for the acceptance of a certain further usage.

Property development risks, be they on greenfield or brownfield sites, will inevitably materialise in financial costs. Brownfield development brings with it added uncertainties in the identification of below surface contamination, and the potential consequences that this presents. These uncertainties need to be considered in a phased approach, with inherent risks at each stage offering different challenges. Mistakes in the calculation of remedial costs for known contamination are compounded by short and long term consequences of unknown and unidentified risks. Secondly, in a time of constantly changing environmental legislation, provision must be incorporated in the remedial strategy in the event that more stringent regulations are imposed. Liability associated with each phase is critical, and a clear statement of responsibilities is crucial if a remediation programme is to be completed to the benefit of all parties.

Management tools and practices have been developed to tackle the most complex sites, as well as risk transfer vehicles and the part they play in the process.

"Risk Management" and "Risk Sharing" as a significant tool led to "Insurance Innovations" a prominent part in finanzing for brownfield development.

For the "Risk Evaluation of Brownfields" there are safety concepts in development. Examples are environmental remediation insurance, stop-loss coverage and pollution legal liability insurance (D. Mehrhoff, 2000).

7 Conclusion

New solutions are needed to acclerate the process of redevelopment of contaminated sites - for sure an interdisciplinary approach. Research topics are defined to finally provide a mechanism for evaluating new approaches and technologies and as a follow up to transfer this to all parties involved.

"Green Brownfields" means a multidisciplinary task to solve for the benefit of our environment and society.

Land-recycling is a market!

References

[1] Thoenes, H., *International verbindliche Bodenkonvention - Chancen für weltweite Altlastensanierung?,* ITVA-Jahrestagung, Kassel, 1998.

[2] Azzam, R., Heinrich, H. & Klapperich, H. (editors), *Kompetenz-Zentrum für interdisziplinäres Flächenrecycling CiF e. V.,* Veröffentlichungen Institut für Geotechnik, Heft 2001-5, Freiberg, 2001.

[3] Klein, J., *Biotechnology,* Vol. 11b, Environmental Processes II, Wiley-VCH Verlag, Germany, 2000.

[4] Genske, D. D., Klapperich, H., Noll, H.-P. & Thamm, B., *Surface confinement technique for derelict industrial sites,* Geoconfine '93, Montpellier, Fance, Balkema, 1993.

[5] Klapperich, H., *Landrecycling - a market,* Recycling '99 - Geneva, Workshop Recycling Derelict Land, 1999.

[6] Klapperich, H. & Azzam, R., *Rehabilitation of Industrial Sites - Innovative Concepts for Geotechnical and Engineering Geological Approaches* GeoEng2000, Melbourne, Australien, 2000.

[7] Klapperich, H. & Azzam, R., *Innovationen im Flächenrecycling der letzten Dekade,* Veröffentlichungen Institut für Geotechnik, Heft 2001-5, Freiberg, 2001.

[8] *Binational US-German Conference "Green Brownfields",* Salt Lake City, 2000, UEF, New York, chairs: H. Klapperich & L. Garczynski.

    [8a] Jacob, D., *Financial and Business Strategies form the Perspective of a Real Estate Developer.*

    [8b] Henry, E., *The Economics of Brownfields.*

    [8c] Katte, H. H., *Economic aspects of reintegration pre-used properties into the real-estate cycle of the Federal Republic of Germany.*

    [8d] Mehrhoff, D., *Securing Liabilities in the German Brownfield Market.*

Section 1:
Case studies

Case history of a "Brownfields" site in Wichita, Kansas USA: innovative approaches to groundwater remediation

R.L. Olsen[1], J. Brown[2], & P. Anderson[3]
[1] CDM, Denver, Colorado USA
[2] Department of Environmental Health, Wichita, Kansas USA
[3] CDM, Kansas City, Missouri USA

Abstract

At the Gilbert-Mosley Site in Wichita, Kansas USA, the groundwater has been contaminated by chlorinated solvents from past industrial activities. Over 3 billion gallons of groundwater have concentrations of tetrachloroethene (PCE) and trichloroethene (TCE) above drinking water standards (maximum contaminant levels [MCLs]). The contamination covers an area of approximately 2,220 acres. To address the Site's environmental conditions, a Corrective Action Decision (CAD) was approved by the Kansas Department of Health and Environment (KDHE) contained several innovative items, including: (1) alternate cleanup levels (ACLs) above MCLs; (2) containment of the contamination migration instead of aquifer restoration; and (3) use of bioremediation to treat the groundwater. Overall the approaches were viewed as potentially more cost effective than conventional remediation methods. Although all anticipated approaches were not implemented, the overall project was very effective in achieving goals and cost significantly less than typical groundwater remediation projects of similar size. The major cost savings resulted from (1) use of ACLs (vs MCLs), which reduced the amount of groundwater requiring cleanup by 40 percent and (2) efficiencies achieved by combining several contaminated plumes from many sources into one treatment system. In addition, the treatment system will include many enhancements resulting in recreational and educational benefits to the citizens of Wichita.

1 Introduction

In 1990, the City of Wichita, Kansas faced a dilemma plaguing cities across the USA. The Central Business District was declining, aggravated by a weakening regional oil and gas industry and a slump in the real estate market. Through public initiative and private investment, the City of Wichita (City) sought to revitalize its downtown. However, plans for revitalization abruptly ceased later in the year as a result of a report from KDHE. The report indicated that the groundwater under downtown Wichita was contaminated with chlorinated solvents. The groundwater plume was more than four miles long and a mile-and-a-half wide, and extended beneath some 8,000 parcels of land, including more than 550 businesses and thousands of residential properties. The area was named the Gilbert-Mosley Site after the intersection of two streets near the center of the area. The knowledge of this groundwater contamination created concern from environmental, public health and safety, and economic perspectives. Faced with their own questions of liability, banks immediately stopped lending to business and home buyers in the area. Without action, property values within the area were predicted to plummet by 40 percent. Unless a solution was developed quickly, the federal government would invoke Superfund, aggravating what already promised to be a long and costly problem.

The City decided to take the initiative and develop a plan to cleanup the Site. A unique partnership between the public and private sectors was established, involving intergovernmental partnerships with local, state, and federal government support, along with participation from the private sector — banks, responsible parties (industry), and the real estate community. The plan's fundamental premise would be the City's acceptance of responsibility for the cleanup of the Gilbert-Mosley Site in exchange for funding commitments from public and private sector partners. With City government in a leadership role, the following actions were taken:

- The State of Kansas and the City of Wichita established an agreement with the State's environmental agency (KDHE) who were acting on behalf of the U.S. Environmental Protection Agency (EPA).
- A primary party, responsible in part for the contamination, signed an agreement with the City to pay for its share of the contamination.
- The City developed a program agreed to by a majority of the local lending institutions to reestablish lending in this area by using a certificate of release from liability issued by the City.
- Authored amendments to Kansas State law that were adopted by the Kansas Legislature to allow for the use of Tax Increment Financing (TIF) for environmental conditions.
- Created a citizen involvement process for community input and awareness of the project.
- Secured a qualified consultant for the job.
- Adopted City Ordinances (institutional controls) to protect citizens from contaminated groundwater and related environmental conditions.

As a result of the above actions, life in the Gilbert-Mosley area returned to normal. Over 4,000 requests for "Certificates of Release" have been received, not only from property owners within the Gilbert-Mosley Site, but from many on the fringes of the Site, suggesting that the "Certificate" is a good method for removing potential liability. Overall, the following have been accomplished:

- Wichita's citizens have been protected from groundwater contamination.
- The City's tax base has been preserved.
- The property values in the Gilbert-Mosley area have been preserved and restored.
- The environment has been protected for future generations.
- To the extent that they can be identified, those responsible for the pollution are paying to clean it up. The remaining costs are being covered by the TIF.

In addition to the above positive socioeconomic benefits, many technical innovations have been implemented. The remainder of this paper describes the technical aspects of the project, including innovative approaches to remediation.

2 Site Hydrology

Figure 1 shows the location of the Gilbert-Mosley Site, the Site boundaries, and the various contamination plumes (A-F). The current Site is approximately 3,850 acres in size, covers an area approximately 4.4 miles long from north to south, and varies in width from 1.1 to 2.1 miles from west to east.

The Site is situated within the Arkansas River floodplain. The topography of the Site has low relief. The Arkansas River is the most prominent surface water feature and flows towards the south near the western border of the Gilbert-Mosley Site and then turns east, south of the Site. Chisholm Creek, which runs between I-135, is present at the eastern boundary of the Site.

The geology of the study area primarily consists of Arkansas River Valley alluvium and associated terrace deposits overlying the Wellington Formation bedrock. The Wellington Formation is primarily a gray and blue shale of Permian Age. The alluvial sediments consist of interbedded gravels, sands, silts, and clays. The alluvial sediments have a thickness of between 24 and 54 feet and averages 30 feet on the Site.

Groundwater depths in the alluvial range from 16 to 20 feet across the Site and average 18 feet below ground surface. Groundwater levels in the alluvial aquifer fluctuate seasonally and may vary as much as 3 feet, depending on the proximity to the Arkansas River, their flow levels, and the amount of precipitation received. General groundwater flow is to the south, with a gradient between 0.0007 and 0.0014 feet/foot across the Site. The hydraulic conductivity ranges between 380 and 707 feet per day. An average groundwater velocity for the Site as determined from the calibrated flow model ranges from 1.2 to 1.7 feet per day.

3 Problem Definition and Remediation Approach

The present boundaries of the Gilbert-Mosley Site and plume extents were developed as a result of a series of site investigations that have been conducted privately and by KDHE since 1986. In January 1991, the City selected CDM to conduct the Remedial Investigation and Feasibility Study (RI/FS) investigation. The Final RI was approved in January 1994 and the Final FS was approved in April 1994. After a public meeting, the Final CAD (similar to EPA Record of Decision) was signed on September 30, 1994.

The groundwater contamination above MCLs covers an area of approximately 2,220 acres (see Figure 1). Over 20 individual sources have been identified. Typical regulatory mandated solutions or Record of Decisions (RODs) for groundwater contamination contain the following components:

Component	Specified Action
Cleanup Level	MCLs
Goal of Remediation	Restoration of aquifer to drinking water quality
Treatment Method	Pump-and-treat

Given the magnitude of the contamination, the large number of sources, and the questionable ability to restore aquifers to MCLs, the City and CDM proposed a new approach. The rationale for the new approach was provided to KDHE in May 1993 in a document titled *Gilbert-Mosley Site Preferred Alternative*. The document addressed each of the typical components of a ROD. The rationale used in 1993 is summarized in the following sections.

Cleanup Levels and the Goal of Remediation: In the early and mid-1990s, the first of a series of documents became available that evaluated the effectiveness of groundwater remediation systems currently in use at a variety of sites ([1], [2], [3]). Conclusions from these reports include:

- In the majority of the cases, the pumping systems were able to achieve containment of the dissolved phase contaminant plume and the extraction systems were effective in reducing the mass of contamination from the aquifer.
- When extraction systems were started up, contaminant concentrations dropped rapidly but then leveled off (tailing effect). The plateau concentrations were above remediation goals (e.g., MCLs).
- Cleanup times and cost were severely underestimated.
- The chemical nature of contaminants and/or the geological conditions of the Site can prevent pump-and-treat systems from restoring aquifers to health-based standards in a relatively short time.

The Comprehensive Environmental Response, Compensation, and Liability Act (CERCLA) requires the attainment of MCLs "where such goals are relevant and appropriate under the circumstances of the release or threatened release." Neither CERCLA or the National Contingency Plan (NCP) require aquifer restoration per se; however, EPA has construed the MCL language to generally obligate restoration of any aquifer that is a potential drinking water source. The NCP does allow for a waiver if the solution is "technically impracticable." Since 1993, guidance has been issued concerning the technical impracticability (TI) of groundwater restoration [4]. The document provides specific guidance concerning the evaluations to be performed to document TI. At Gilbert-Mosley, large areas exist with low concentrations of contaminants. These levels are above MCLs but below the concentrations typically achieved during pump-and-treat remediation. Given the historical record of sites not achieving MCLs and the low concentrations observed, restoration of the Gilbert-Mosley Site to MCLs may be technically infeasible.

Overall, the emphasis of any groundwater remediation should be on containment (vs restoration), institutional controls, realistic goals, natural attenuation (if applicable), and frequent re-evaluation of remediation goals. At the Gilbert-Mosley Site, a unique combination of approaches to remediation goals was proposed in 1993. First is the recognition that aquifer restoration to MCLs is not achievable in a reasonable time frame. To design a system for the sole purpose of showing an aquifer may not be restored to MCLs (i.e., to prove technical impractability) is not a cost-effective approach. The alternative approach, to contain the contamination at higher levels and implement higher remediation goals with frequent evaluation, is cost-effective. In fact, the goal proposed (10^{-4} additional cancer incidents) falls within EPA's range of acceptable risk levels. To further minimize any risk to human health, strict institutional controls and public education will be implemented to prevent future groundwater usage in all areas with contaminant concentrations above MCLs. Remediation to MCLs is also protective of human health, but it is not achievable and it is not cost-effective.

Method of Treatment: The previous reports clearly document the ineffectiveness of conventional pump-and-treat technology to cleanup contaminated groundwater to MCLs. Besides a basic change in the goal of remediation and the level of cleanup (containment vs restoration, and 10^{-4} goals vs MCLs), better treatment technologies are also needed. These technologies may include: Soil vapor extraction, air sparging, addition of surfactants or co-solvents (i.e., chemical enhancements), *in situ* chemical oxidation, monitored natural attenuation (MNA), reactive barrier walls (e.g., reductive iron), and *in situ* bioremediation.

4 KDHE and EPA decision

After reviewing the RI data and the technical evaluations in the 1993 Preferred Alternatives (summarized above) and receiving public comments, KDHE (with EPA's approval) issued a CAD, which is equivalent to an EPA ROD. The CAD contained the following components:

Institutional Controls: Establish institutional controls within the defined boundaries of the Gilbert-Mosley Site. The City of Wichita staff must propose an ordinance to the City Council to prohibit the connection of newly-constructed private water wells for private or public drinking water purposes. In addition, a public educational program should be initiated to discourage the use of groundwater contaminated above the MCLs within the Site. These items have been implemented by the City.

Hydraulic Containment: Establish hydraulic containment to prevent further migration of contaminated groundwater. Groundwater contaminated above KDHE's ACLs would be targeted for containment. Any recovered groundwater would be treated to MCLs for the contaminants of concern. Hydraulic containment could be terminated once the ACLs have been achieved and sustained over a one-year period.

Monitoring: Establish compliance monitoring wells at the zero line (i.e., the area where groundwater contamination is below MCLs) to monitor for the chemicals of concern on a quarterly basis or other frequency as determined by KDHE. If any one of the compliance monitoring wells exceed the federal MCLs, additional remediation would be required. Long-term monitoring would be required at the compliance and selected monitoring wells for a minimum period of 10 years of annual monitoring following termination of hydraulic containment.

Source Control: Individual source control activities must be established at all identified source areas to eliminate and/or reduce the toxicity, mobility, and volume of waste/contaminants at the Site. Source controls will be determined on an individual basis following an appropriate source investigation. Source control has been implemented by Coleman (in 1993) and the City (at one site in 1998).

Bioremediation Demonstration: A microcosm study and a field demonstration would be performed at the Site to demonstrate the efficiency and economics of microbiological enhancement. The pilot demonstration was completed in 1995.

The CAD describes KDHE's selection of ACLs at the Site. The levels proposed in the Preferred Alternative (10^{-4}) were modified based upon consideration of appropriate factors including: exposure factors, uncertainty factors, and concerns regarding cumulative effects of multiple contaminants. The ACLs include chemical-specific 10^{-5} excess carcinogenic risk concentrations, or federal MCLs, whichever are greater, to address the uncertainties associated with cumulative risk factors. KDHE's ACLs, the 10^{-5} chemical-specific risk levels, and the MCLs are provided in Table 1.

Table 1: Comparison of MCLs and KDHE's ACLs (μg/L)

Contaminant	MCLs	10^{-5} Chemical-Specific Risk Levels	KDHE's ACLs
Trichloroethene	5.00	21.00	21.00
Tetrachloroethene	5.00	14.00	14.00
1,2-Dichloroethene*	70.00	36.50	70.00
Vinyl Chloride	2.00	0.25	2.00

* Not a carcinogen (based on a Hazard Index of 1.0).

5 Application of the CAD

The CAD addresses the three components of concern evaluated in the Preferred Alternative document. The application of the CAD results in the following modifications to a typical ROD.

Cleanup Level: ACLs were selected versus MCLs for the cleanup criteria. Even though KDHE selected a 10^{-5} level (vs the 10^{-4} recommended), the area of contamination needing to be addressed was reduced from 2,220 acres to 1,350 acres (see Figure 1). The volume of contaminated water requiring treatment was reduced from 3 billion gallons to 1.8 billion gallons. This is almost a 40 percent reduction.

Goal or Results of Remediation: Containment was selected versus aquifer restoration as the ultimate goal of the remediation. As a result, remediation systems can be installed at the leading edges of the plumes versus within the plumes. The systems will also be installed near the ACL limits instead of the MCL limits.

Cleanup Method: Innovative treatment methods vs conventional pump-and-treat methods were selected by the City as the potential treatment methods. *In situ* bioremediation or *in situ* reductive iron walls were initially selected for groundwater contaminated with TCE, DCE, and VC. The innovative methods allow for flexibility in the future if newer and better technologies become available.

6 Progress Since the CAD

Since signing of the CAD in 1994, CDM, the City, and KDHE have worked to:
- Provide better delineation of the ACL extents of the groundwater plumes
- Conduct pilot scale studies to evaluate *in situ* bioremediation
- Monitor the downgradient extent of the groundwater plumes
- Better identify and characterize the source areas within the Site
- Collect data necessary for remedial design and groundwater modeling efforts
- Conduct groundwater modeling for the remedial design
- Perform preliminary designs for downgradient groundwater remediation systems

After collection of pre-design data in 1995 (groundwater samples to define the ACL extent and pilot plant evaluation of *in situ* bioremediation), a draft Preliminary Design Report was completed in March 1996. The report evaluated two remedial alternatives to establish hydraulic containment. These two alternatives were a pump-and-treat option and an innovative alternative option, which consisted of reactive iron walls and *in situ* bioremediation. The innovative options were evaluated in order to provide alternatives with less operation and maintenance and potential overall cost reductions. After input from a Citizens Technical Advisory Committee and KDHE, the City decided to implement full-scale demonstrations of the reactive iron wall and *in situ* bioremediation technologies at Plumes D and C, respectively (Figure 1).

An investigation program to acquire the data needed for the design of the remediation systems was conducted in July through September 1996 and was reported to KDHE in meetings in October 1996. Based on results of the 1996 investigation programs, KDHE concluded that hydraulic containment of Plume D was already in place via Chisholm Creek (Figure 1). Since the requirement of the CAD that a remediation system "establish hydraulic containment to prevent further migration of contaminated groundwater..." was met, no remediation of the downgradient end of Plume D was considered necessary by KDHE. As a result, the implementation of the full-scale demonstration of the reactive iron wall technology at Plume D and *in situ* bioremediation on Plume C was not completed.

In order to proceed with remediation on the other plumes, the City and CDM submitted Plume Remediation Investigation Work Plans for each of the other plumes (A, B, E, and F) to KDHE in 1996 and 1997. Investigations associated with these work plans were conducted from October 1997 to March 1998. Based on the new data collected, the City issued a RI/FS addendum in February 1999 for Downgradient Plume Remediation. The FS addendum evaluated four technologies in detail: monitored natural attenuation, pump-and-treat, reactive iron walls, and *in situ* bioremediation. The four technologies were modified by combining them into the following alternatives:

- Alternative 1   Monitored Natural Attenuation (MNA)
- Alternative 2a   Pump-and-Treat Downgradient
- Alternative 2b   Enhanced Pump-and-Treat
- Alternative 3a   Iron Walls and Downgradient Pump-and-Treat
- Alternative 3b   Iron Walls and Monitored Natural Attenuation
- Alternative 4   *In situ* Bioremediation

A summary of the evaluation and estimated costs for a combined treatment system for Plumes A, B, and E are provided in Table 2. Based on input from a major responsible party, a Citizens Technical Advisory Committee, and KDHE, Alternative 2a, Enhanced Pump-and-Treat, was selected. The major reasons for this selection follows:

- MNA was not effective in meeting the requirements of the CAD and had extremely long cleanup times. Although PCE and TCE do degrade anaerobically in the site groundwater, the degradation rates are very slow and the degradation does not proceed beyond *cis*-1,2-DCE. These phenomena exist due to the limited electron donors (carbon) and excess electron acceptors (sulfate).
- Iron walls and *in situ* bioremediation are new technologies that lack long-term performance data. The technologies are also the most expensive.
- Pump-and-treat at only the downgradient end of the plume allows lateral expansion ("smearing") of the ACL extent and takes a long time.
- Enhanced pump-and-treat includes additional pumping wells in the plume to prevent lateral smearing and decrease cleanup times.

The enhanced pump-and-treat alternative was approved by KDHE in October 1999. After submittal of preliminary, intermediate, and final design reports,

KDHE approved the final design in October 2000. The Remedial Action Work Plan was approved in March 2001 and construction commenced in April 2001. The treatment system will start treating water in June 2002. A summary of the treatment system components follow:

- Thirteen extraction wells
- 5.6 miles of high-density underground pipe
- One treatment building with a hydraulic venturi and stripper system
- Design flow rate of 860 gallons per minute (gpm) (maximum of 1,155 gpm)

The locations of the treatment building, extraction wells, and underground piping are shown on Figure 1. The treated effluent water will be discharged through a series of ponds and meandering creeks to enhance the park setting. The treatment building will be architecturally enhanced to serve as a display and education center. Features will include an educational wing and aquarium. The costs for the Plume ABE basic treatment system (without enhancements) follow:

- Capital Costs: $5,300,000
- Average Annual O&M $213,000
- Net Present Value $13,800,000

Net present value is for the projected clean-up time of 60 years plus 10 years of post remediation monitoring.

7 Conclusions

The unique approach by the City of Wichita to this "Brownfields" type project have resulted in many benefits, including restoration of property values and redevelopment of portions of the Gilbert-Mosley Site. Several innovative technical approaches to remediate the groundwater at the Gilbert-Mosley Site were also originally anticipated. A summary of approaches actually implemented follow:

- The use of ACLs (vs MCLs) has reduced the amount of groundwater requiring cleanup by 40 percent. Although about 10 percent less than originally anticipated, the cost savings have been significant (millions of dollars).
- Containment at the downgradient end of the plumes only was shown to be ineffective in completely controlling contaminant migration. Additional extraction wells within the plumes were necessary and helped reduce cleanup times significantly.
- Use of innovative remediation techniques (*in situ* bioremediation and iron walls) were determined to be more expensive than pump-and-treat alternatives. In addition, the responsible parties favored more conventional technology versus innovative technologies with no long-term performance history.

Overall, the approaches and treatment methods implemented have been the most cost-effective available. In addition, the costs have been significantly less than typical groundwater remediation projects of similar size. The cost savings have resulted from use of ACLs and combination of several plumes from many sources into one treatment system. In addition, the enhancements to the treatment system will result in increased benefits to the citizens of the City including recreational and education facilities. The project has been a true "Brownfields" success story that serves as a model for future sites.

Table 2: Plume ABE Remediation, Comparison Summary of Remediation Alternatives

Alternative	Cleanup Time (Years)	Capital Cost (Millions)	Net Percent Value (Millions)	Advantages	Disadvantages
1. Monitored Natural Attenuation	Plume A: 90 Plume B: 130 Plume E: 130	$0.08/ $0.30	$7.45/ $12.1	-Very Low capital cost -Easy to implement -Possible low Net Present Value -No Surface Structures	-Very long cleanup time -Does not contain high concentrations -Permits significant discharges to surface water -Does not meet CAD
2a. Pump-and-Treat Downgradient	Plume A: 40 Plume B: 80 Plume E: 100	$2.35	$10.1	-Low capital cost -Easy to implement -Lowest Net Present Value -Effective Containment	-Long cleanup time -Discharges VOCs to air -High O&M cost -Effectiveness may decline over time -Lateral expansion of the ACL boundary will occur
2b. Enhanced Pump-and-Treat	Plume A: 40 Plume B: 40 Plume E: 80	$3.47	$11.6	-Shortest cleanup time for Plumes A and B -Low capital cost -Easy to implement -Effective containment	-Discharges more VOCs to air -High O&M cost -Effectiveness may decline over time -Moderate Net Present Value
3a. Iron Walls with Downgradient Pump-and-Treat	Plume A: 50 Plume B: 50 Plume E: 70	$10.1	$14.6	-Shortest containment time for Plume E and overall -Effective containment -Most VOCs destroyed in situ -Moderate O&M Cost	-High capital cost -Lack of demonstrated long-term performance (iron wall) -More disruption during construction (iron wall) -Discharges some VOCs to air (pump-and-treat) -High Net Present Value
3b. Iron Walls with Monitored Natural Attenuation	Plume A: 50 Plume B: 50 Plume E: 70	$8.5	$10.8	-Shortest cleanup time for Plume E and overall -Containment of high concentrations -VOCs destroyed in situ -Low O&M cost -No surface structures -No discharge of VOCs to air	-High capital cost -Disruption during construction -Lack of demonstrated long-term performance -Moderate Net Present Value -Does not contain all contamination above ACLs
4. In situ Bioremediation	Plume A: 50 Plume B: 50 Plume E: 70	$10.5	$29.6	-Shortest cleanup time for Plume E and overall -Effective containment -Most VOCs destroyed in situ	-High capital cost -Very high O&M cost -Very high Net Present Value -Disruption during construction -Lack of demonstrated performance

References

[1] C.B. Doty and C.C. Travis. 1991. The Effectiveness of Groundwater Pumping as a Restoration Technology, Oak Ridge National Laboratory. ORNL/TN-11866. May.

[2] EPA. 1992. Evaluation of Groundwater Extraction Remedies: Phase II, Publication 9355.4-05 (Volume 1, Summary Report) and 9355.4-05A (Volume 2, Case Studies). February.

[3] National Research Council. 1994. Alternatives for Ground Water Cleanup. National Academy Press. Washington, D.C.

[4] OSWER. 1993. Guidance for Evaluating Technical Impracticability of Ground-Water Restoration. Directive 9234.2-25.

Monitoring the effectiveness of remediation: A case study from Homebush Bay, Sydney, Australia

E. Laginestra
Sydney Olympic Park Authority, Australia

Abstract

Sydney Olympic Park, the main venue of the 2000 Summer Olympic Games, contains world class sporting facilities, residential areas, a commercial centre and extensive parklands. It was previously 760 hectares of degraded land that required extensive restoration and remediation.

Remediation of the site commenced in 1992, well in advance of the US Brownfields initiative, and provides useful information for other contaminated land managers in the selection of clean-up options, environmental monitoring and assessment, and adaptive management success.

Treatment of the estimated 9,000,000 cubic metres of waste was carried out in-situ, as a preferable alternative to off-site transfer. This has necessitated ongoing monitoring to ensure the remediation options chosen remain effective in protecting environmental health and safety. The monitoring program has provided baseline data and management tools for long term assessment.

Compilation of historic site data provided the necessary context for new monitoring programs, initiated to give a more complete environmental picture, which included leachate toxicity monitoring, sediment quality triad studies, sediment core analysis, bioaccumulation assessment and bioremediation studies. Promising initial results indicate attenuation of leachate toxicity, undetectable impacts on the surrounding environment due to site influences and identification of optimum conditions for bioremediation. The monitoring has also suggested streamlining some monitoring and highlighted areas of potential concern, assisting in the prioritisation of resource allocation.

Introduction

Following a successful bid to host the 2000 Summer Olympic Games, work commenced in earnest to rehabilitate 760 hectares of degraded and contaminated land at Homebush Bay, located in Sydney's geographic centre on the southern shore of the Sydney Harbour Estuary. The estimated nine million cubic metres of waste had to be treated within seven years to provide a suitable site for a showcase of "green" development and Olympic goodwill. The remediation was arguably the largest clean up undertaken in Australia (NSW EPA [1]).

The remediation process was similar to other Brownfield projects in that monitoring, hazard and risk assessments were undertaken, clean-up options were assessed and remediation carried out. However, the clean up at the the Sydney Olympic site was distinctive in two ways:

1. due to long term commitments made by the NSW State Government to the public regarding the use of the land after the Olympic Games, programs were set in place to assess the long term effectiveness of the remediation options chosen for the site, and
2. contaminated material was not removed, but consolidated within the site for future management. This provided a benefit by not re-locating a pollution problem but dealing with it in-situ.

The NSW Government assigned AUD$12 million for an "Enhanced Remediation Strategy", a long term monitoring and management program, which included:

*   The establishment of an environmental reference group, including representatives from the community, scientific experts and environment groups,
*   Biological monitoring, which would go beyond legislative requirements,
*   A more coordinated "catchment" approach to remediation, compiling relevant data and corporate memory in a GIS-based decision support system,
*   The introduction of an environmental sciences education program that would involve primary, secondary and tertiary students in the monitoring programs, and
*   Documenting the experiences and technologies employed for the benefit of industry involved in remediation elsewhere (Knight [2]).

This enhanced remediation strategy outline was the basis of the framework put in place by the Olympic Coordination Authority (OCA) to enable management of the site after the Olympic Games. Remediation works were well underway at this time, and although the community had been involved in the selection of remediation options during the early stages of work, there remained an element of mistrust about the ability of the government to manage the site after the Games were over and the excitement, and possibly the funding, died down. The community needed re-assurance that resources would be available to ensure ecological sustainability and long-term site safety.

A long-term remediation monitoring program would not only provide information regarding the ongoing effectiveness of the safety systems in place but provide ongoing lessons from the field. Lessons learned from case studies have helped shape government policies (Pepper [3]) so continued learning from experience, a form of adaptive management if well documented and assessed, should allow improvement in the state of the environment and reduce the repetition of mistakes. This long term approach is consistent with the United Nations Environment Program in sustainably managing world resources (UNDP [4]). The fundamentals of an ecosystem approach include the integration of multiple issues, systemic management and a longer term view, working across a variety of spatial and temporal scales.

Background

The Sydney Olympic Park site was formerly degraded land affected by past site practices including agricultural, industrial and waste dumping activities during the 19th and 20th centuries. The type of contamination included petroleum waste, industrial rubble, asbestos and flyash, unexploded ordnance, putrescible municipal waste, abattoir discharge and dumped or spilled chemical wastes.

By the end of the 1980s most of the 760 hectare site (now largely Government owned) was unsuitable for most urban uses and was a contamination source for Sydney's waterways. At this time the site was identified as a possible location for major sporting facilities and a series of Masterplans were produced, incorporating the potential for Olympic development which would help resource the cleanup and development costs. The contamination of the land was recognised as a possible constraint to development so a detailed investigation program was instigated in 1990 to estimate clean up costs. The initial cost estimates cast a cloud over the economic viability and proposal to bid for an Olympic games, nor did it include any ecological or landscaping costs (Pym [5]). A rethink of the proposed remediation works was required.

The consideration of the end-use was an important aspect of the clean-up method. Most Brownfields sites clean up degraded industrial land for use as industrial land. At the Sydney site, the final end-uses desired included residential, commercial and recreational uses.

Remediation activities

Ultimately, the strategy adopted for remediation of Sydney Olympic Park was to manage the risk by blocking exposure pathways. To determine the best options for clean up at the site a good understanding of the site's conditions, specific characteristics, contamination type and the proposed end-uses was required.

There are a number of frameworks outlining what should be included in an ideal clean-up project (US EPA [6], Smartgrowth Network [7], ANZECC & NH&MRC [8], NEPC [9]). All the frameworks agree on determining the hazard based on site specific information, using a coordinated approach and community involvement. Feedback is required from long-term projects to add the requirements needed for frameworks for long-term monitoring of remediation projects where contamination remains in place.

At Homebush Bay, targeted soil, surface water and groundwater studies were initiated in 1991 and migration pathways for volatiles and liquids determined. While the final end-uses affect the selection of clean-up methods in an area, the types of treatment options available for the site were constrained by the variety and haphazard mixtures of the wastes. A number of treatment options were proposed and the community was involved in discussion and selection of the options and kept informed of progress (Pym [5]).

The initial remediation plan involved stripping the topsoil from hectares of contaminated land and containing it in a huge pit created by the former brickworks, with the remaining landfill areas contained by the construction of barrier walls, perimeter drains, capping and installation of gas control systems. This was abandoned following the discovery of a threatened frog species in the brickpit. The final model settled upon retained the strategy of "move and contain", but with significant modifications. As the site was largely owned by the Government, it was treated as one area and did not require approvals to move contamination within the area, minimising approval time and transport problems.

The options selected for waste treatment at the site were consolidation, bioremediation and thermal desorption. Mixed and unknown wastes from waterways plus contaminated soils and sediments were consolidated into four main containment mounds, with capping and perimeter drainage installed for these and five other waste areas. Where the wastes were either known or could be separated they were treated, either by bioremediation using indigenous microorganisms to breakdown petroleum contamination, or by thermal desorption plus catalytic conversion for scheduled chemicals. Two containment methods were used:

1. a totally enclosed "dry" landfill contained by a double layer HDPE membrane and covered with a carpark; and
2. an isolated, but not enclosed, "wet" landfill, where drains were installed on the downstream side to collect contaminated groundwater and leachate, so blocking the migration pathway to surfaces or waterbodies.

The consolidated mounds were shaped and landscaped with a mounded shape and semi-permeable capping allowing air-flow within the mound to enhance natural attenuation of toxicity over time. The surface capping and landscaping provided a zone where volatile gases appeared to be broken down by bacteria and taken up by the vegetation. Stormwater was redirected around the mounds and away from the cap to reduce inflow. Monitoring of leachate toxicity, system integrity, mound and landscaping cover was required to ensure the effectiveness of this remediation strategy.

The topography and the groundwater flow at the site have largely been altered due to reclamation, excavation and finally, the remediation. Shallow groundwater that previously flowed towards the waterways has been caught in leachate drains or re-directed past geotextile or clay barriers. The landform now consists of several hills and viewing mounds, re-aligned creeks and new water features. Approximately 460 hectares of the site are parklands, containing remnant forest and wetlands, rehabilitated grasslands and wetlands, and recreational areas. The majority of the containment mounds are located in the parklands. The parklands are significant in terms of their size, the presence of endangered, locally rare and internationally protected migratory species and the range and connectivity of different habitat types.

Post Remediation Monitoring

Funding for the Enhanced Remediation Strategy monitoring lasted 18 months. To collect the site information and prepare the groundwork for future monitoring within the set time frame, a staged approach was taken (Laginestra [10]). Important steps included defining the program scope, identifying contaminant sources, establishing an independent review team, determining the site functions, setting up the data collection system to compile historic monitoring data and then implementing the new monitoring programs.

After discussion with managers and scientific specialists, a list of questions likely to be asked in the future were developed:

- Is the site safe?
- Have the remediation actions taken to improve the site been effective, and do they remain effective?
- Are the ecosystems improving in health and providing suitable habitat for desired species/communities?
- If there are adverse impacts observed, are they caused by activities on-site?
- Are the adverse impacts observed changing over time, and do the changes enhance the environment?
- What are the issues of concern at various site locations and is current monitoring adequate to address these issues and to detect change?

To be able to answer these questions and ensure protection of the environment, the new monitoring programs had to determine:

- the current status of the site and surrounding ecosystems,
- the possible hazards posed by the site,
- the site risks and management controls in place to minimise these risks,
- the success of the innovative remedial technologies,
- the suitability and success of the maintenance monitoring regime,
- the change in toxicity of leachate over time, and
- an estimate of when the contamination may reduce to an acceptable level.

Data compilation

To assess the site status as a whole and identify data gaps, all monitoring data from different Authority divisions were collected, sorted by media (soil, groundwater, surface water, sediment and air) and spatially identified before being entered into a database. Geographic location and other metadata are essential for managing data. They allow managers to layer data in a GIS system (to identify patterns or conflicts), assess the quality of the data and view trends by area or media.

It is important to understand the site history when preparing future management plans. Variations in site condition can be explored and often explained by knowing past land-uses by location, so unnecessary or environmentally harmful management actions can be avoided. Consequently, the management of data and capture of corporate memory are essential elements in effective long term monitoring programs (Hudson *et al.* [11]). Using GIS techniques, number of maps were prepared for use as management tools. These included:

- types of vegetation/habitat,
- vegetation changes over time (from 1930 to 2000),
- monitoring site locations, historical landuse/ownership,
- groundwater flow direction,
- surface water catchments,
- land use constraints (leachate mounds and drainage systems, protected/ endangered species, public safety risks, horticultural chemical usage), and
- catchment pollution sources.

Leachate monitoring

The new monitoring programs implemented to help assess the environmental health status included toxicity testing alongside the chemical analysis and biological monitoring. Toxicity testing was used to assess the biological impact of chemicals and mixtures in the leachate and whether any impacts were observed in the surrounding environment. Toxicity testing is not mandatory in Australia, but as contaminated site assessment and water quality guidelines use a risk based approach, they provide invaluable risk management information.

The first stage of leachate monitoring tested seven analysis techniques to provide information on the most sensitive tests for use in long-term monitoring and to give an indication of the relative toxicity of different leachates. Estimates of toxicity due to ammonia were also calculated from toxic unit charts for each method (CSIRO [12]). Further toxicity testing with the four most sensitive methods identified major compounds of concern using Toxicity Identification and Evaluation (TIE) techniques. This is essential information for the assessment of leachate treatment alternatives by management. Chemical analyte concentrations had generally decreased from the pilot study (SKM [13]), however, these results need to be assessed with caution as different laboratories were involved in the second round of tests. The third round of testing will commence in April 2002.

Site environmental status monitoring

Toxicity tests were also used in a weight-of-evidence approach for assessing possible adverse impact on sediment in the surrounding waterways and on petroleum contaminated soils undergoing bioremediation.

A sediment quality triad (SQT) project (the use of toxicity tests, chemical analyses and benthic monitoring as described in Chapman [14]) was undertaken to assess impact in the surrounding waterways. Concurrently, edible fish species were caught and analysed for chemical concentrations in muscle and liver tissues to assess bioaccumulation. The combined results indicated that the impact observed in the samples taken from the site and surrounding areas did not appear to be statistically different from other sites in Sydney Harbour (EVS [15], The Ecology Lab [16]). Again, results require careful evaluation due to problems with the experimental design (EVS [15]). The SQT project had not been set up solely for OCA, but was part of another Sydney Harbour project. Valuable lessons regarding the preparation of monitoring programs were learnt from this project, and led to the refinement of monitoring program quality checklists (Statzenko and Laginestra [17]). Sediment cores were also taken from selected sampling sites to provide information on the history of contamination and provide background chemical levels. As not all the samples provided background chemical levels, a bathymetric survey of the main waterway was conducted to provide information for future sampling sites. Despite the drawbacks, the cores did provide supporting information on the health of the site's waterways and changes in sediment chemical levels over time.

Extensive monitoring has been undertaken at the petroleum contaminated site of Wilson Park. As bioremediation has been found to be successful in other petroleum contaminated sites (Maier *et al.* [18]), it was chosen as the treatment method at Wilson Park, which had not been contaminated with other pollutants. Trials were undertaken to establish if the indigenous population of microorganisms that had developed at the site were breaking down the petroleum wastes. The trials indicated that the indigenous microorganisms were involved in attenuation and that benzene was being degraded in the treatment ponds within five days. The microbial community involved in the degradation has been characterised using DNA comparisons (Hanitro [19]). This project also determined the optimum site conditions required for microbial degradation to occur. Toxicity testing and chemical analyses of the soil, elutriate washes and the treatment pond water and sediment were undertaken to provide baseline information on contamination levels. From these results, a chemical concentration trigger level has been determined for toxicity re-analysis.

Integration with other programs

As part of the Enhanced Remediation Strategy, three other programs were initiated alongside the Biological Monitoring Program. These were the Databank Program for managing the data and harnessing corporate memory, the Education Program to provide learning programs and tools for schools and the community, and the Advocacy Program, which arranged seminars and managed the environmental reference group. The programs interrelated closely, as the education program required information from the monitoring program, with most data being managed through the databank. Information gained from the Enhanced Remediation Strategy has been prepared for release through a number of methods including educational CDs, seminars, school curriculum courses (developed with NSW Department of Education and Training) and adult education short courses. Other partnerships are being sought to develop the research program, utilise the databank and extend the education program.

The importance of input from the community group to the success of the programs cannot be overemphasised. Group members were paid to attend and were provided with meeting protocols. The program managers reported on program status at each meeting and results were discussed. The serious tensions encountered at the beginning eventually resolved to trust through sharing of information and strong management support. For a potentially contentious site very little controversy arose.

Management actions based on monitoring and assessment

The Sydney Olympic Park Authority has already used data obtained from the post remediation monitoring program. The site-specific information and land-use constraints provided valuable information for preparing the Millennium Parklands Plan of Management (as required under the SOPA Bill [20]). The Plan of Management must consider current and long-term usage at each site location. Tender contracts for landscape managers include the monitoring regime and reporting requirements based on monitoring information. It is hoped that as contractor monitoring results are delivered to managers, issues or emerging trends by site can be highlighted and fed back into the research program. Information from the bioremediation work regarding optimal site conditions has been incorporated into site procedures.

A post remediation reference manual has been prepared for the site as a base line document for site managers. It compiles, by location, information on history, ecology, attributes and uses, issues, recent monitoring, monitoring gaps and issues for consideration. To prepare this document, remediation monitoring was assessed alongside the compliance monitoring. Monitoring required by legislation includes water quality monitoring, leachate chemical assessment, maintenance programs (including system integrity assessment), and endangered flora and fauna monitoring (birds, frogs and wetlands). Recommendations and issues for consideration arising out of the combined monitoring assessment form the second stage of preparing a long-term management plan, which is to determine risk priorities and further streamline the programs.

Future management issues

Sustainable management decisions at a remediated site cannot be purely political expediencies if we are to manage in a truly ecologically sustainable way. The development of future research monitoring has been based on our assessment of the site monitoring results. Areas proposed for next year's research include testing for endocrine disruption, leachate re-use options, efficacy risk assessment (determining residual risk or identifying new risks), hydrological studies (including sediment monitoring and pollution load modeling), monitoring after events and studying the effectiveness of landscape monitoring regimes.

Frameworks under development to enhance site management capabilities include decision support systems, adaptive management and streamlining feedback mechanisms, and identification and management of confounding effects.

Conclusions

A number of conceptual problems arise where contamination is left in place with the expectation that future land use and natural degradation will protect the environment (Simon [21]). Short-term monitoring frameworks can usually be provided, but where the site could pose a risk for decades, and there is potential for site use or ownership changes, we need to ensure continued protection of the environment. Contaminated land managers need to learn from other site examples about the issues and advantages of different methods, and the effectiveness of controls over the long-term.

The State Government made a commitment to manage the restored Sydney Olympic Park site for the long-term. The remediation strategy chosen at the site has so far appeared to be successful, judged against other Brownfields case studies and from our monitoring results. Although the monitoring data, amassed since 1992, was not collected for assessment of long-term effectiveness of the remediation options, it is now being compiled and managed to provide baseline data for future questions regarding changes over time. The assessment of historic data with the addition of the weight-of-evidence monitoring by site function is starting to highlight successes and define areas to concentrate on. Natural attenuation appears to be underway in the leachate mounds, but modifications have had to be made to prevent erosion. Bioremediation appears to be successful, but requires active maintenance in the treatment ponds.

Elements highlighted as contributing to a successful clean-up (Pepper [3]) also present at the Sydney Olympic Park site include:
- the presence of a proactive authority who coordinated the different levels of government and consolidated the project management teams under one roof,
- project leadership driven by key individuals with backing from senior management,
- Strong community participation,
- Use of innovative remedial technologies,
- Financing advantages, including prime location and piggy-backing on a major public works project

By not transferring the contamination problem and receiving funding from the government to monitor the long-term effectiveness, without fear of changes to landuse or transfer of ownership, valuable information can be obtained for assessment by Brownfield site managers.

References

[1] NSW Environment Protection Authority, *State of the Environment Report 1997*, www.epa.nsw.gov.au/soe2000/cl, 2000

[2]   Knight, The Hon. M., Minister for the Olympics Long Term Monitoring of Remediation Program, *News Release*, March 24, 1998

[3]   Pepper, E.M. *Lessons from the Field, Unlocking Economic Potential with an environmental key*, Northeast Midwest Institute, January 1997

[4]   United Nations Development Programme, United Nations Environment Programme, World Bank and World Resources Institute (UNDP), World Resources 2000-2001 *People and Ecosystems, The fraying web of life*, World Resources Institute, Washington D.C. April 2000

[5]   Pym, J., From Liability to asset: Homebush Bay Remediation 1989 – 2000, *Report for the Olympic Coordination Authority* June 2001

[6]   US EPA A Sustainable Brownfield Model Framework, *EPA500-R-99-001*, January 1999

[7]   Smartgrowth Network, An Integrated approach for Brownfield redevelopment, A priority setting tool, smartgrowth.org/library/brownfields_tool/brownfields _priority_set.html, September, 1996

[8]   Australian and New Zealand Environment and Conservation Council and National Health and Medical Research Council (ANZECC), Australian and New Zealand Guidelines for the assessment and management of contaminated sites, January 1992

[9]   National Environment Protection Committee (NEPC), National Environment Protection (Assessment of Site Contamination) Measure 1999, www.nepc.gov.au, 1999

[10]   Laginestra, E. Developing a long term monitoring plan for the Homebush Bay Olympic Site, *Australian Journal of Ecotoxicology*, in press 2002

[11]   Hudson, J., Hughes, K. and Daffern, P. Harnessing Corporate Memory for the environmentally responsible management of remediated lands, *Proceedings from the 28th annual conference AURISA*, 20-24 November 2000, www.sopa.nsw.gov.au/ecology/science

[12]   CSIRO, Centre for Advanced Analytical Chemistry, Toxicity Assessment of Leachates from Homebush Bay Landfills, *Report ET/IR341R*, Prepared for the Olympic Coordination Authority, November 2000

[13]   Sinclair Knight Merz, Ongoing Toxicity Monitoring of Leachate, *Final Report for Olympic Coordination Authority*, September 2001

[14]   Chapman, P.M. The sediment quality triad approach to determining pollution-induced degradation, *The Science of the Total Environment*, 97/98, 815-825, 1990

[15]   EVS Environment Consultants, Sediment Quality Triad Assessment, *Final Report for Sydney Olympic Park Authority*, March 2002

[16]   The Ecology Lab, Bioaccumulation of Contaminants in fish, Newington Wetlands, Haslams Creek and reference locations, *Report for the Olympic Coordination Authority*, March 2001

[17]   Statzenko, A. and Laginestra, E. Homebush Bay Monitoring Methodology Review, Vol. 2, www.sopa.nsw.gov.au/ecology/science, October 2001

[18]   Maier, R.M., Pepper, I.L. and Gerba, C.P. *Environmental Microbiology*, Academic Press, 2000

[19]   Hanitro Pty Ltd, The demonstration of Bioremediation processes at Wilson Park, Silverwater, *Reports for Olympic Coordination Authority*, October 2000
[20]   Sydney Olympic Park Authority Bill, www.parliament.nsw.gov.au 2001
[21]   Simon, S.E. Editor's Perspective: The effectiveness of nonpermanent remedies, *Remediation*, Vol 12, #1, 1-4, Winter 2001

Redevelopment potential of landfills. A case study of six New Jersey projects

J.B. Wiley, III & B.Asadi
Sadat Associates, Inc, United States of America.

Abstract

The paper will compare and contrast the experience of six landfill redevelopment projects engineered by Sadat Associates, Inc., where the authors, Messrs. Wiley and Assadi, have each worked for over 14 years.

The projects include one community college, the largest mall in the region, an office development, an 18-hole golf course and two housing developments. A total of 378 acres of waste have been properly closed and redeveloped in these projects. The construction value of the projects currently exceeds $500 million, and when completed, will exceed $1 billion.

Some of the projects involved constructing major building and site improvements, such as parking areas on top of old waste. Some projects involved waste consolidation/ relocation on site. Two of the projects involved beneficial use of over 7 million cubic yards of recycled materials and stabilized, contaminated dredge as part of the site development.

The paper will provide a comparison of the study, design and construction process at these six sites and will address relevant environmental and geotechnical factors. The paper will also address regulatory issues, including site contamination/remediation, and other environmental permit needs. In addition, the paper will explore the financial incentives and economic factors that made the projects feasible as well as institutional arrangements and community relations aspects.

The paper provides an evaluation of the many factors that make landfill redevelopment feasible in order to encourage practitioners in the field to consider the widest range of options for end uses of closed landfills.

The main factors that affect landfill redevelopment are: size of site, degree of contamination, type of waste, depth of waste, location, wetlands and open water bodies, use of recyclable materials for remediation and development purposes, land value, willing developer, regulatory agency policy/rules, engineering solutions, and financial initiatives.

1 Introduction

1.1 General

This paper provides a summary of technical, regulatory, institutional and economic aspects of redevelopment of six former landfill sites in New Jersey. These projects were permitted and designed by SAI for various clients over a period from 1988 to the present. The end uses of the sites range from commercial/retail to institutional and residential uses. In addition to compliance with landfill closure requirements, several of the sites (which had a history of industrial waste disposal) were characterized in accordance with site remediation rules and received (or will receive) a No Further Action Determination under a State remediation program (N.J.A.C. 7:26B).

1.2 Background

New Jersey has a legacy of improperly closed landfills. The New Jersey Department of Environmental Protection (NJDEP had historically) registered 387 landfills, most of which are currently closed. The State estimates that the total number of landfills registered and unregistered) may approach 600. Most of the unregistered landfills were never properly closed. A handful of those landfills were properly closed and received a Closure and Post Closure Plan Approval pursuant to the Amended State Solid Waste Management Act of 1975 and/or the Sanitary Landfill Facility Closure and Contingency Fund Act of 1992. Hundreds of registered landfills were never properly closed because the owners lacked the resources to conform with regulatory closure requirements. While the Sanitary Landfill Facility Closure and Contingency Fund Act provides a revenue source through a tax on operating landfills, the State has historically not utilized these funds for closure of abandoned landfills and reserved the public funds for emergency actions, such as extinguishing a landfill fire or remediating methane migration. The State has also used the Spill Fund (Spill Compensation and Control Act) to study and remediate landfill sites which had a clear record of industrial waste disposal. Another group of large landfills were closed utilizing the Superfund. Nevertheless, hundreds of essentially orphaned landfill sites remain a problem with no plan for remediation being pursued either by current owners or the State. At one time, the NJDEP had worked on a Statewide Landfill Closure Plan. In the late 1980s, this plan estimated the unfunded capital costs to implement a program for closure of the registered landfills would have exceeded $1.2 billion. Clearly the current capital costs of closure of registered and unregistered landfills would probably exceed $2.0 billion. No coherent program to provide public funding for closure for the large number of unclosed sites has been developed.

The authors first became familiar with the problem of improperly closed landfills when they worked at the NJDEP. Mr. Wiley's last government position was Deputy Director of the Division of Solid Waste Management, where he had also served as Assistant Director for Planning. Mr. Assadi served as a Review Engineer in the same division in the Landfill Bureau.

Sadat Associates, Inc. is a 55 person environmental engineering firm which specializes in site investigation and remediation, and has developed unique expertise in landfill redevelopment.

While the case studies of landfill redevelopment covered herein are for New Jersey sites, the technical issues are relevant to sites in other jurisdictions. Environmental conditions in New Jersey are very sensitive (the state has the highest population density in the country). Therefore, the case studies represent a model for both redevelopment potential of old landfills and proper remediation to meet the strictest environmental standards likely to be encountered in any jurisdiction.

2 Project summary descriptions

2.1 Overview

The six projects included a wide range of and uses including residential, commercial and institutional uses. Table 1 summarizes the key features of the projects including name/location, waste acreage, end use, remediation techniques, remedial costs, value of land uses constructed and beneficial use of recyclables.

2.1.1 Location and environmental setting
Most of the projects were located in the central/northern part of New Jersey where land is scarce and real estate values can range from $100-300,000 /acre and more for developable land. One project was located along the Southern Jersey shore in an area of high real estate value.

2.1.2 Waste acreage/thickness
The projects ranged from 12 to 165 acres of waste. In all cases the waste was identified as municipal solid waste (MSW). However, both the Elizabeth and Bayonne sites had a history of industrial waste disposal as part of the MSW disposal operations. Waste thickness ranged from 10 to 30 feet.

2.1.3 Previous contamination
Previous contamination at the sites included typical parameters associated with general municipal solid waste (MSW). Sites such as Ashbrook Farms, Federal Business Centers, Wanaque/Passaic Co. Community College and North Wildwood had some heavy metals such as Arsenic, Lead or Beryllium above the State's industrial cleanup criteria. However, these parameters are related to typical urban soils and/or coal ash/and are not unexpected in MSW. Two of the sites, Elizabeth and Bayonne, had a history of industrial dumping in portions of the sites. As a result, parameters such as petroleum, hydrocarbons and PCBs above the State's 2 PPM Nonresidential Cleanup Criterion, were encountered on portions of those sites. These sites required more intensive sampling to delineate Areas of Concern (AOC) within the landfills and to design limited removal of hazardous waste drums.

Groundwater was a potential concern at all of the sites because they were not designed as contained facilities with liners or leachate collection systems. In the case of all of the landfills, except Wanaque, the sites were located on naturally impermeable formations (either clay or meadow mat). SAI conducted detailed groundwater studies for the Bayonne and Elizabeth sites (which were large and had a history of receiving industrial wastes) to confirm that the aquifer beneath the meadow mat had not been contaminated. For Ashbrook and Federal Business Centers, the landfills were relatively small and were sited on deep natural clay. Therefore, there was no concern of deeper aquifer contamination. In the case of Wanaque, a number of monitoring wells were placed in the rock aquifer. While there was slight landfill-related contamination (i.e. ammonia and TDS), the aquifer had a very low yield. No groundwater remediation was required.

2.1.4 End uses

The end uses varied widely. In an earlier project, such as Ashbrook Farms, the waste was moved, consolidated and capped so that the end use could be built on remediated areas of the site. The same occurred in the case of Federal Business Centers, except that parking areas were constructed over the re-encapsulated waste. The most recent project, North Wildwood, involves construction of elevated residential units on pile structures over the capped waste.

In some cases, the project owner had a well identified end use at the start of the project (North Wildwood, Bayonne). In the four other cases, a lengthy process of exploring alternatives for end use took place before the final use was selected. In three cases (Passaic, Ashbrook and Federal Business), the landfill was part of a larger property that was owned or acquired to develop the usable land. The landfill portion was an impediment to proper and complete use of the property.

2.1.5 Remediation techniques

A wide variety of remediation techniques were utilized. In the simplest case, waste was capped in place with one foot of silty, clayey material and one and one-half feet soil cover. In the most complex case, a slurry wall/sheet pile wall was used to contain leachate from outflow from the site and an interior leachate system was installed. The degree of capping, containment and leachate collection was influenced by the underlying geology, leachate strength and site specific cap design.

2.1.6 Remedial costs

Remedial costs ranged from $0.3 to $11 million. The cost per acre of closure/ remediation ranged from $10,000 to $100,000. In addition to remedial costs, additional costs were incurred for the Elizabeth and Passaic projects for improving the geotechnical conditions of the waste to make it possible to build parking areas on the closed waste.

2.1.7 Value of end uses

The value of the constructed end uses varied from $3 million to over $500 million. The value of the end uses reflected in Table 1 only represents the approximate investment costs for building the end use and does not include the value of the property after remediation. Except for Passaic, all the projects were undertaken by private owners/developers. It is understood that in all the cases, the development project was profitable. That is the value of the property after remediation/approvals for development exceeded the property acquisition costs, all remediation costs, all soft costs and carrying costs.

Table 1: Summary of Six Landfill Redevelopment Projects

Project Name	Location	Owner/ Developer	Acres	End Use Type	Remediation Technique	Value of Development	Use of Recyclables/ Dredge
Jersey Gardens Mall	Elizabeth, NJ	OENJ Corporation	166	Mall, Hotels, Commercial, Ferry Service	L,V.C,H,G	$700M Constructed $300M Planned	2.5 MCY
Bayonne Golf Course	Bayonne, NJ	OENJ-Cherokee Corporation	120	Golf Course	L,S,C,H,G	$10M Planned	5.0MCY+
Seaboard Point Resort	North Wildwood, NJ	Seaboard Development	12	Residential Condominiums	C, G	$50M+	50,000 cy
Passaic County Comm. College	Wanaque, NJ	Passaic County Community College	12	Community College	L,C,W,G	$10M	--
Ashbrook Farm	Edison, NJ	W&F Developers	30	Residential	W,C,L,G	$3M	--
Federal Business Center	Woodbridge, NJ	Federal Business Centers	38	Office/ Warehouse Development	W,C,L,G	$9M	--
Totals			378			$1 Billion+	7.5 MCY+

L = Leachate Collection & Treatment   C = Capping   S = Slurry Wall   H = Hazardous Waste Removal
V = Vertical Membrane Wall   W = Waste Relocation   G = Landfill Gas Controls

2.1.8 Associated permits
In addition to a Landfill Disruption and Closure/Post Closure Care Plan, each of the sites required a variety of other State environmental permits.

Three of the sites (Elizabeth, Bayonne and North Wildwood), included preparation of a Remedial Investigation/Remedial Action Workplan (RI/RAWP) under the State of New Jersey Industrial Site Remediation Act (ISRA, N.J.S.A. 13:1K). In order to receive a letter of No Further Action (NFA), which addresses remediation of past releases of hazardous materials, it is necessary to complete the RI/RAWP. While MSW landfills are not required generally to undergo the ISRA process, in the case of sites with a listing of hazardous materials and/or sensitive end use issues, it is prudent for the site developer to receive an NFA. This also assists the developer in obtaining bank financing. Finally, compliance with ISRA is a pre-requisite for financial incentives available under the State's Brownfields Redevelopment Act.

2.1.9 Beneficial uses of recyclables

Two of the projects (Elizabeth and Bayonne) utilized large amounts of recyclable materials to prepare the land for end use development. In Elizabeth, approximately 2.2 mcy of contaminated dredge spoils and recyclables were used to re-contour the site for development and surcharge the old waste to provide adequate foundation conditions. In the case of Bayonne, over 5 mcy of contaminated dredge spoils and recyclable are being used to create a rolling topography for use in golf course construction.

3 End use planning approaches

3.1 Historical overview

MSW landfills have been redeveloped historically without necessarily following the technical or regulatory approaches described in this paper. The Wanaque Landfill (now redeveloped as Passaic County Community College) had a Vo-Tech School built within the waste area after only the waste under the building footprint was relocated. The Vo-Tech School project was abandoned around 1985 before construction of the building was completed due to reports of methane migration as well as state enforcement action for lack of proper state permits to construct on the landfill.

A new era of proper planning design and remediation to allow landfill redevelopment was pioneered in the late 1980s in New Jersey by SAI, when it undertook the first projects described in this paper.

3.2 Planning approaches

Some of SAI's early projects were the result of owners "discovering" old MSW on their properties in the process of developing other portions of their sites. In these cases, SAI was retained to determine how to best "work around" the old waste and still maximize the use of the property. The solution in two cases was to move and consolidate the waste on site into a smaller controlled area with leachate collection on natural clay and capping with clay or development.

In other projects, the owners started their activities with the full knowledge that the MSW was a major constraint in use of the site.

In these cases, in order to justify the costs of remediation, it was necessary to consider a variety of alternative end uses. In the case of Elizabeth Landfill, several years of alternative use analysis took place in which a real estate expert (Eli Cohen Realty of Paramus, NJ) explored the real estate value of a wide range of uses. Some of the uses considered included container storage, car staging for imported vehicles arriving by ship, rail yard expansion and various types of retail. The alternative finally chosen to begin development was a 100 acre regional mall with a 1.2M ft^2 two story mall (largest in the region). Four smaller parcels in the 100 acres were reserved for other commercial uses. Eventually four hotels were sited and are being completed on the original 100 acres. An additional 25 acres "waterfront" parcel has been approved for office development, a ferry service and a marina. At another site (North Wildwood), the original plan was development of a minor league baseball stadium. After this project was abandoned for financial reasons, a new developer acquired the site with the objective of developing five story, high rise condominiums. The Wanaque Landfill, where the Vo-Tech School project was abandoned in the mid-1980s, was evaluated by the owners by default (Passaic County Board of Chosen Freeholders) in the late 1980s. After many alternatives were considered for end use, ranging from open space to public works garage, the alternative of redevelopment as a satellite community college was selected.

As more experience has been gathered and successful projects completed, developers are now acquiring old landfill sites with the express intention of redevelopment.

4 Regulatory issues

New Jersey has had statutes and regulations related to landfill closure since 1970. The Solid Waste Management Act of 1970) and regulations N.J.A.C. 7:26 et seq. require approval by NJDEP for "disruption" of a landfill. The Landfill Closure and Contingency Fund Act of 1987 requires landfills closed after January 1, 1982 to have a Closure and Post Closure Plan and Financial Plan to ensure proper closure. Therefore, a long standing program has been in effect, whereby NJDEP must approve landfill closure and development.

In the mid-1980s, the NJ Legislature passed the ECRA law which required NJDEP approval of property transactions involving certain prior industrial uses, MSW landfills were not included. When the ECRA law was amended by the ISRA Law in 1993, discharge of hazardous materials from any property became the subject of additional regulation. This resulted in a situation in which old MSW landfills became the subject of state cleanup requirements, assuming they had received hazardous substances, which required control /remediation.

ISRA and the State Spill Fund Act was amended in 1998 under the Brownfields Act, which offered liability protection for innocent purchasers in the form of a covenant not to sue. Financial incentives for remediation were also provided in the 1998 Brownfields Law, including provisions for grants to municipalities to study cleanup needs, up to $2 million in grants to municipalities for actual cleanup and loans for private innocent purchasers. Also included is a 75% state tax credit program for remediation by private innocent purchasers.

After the ISRA, it became necessary for landfill redevelopment projects to explicitly address contamination of soil and migration of hazardous materials to receiving media. Prior to ISRA, the issue of migration was addressed under the Solid Waste Management Act and the State Water Quality Management Act. ISRA has extensive rules regarding site investigation which require that the waste itself be characterized through Priority Pollutant +40 testing.

5 Technical issues

The redevelopment of old landfills involves a multitude of technical factors: remedial investigations, natural constraints, contamination, safety, and serviceability. Each site may have some of all of these factors with varying degrees of difficulty.

5.1 Remedial investigations

Remedial investigations at old landfills that are considered for redevelopment usually take a different approach from those conducted for landfills that are investigated to be remediated without being considered for redevelopment. The limited or intensive use of the redeveloped sites by human beings raises the concerns of the regulator, the developer/owner, and the professionals involved in the contaminated site redevelopment to a much higher level. Whence a new and more important objective of the remedial investigations becomes making sure that the health and safety of the short and long-term users of the redeveloped site are protected. This new objective requires more extensive investigations. This means additional cost to the owner/developer.

In two of the six cases presented in this paper (Elizabeth & Bayonne), SAI had to develop sampling plans with the cooperation of the New Jersey Department of Environmental Protection (NJDEP) to address human exposure, and characterize onsite contamination. Geostatistical models were employed to optimize the number and the location of samples to be collected and analyzed. In certain cases, the sampling had to be biased towards areas of concern where easily discernable contamination was encountered, or where sensitive environmental resource was present (wetlands, mudflats...).

In some cases, remedial investigations were limited to collecting information about the degree of waste decomposition using test programs, the presence of landfill gas, the strength of landfill leachate, and collecting few samples for chemical analysis. In other cases, extensive investigation of the waste, groundwater (deep and shallow), surface water, sediment, landfill gas, tidal influence and the impact on the ecological receptors.

5.2 Natural constraints

In general, site development is challenged by natural constraints: wetlands, open water bodies, streams, irregular topography. At old landfills, this factor is compounded by the presence of contamination that exacerbates the impact on development.

In three of the cases (Elizabeth, Bayonne & North Wildwood), wetlands were present at the site. These wetlands were for the most part degraded; however, they provided habitats for animals or birds. In order to develop these sites, the impact of the site redevelopment on the wetlands had to be addressed. In one case (North Wildwood), the wetlands were completely avoided, but the wetlands transition area was used for development related permitted uses such as storm water basins. In another case (Bayonne), the existing wetlands were partially filled, and higher quality mitigated wetlands were created. The filling of existing wetlands was necessary for the remediation and the redevelopment of the site. The remediation of the site included the containment of the onsite waste, which in some areas existed in wetlands. In the case of the Elizabeth site, the wetlands were filled in the process of installing a ten-foot RCP to pipe a one mile long ditch that bisected the site. Previously the ditch acted as a conduit for leachate from the landfill allowing it to discharge into the Newark Bay. Filling of the wetlands was not only required for the proper remediation of the environmental impact, but also enhanced the potential for site redevelopment.

The landfill redevelopment in North Wildwood has been impacted by the extent of the wetlands' transition area and the presence of a bird habitat. Discussions with the NJDEP are underway to mitigate this impact.

Another natural restriction to site redevelopment are drainage ditches. In two of the six sites (Elizabeth and Bayonne), large drainage ditches had to be filled to accommodate redevelopment. However, they would not have been permitted to be filled without the special justification provided by the need to prevent leachate from discharging into the surface water bodies of the state. The main justification for piping these ditches was preventing contaminants (leachate or contaminated sediments) from being transported via these ditches to the surface water bodies of the State.

5.3 Contamination

Contamination at old landfills varies from one site to another. There are common contaminant that are present at most landfills at varying levels and are considered landfill indicators: BOD, COD, Ammonia, some heavy metals, and chlorinated organic compounds. However, other contaminants are encountered at old landfills due to uncontrolled dumping. This especially true for old landfills, which are the majority of landfill sites considered for redevelopment. These contaminants include PCBs, tar, paint sludge, waste oils, drummed industrial waste, medical waste, and others. At the two larger sites (Elizabeth and Bayonne) of the sites under study, there were several examples of these types of contaminants, which required special types of treatment. Some required removal and offsite disposal: drums (Bayonne), and paint sludge and tar (Elizabeth). Some required in-situ treatment: oil sludge (Bayonne). Some required complete encapsulation due to the large volume and high cost of removal and disposal: PCB contaminated waste (Elizabeth).

5.4 Remedial action

Remedial action at old landfills normally includes capping of the waste, managing landfill leachate and gas, and monitoring the impact on the environment.

For the six cases that are considered here, some or all of the landfill remediation elements were implemented to prepare the sites for redevelopment. The least remedial action involved: capping, landfill gas management, and maintenance and monitoring program (North Wildwood). The earliest remediated site (Edison) involved: partial relocation of waste to accommodate redevelopment, capping the relocated waste with one foot of clay and one and a half foot of soil to promote vegetative growth, landfill gas control and venting system, leachate collection and disposal system. The second earliest remediated site (Woodbridge) involved: excavation and relocation of the entire waste, the construction of a state-of-the-art small landfill on a two acre lot (clay liner, leachate collection and disposal system, passive gas venting system). Thirty six acres of this thirty eight acre site became available for development after implementing the remedial action/closure. The closed and capped two acre new landfill has also been used for parking purposes.

The waste relocation and consolidation approach was later enhanced. During the implementation of a test pit program at the Wanaque landfill site, it was observed that the majority of the waste (70%) had decomposed and was reduced to soil like material. The excavation and relocation was augmented with a screening process prior to relocation and encapsulation. The material passing the screen was tested and found suitable for use as backfill for redeveloping portions of the site.

At the three more recent remediated landfill sites, the costliest element of the remediation, capping, was incorporated into the redevelopment. The two main purposes of capping are: creating a physical barrier between the waste and the environment, and reduce the potential for precipitation entering the waste fill and producing more leachate. Therefore, buildings, paved roadways and parking areas were used for achieving these two objectives (Elizabeth and North Wildwood). In Bayonne, the fill material for shaping the site to accommodate the construction of a golf course was also used as a cap, using low permeability recyclable materials such as stabilized dredge material.

5.5 Safety

One of the main concerns of redeveloping landfill sites is the safety of the eventual users of the sites: residents, shoppers, workers, or golfers.

Providing sufficient physical barrier (cap) between the waste and the users, coupled with institutional controls, proved to be an effective means of protection against human exposure to onsite contaminants. To enhance this protection, SAI with the direction of NJDEP, made it a practice to design backfill for all utility trenches with clean fill to prevent the exposure of maintenance crews to any contaminants at the landfill. Also SAI developed operating procedures for maintenance crews for areas where the potential for exposure still exists, such as leachate pump stations or valve chambers.

The second biggest concern at landfill sites is landfill gas, due to its explosive potential at certain concentrations. At each of the six sites under study, a landfill gas management system was installed or is planned to be installed. These systems are at varying degrees of sophistication depending on the level of exposure and the levels of methane gas encountered during site investigations. The system range from simple passive gas vents (Woodbridge), to synthetic barrier wall with venting trench and passive vents (Wanaque & Edison), to the more advanced active venting system under commercial or residential buildings (Elizabeth and North Wildwood). For residential development on landfills, the regulatory agencies have been more stringent. In the earlier residential development (Edison), the NJDEP allowed the construction of the houses after the waste was relocated and a passive gas barrier was installed between the housing development and the waste. In the more recent residential development, the NJDEP permitted the construction of the condominium building on top of the waste, but with an open-air first floor used for parking and an active gas venting system under the first floor.

In commercial developments, the NJDEP permitted the construction of buildings with an active gas venting system under the first floor, with explosive gas sensors in the first floor.

5.6 Stability

Another safety concern at landfills is stability of the developed structures. Stability can be compromised by differential settlement in the waste fill and the underlying soils. This can be controlled by engineered improvement of the characteristics of the waste fill and the underlying soils. For the Elizabeth site, fill surcharge and deep dynamic compaction were employed to improve the characteristics of the waste fill and the underlying soil to accommodate the construction of roadways and parking lots. However, the mall building and the hotels were built on pile foundations. The waste and underlying soft soil layer could not have been improved to accommodate the loads from these structures in time for the opening of the mall. The additional cost of pile foundation was justified by having high returns from development. Similarly, at the North Wildwood site, the waste will be surcharged with fill to mitigate the potential for differential settlement in the waste and underlying soft soils in the roadways, parking areas, and the landscape areas, while the high rise residential buildings will be constructed on piles.

At the Bayonne site, the clubhouse for the golf course is located in an area that was originally at much lower grade. Therefore, imported fill to bring the site to grade at that location to final grade needed to be select fill, to provide a competent foundation. The clubhouse location will also be surcharged to mitigate future settlement in the thin layer of waste and underlying organic clay layer. The small thickness of the waste layer and the less aggressive schedule for constructing the golf course, due to the need for bringing large volume of recycled fill, permitted implementing geotechnical improvements with a less expensive foundation solution.

Like any other sloped fill structure, redeveloped landfills should be analyzed for slope stability whenever the conditions warrant. At the Bayonne site, the stability of the side slopes was analyzed to ensure against failure. In certain areas of the site, the proposed slopes could have proven unsafe to accommodate the proposed golf course. Engineered retaining walls (sheet piling) had to be installed to prevent unsafe conditions from developing. At the North Wildwood site, the side slopes face the Atlantic Ocean and have to be protected against wave action. The slopes are designed to be planted with types of vegetation that provide protection against erosion.

5.7 Serviceability

In addition to the above challenges, redeveloping old landfills does have its limitations: potential settlement due to decomposition and consolidation of the waste and the underlying soils, and the non-homogeneity of the waste causing differential settlement as discussed above. The impact of differential settlement on the stability of structures is one effect of potential settlement, serviceability is another. Settlement may cause the surfaces of roadways and parking areas to be uneven, or it may cause damage to utility lines. In one case (Elizabeth), the pavement settled evenly, but the lighting poles were constructed on piles and which did not settle. This caused the base of the pole to project above the pavement. Such an occurrence can be problematic if the differential settlement is high. Luckily this has not been the case, since the waste was dynamically compacted or preloaded prior to development. In order to avoid settlement in the different imported recyclable materials, SAI developed a laboratory and field-testing Protocol (Protocol for Review and Certification of Recyclable Material at the OENJ Elizabeth Site, copyright 1995, Princeton Recycling Technologies, Inc.) for all imported recyclable materials. The laboratory testing Protocol focused on providing on the density and strength of the recyclable material. These parameters were then verified in the field by an SAI representative for the actual material imported to the site.

SAI monitored the compaction of the different recyclable materials, and used a nuclear density instrument to check the field density. This became part of the record and was shared with the geotechnical engineer for the mall developer.

6 Economic/financial Issues

6.1 Economics of remediation/closure

As mentioned earlier, the remediation and proper closure of old landfills is a costly proposition. This is perhaps one reason why there are many landfills that have not been properly closed. Redeveloping old landfills provides the resources for implementing proper closure and rendering uncontrolled old landfill sites valuable, job generating and tax paying. Incentives described elsewhere in this paper helped create a success story for developers and the State and local governments. However, landfill redevelopment projects have to make economic sense to developers to embark on them. In four out of the five privately owned projects (those are the ones that have been completed), the developers were able to achieve reasonable profits.

At the Woodbridge site, the developer spent approximately $200,000 in engineering and construction oversight costs, and approximately $800,000 in construction costs. The end results were: the site no longer had scattered waste, and thirty six acres of developable property became available for development with a value of $9M at the time of completion.

At the Elizabeth site, the site was uncontrolled contaminated dump with a lot of surficial dumping, overgrown vegetation, and a large open ditch that bisected the site. The value of the land was less than $100,000/acre. However the location was highly attractive (next to the New Jersey Turnpike and near Newark Airport), which attracted a Danish redevelopment company to pursue developing the site. The Danish company spent approximately $3M in engineering and construction oversight costs and $8M in construction costs to remediate the site without applying a cap. The redevelopment of the site required importing approximately 2.2 million cubic yards of fill. This would have cost the developer approximately $13M. Instead, the developer, with the assistance of SAI, developed a program for using recyclable materials as fill to accommodate the development and cap the landfill. SAI also developed a closure plan that incorporated the capping into the development plan. The buildings, the roadway pavement and the parking lot pavement became part of the cap. Only in landscape areas did the developer have to install a two-foot soil cap, which was also part of the development cost. The value of the property increased to $400,000/acre for 100 acres of upland, and to $1M/acre for eighteen acres of waterfront upland parcel.

The Bayonne site has 120 acres including an old municipal dump and a contaminated former industrial site. The remediation cost of this site is estimated at $3M in engineering and construction oversight costs and $11M in construction costs. For the site to be developed as an 18-hole golf course, which is the intended use of the site, 6 million cubic yards of fill material were needed. This fill would have cost $30M. Instead, SAI developed a program with the NJDEP for accepting recyclable materials as fill. To date, the site has received approximately 4 million cubic yards of recyclable materials. The contaminated site has been transformed into a valuable piece of property in the middle of urban area.

The site in North Wildwood is a twelve-acre old municipal landfill that is being developed into a high-end apartment complex. The unremediated site would have little to no value. The remediation cost is limited to the installation of gas venting systems for the different buildings. The estimated cost of construction for these systems is $300,000, and the estimated engineering and construction oversight costs for obtaining the necessary approvals are $300,000-$500,000.

The cost of remediation/closure for the sites under study can be plotted against the acreage of the sites. The curve generated presents the trend of the cost of remediation/closure versus the acreage of the landfill site. It is clear that the larger the site, the higher the cost of the remediation/closure. The best fitting curve of the available data is linear (see Figure 1, next page). However, inspection of the data points indicates that costs per acre appear to reach a plateau at landfill sites of the size 130 acres or more. This trend makes sense since the larger sites are more likely to have more difficult natural constraints, more illegal dumping, more hazardous waste to deal with, which drive the cost higher. The Wanaque site cost per acre probably an outlier since the excavation, screening and re-deposition of waste increased the cost of remediating the site and preparing for closure and development. This site topography, geology and size required such a solution for dealing with the on site waste, and caused the cost to be higher than the other sites evaluated in this paper.

The economics of redeveloping old landfill sites is impacted by many factors: type of development, intensity of development, schedule, site conditions and environmental setting. These factors are inter-related. For example, a shorter schedule may not allow for a less expensive option for resolving foundation issues. As mentioned earlier, the opening of a mall and the loading from the mall building required the use of pile foundation (more expensive), instead of pre-loading which would have taken much longer and would not have supported an aggressive development plan. At the Ashbrook Farms site, the development intensity and type were low and allowed for an inexpensive solution. The housing units were built on a small portion of the landfill, which allowed for waste relocation from the developed area to the undeveloped area. The residential units were low rise, which did not require expensive foundation improvements. At the North Wildwood site, the intensity of the development does not allow for waste relocation, and will require expensive foundation and landfill gas management solutions.

The additional costs for redeveloping landfills are partially compensated by the relatively low cost of land before development. However, this low cost of land usually is not sufficient to compensate for the high cost of remediation/closure. This prompted the legislature in the State of New Jersey to develop incentives for redeveloping contaminated sites, including landfill sites as discussed elsewhere in this paper.

6.2 Financial incentives

A number of significant financial incentives are available in New Jersey for landfill redevelopment.

In 1996, the Gormley Bill (Senate Bill 294 " Municipal Landfill Site, Closure, Remediation and Redevelopment") was enacted. That bill provided up to 75% in state tax credits for remediation costs for qualified landfill redevelopment projects. Typical state taxes that can be defrayed through the former bill include a six percent sales tax. Redevelopment projects that involve retail development provide an excellent opportunity for setting aside State taxes into a State administered fund to allow the developer to recoup the costs of remediation. Finally, special legislation has been used in the case of the Elizabeth Landfill to utilize the framework of the Gormley Bill to provide a funding source to pay for major infrastructure improvements. For the Elizabeth Landfill redevelopment project, the 3 percent sales tax from the mall project was allowed to be set aside to repay approximately $80M in roadway improvements needed to provide a four lane connector road from a toll plaza on the New Jersey Turnpike directly to the mall property.

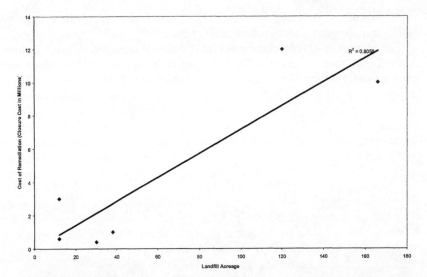

Figure 1: Landfill acreage vs. cost of remediation (closure cost in $millions)

7 Community relations issues

The landfill closure and redevelopment program in New Jersey does not have a required community involvement component in state level permitting except insofar as certain state permits, such as Upland Waterfront Development (similar to a Coastal Zone Management Act approval) have prescribed public notice/hearing provisions.

Community input to the planning for site redevelopment for the projects discussed herein resulted either from the project developer's desire to involve local officials or from required local land use approvals for site development.

In the case of the Elizabeth Project, the developer took advantage of strong interest from the Regional Planning Association (RPA) to support the project. RPA assisted the developer in coordinating local, regional and state permits needed for the project. The involvement of RPA was helpful in obtaining local and legislative support needed to gain special legislation to provide financing for the transportation infrastructure needed for the project.

City of Hopewell, Virginia – learning to deal with its industrial legacy

E. Bogdan
Earth Tech, Inc., USA

Abstract

The City of Hopewell is a perfect example of an American "Brownfields community." E. I. DuPont de Nemours purchased 1,800 acres along the James River in 1915 for the production of gunpowder. Over 25,000 employees aided in the manufacture of munitions for World War I. That plant was one of the first of many to be located in Hopewell, a community isolated from Virginia's capitol, Richmond. Unbridled growth and virtually unregulated chemical waste disposal led to contamination of Hopewell's air, water, and soil.

Many industrial properties in Hopewell now lay "idle, abandoned, or underutilized." Most are either contaminated or have been affected by nearby contamination. The resultant blight and threat to human health and the environment have thwarted past community redevelopment efforts.

The City of Hopewell has now tied its future directly to the cleanup and redevelopment of the Exeter property, located in close proximity to downtown, and within view of the old DuPont property. Cleanup will lead to reuse as an area shopping mall, but more significantly will foster refurbishment of Hopewell's downtown area and future cleanup and redevelopment of other nearby industrial "Brownfields" properties. Hopewell is looking to its industrial past for the promise of its economic future.

1 Introduction

Hopewell's 20th century industrial history makes it a perfect 21st century candidate for the designation "Brownfields community." To date, 15% of the city's 11.3 square miles has been targeted for Brownfields redevelopment. The Exeter property is the first of many such properties projected to be remediated, redeveloped, and returned to the city's tax base. Successful redevelopment of Exeter will provide the needed impetus for the conversion of Hopewell's industrial legacy into a modern, community-wide revitalization program.

Hopewell did not incorporate until 1916, but its forbears date back to 1613 when Governor Thomas Dale began a settlement at Charles City Point, near the confluence of the James and Appomattox Rivers. The Reverend Patrick Copeland opened the doors to America's first free public school nearby in 1621. In 1622, however, the settlement was nearly destroyed in an attack. Fortunately, the ship "Hopewell" arrived from Jamestown with 70 men in time to drive off the Native American invaders and rescue the survivors. King Charles rewarded the ship's captain, Francis Eppes, with a large tract of land that he named "Hopewell Farms."

Both the James and Appomattox Rivers were significant sources of water transport (with City Point at their confluence) during the 19th century. City Point (the future Hopewell) served as the deepwater port for the cities of Petersburg and Richmond prior to the Civil War. In 1884 the United States Congress authorized improvement of the James River up to City Point. Shipping and trade were thus added to the city's predominantly agricultural base.

Hopewell and its immediate surroundings remained predominantly rural through the remainder of the 19th century, lying outside of the influence of Richmond and Petersburg. Commercialization and industrialization had not yet taken hold, leaving the environment relatively unscarred. With the advent of the 20th century, however, Hopewell's environment was to be affected dramatically, as were scores of other communities throughout the United States.

One century later, the once pristine environment has become a community burdened with heavily contaminated properties, an ailing downtown, and an eroding economic base. The Exeter property lies within an industrial corridor with many other idle, abandoned, or underutilized sites. Examining the city's 100-year path toward becoming a Brownfields community will help us understand the process chosen for transformation of Hopewell from a locality in decline into a revitalized and rejuvenated city.

2   Industrialization of Hopewell

Prior to formal incorporation as a city in 1916, Hopewell had, like many other rural communities throughout the United States, become the target of a "policy of isolation" adopted by major manufacturers. Siting criteria for chemical manufacturing facilities during the first three decades of the 20th century specified separation from major population centers, minimizing public and municipal opposition. In 1912, DuPont purchased a portion of Hopewell Farms (including the future 43-acre Exeter parcel) to build and operate a dynamite plant. Perhaps of greater significance, DuPont also purchased the rights to pollute neighboring properties. It drew up contracts with adjacent landowners that included disclaimers for property damage caused by the processes used at the plant, beginning decades of unbridled industrial growth and virtually uncontrolled pollution.

World War I broke out soon after the dynamite plant began production. DuPont purchased an additional 1,700 acres from Hopewell Farms, converted the small dynamite facility into a 14-unit, 1.5-million pound/day gun cotton plant, and began construction of new mills. City Point's population swelled from 200 to over 30,000 almost overnight. DuPont attempted to accommodate its workers via construction of three nearby housing developments, rentals, and the DuPont Hotel. Similar to scores of other American communities of that era, Hopewell became a company town with "the company store," company housing, company recreation facilities, and a company hospital.

Hopewell's rapid growth vastly exceeded the resources of local law enforcement. An official police department was not established until after the city was incorporated. Corruption of every kind abounded, protected by a so-called "businessman's uplift organization." The city quickly gained a reputation as the "toughest town north of Hell."

The gun cotton facility closed within one month of Armistice in 1919, and the city's population dropped to a level below 1,500 within one year. The people were gone, but the impacts from uncontrolled contamination remained. For example, scores of munitions casings and associated debris were discovered on the Exeter property during Earth Tech's site investigation in 2001.

Industrialization of Hopewell continued immediately after the war. Tubize Artificial Silk Company began construction of its textile building on the western sector of the site in 1920 and began operations with over 3,000 employees shortly afterward. Other companies operating manufacturing facilities in the city during the 1920s included Hummel Ross Fiber, Hopewell Insulation & Mfg., Special Products, Hopewell Sheet Metal Works, Imperial Bagging, Hopewell Trunk & Bag, Allied Chemical & Dye, and Hercules Powder. All opened virtually without restrictions on discharges to adjoining water bodies, exhausts to the atmosphere, or disposal on land.

Tubize Artificial Silk Company vacated the site in 1935, leaving it abandoned until 1946 when the Celanese Corporation became the new tenant. Firestone Fibers & Textiles purchased the property in 1965 for the production of polyester fibers for use in tires. The property was purchased by Perry Realty in 1983, subsequently by Exeter in 1985, and has been abandoned ever since. Other industrial properties throughout the city likewise have had multiple tenants and have been subjected to environmental neglect.

3 Superfund action – 1991

The United States Environmental Protection Agency (USEPA) was formed in 1969 as a result of heightened public environmental awareness and activism. It took another 11 years for the U.S. Congress to enact the Comprehensive Response Compensation Liability Act, otherwise known as CERCLA or Superfund. Finally in 1980, a federal trust fund was established for the cleanup of contaminated properties.

CERCLA also mandated that responsible parties be held liable for cleanup. Unfortunately, only a handful of significantly contaminated properties were remediated during the 1980s. Most cases were delayed in the courts to determine responsible parties and levels of liability. The 1990s saw more sites remediated as court issues were resolved. It was only as recently as 2001, however, when the U.S. Congress passed and President Bush signed new legislation mitigating the liability issue.

The Exeter property provides a perfect example of century-long environmental degradation. The first evidence of environmental neglect at Exeter came with USEPA's Superfund action in 1991. Fire on the property ravaged portions of the Sawtooth Building and outlying structures. Inspection of the fire damage revealed large quantities of loose asbestos-containing materials (ACM) and exposed polychlorinated biphenyls (PCBs). The potential for airborne dissemination to neighboring properties and to the city's downtown created an imminent threat to human health. The Superfund action was limited to removal of the exposed asbestos and PCBs. Other concerns arising from 70 years of industrial activities on site would not be addressed until 2000, the year the City of Hopewell activated its progressive Brownfields redevelopment program.

4   The Exeter property and Brownfields redevelopment

Many industrial properties along Hopewell's Route 10 corridor lie idle, abandoned, or underutilized. Population figures have been either steady at approximately 20,000 or declining since the 1980s. In 2000 the city administration recognized that past failures at economic revitalization stemmed from several factors:

- Blight and environmental hazards presented by the Exeter property and other properties near the city's central business district;
- Deterioration of the city's central business district;
- Economic deterioration of other properties, including the city's only shopping mall;
- Declining or stagnant residential real estate values; and
- Lack of surrounding employment opportunities to support a population influx.

Perhaps as important, the City of Hopewell's charter prevents the city from annexation of additional land. Its only true hope for economic revitalization is tied to the return of its idle and abandoned industrial properties to the city's tax base. All other properties within its municipal boundaries have been developed.

The Exeter property has become the prime target for renewal because it is located only three blocks from and within view of the central business district. Successful redevelopment of the property will, if the city's approach is proven correct, act as the incubator for rejuvenation of the downtown area and redevelopment of other abandoned industrial properties. Conversion from blight to beautification will instill a renewed civic pride. We will now examine in detail the Exeter property, the contaminants generated, and ongoing remediation operations.

5   Comprehensive site assessment

The City of Hopewell is operating under an "Agreement and Covenant Not to Sue" among USEPA, the city, and a third-party developer. Pursuant to the Agreement, the city is proceeding in good faith to remove all remaining asbestos and demolish all site structures. The city administration had the foresight to allocate millions of dollars from its 2001 budget toward site cleanup, knowing that completion would require additional funds. Entry into the Brownfields arena and receipt of future federal and state grants and loans would be their future ticket to success.

Earth Tech, through the standard competitive bidding process, was awarded the Environmental Consultant and Site Remediation Management contract in July of 2001.

Earth Tech first conducted a comprehensive, but non-intrusive, site assessment. Through prior knowledge of the site's history and the industrial processes conducted on site, the list of known chemical substances was determined to include asbestos, lead and other heavy metals, PCBs, solvents, petroleum products, volatile organic compounds (VOCs), and particulates.

A building-by-building investigation yielded the following observations.

5.1   North Building

This building had been used for shipping and receiving. More than one hundred fluorescent lighting PCB ballasts were identified and tagged for removal. One section with unstable and partially collapsed roofing was targeted for shoring or dismantling. Other environmental/safety concerns included ACM-laden pipe runs and dismantled heating oil tanks.

5.2   Nylon Building

During identification and quantification of ACM for ultimate removal, the Nylon Building was quickly identified as "an accident waiting to happen." Several large storage tanks and vessels had been removed from the five-story building, leaving gaping holes in the steel-plate flooring. Many multi-floor, vertical pipe runs were also removed, leaving smaller trip/fall hazards. One storage room containing several hundred munitions casings, and a nylon synthesis process area containing two boilers marked as a "Radiation Hazard," were isolated pending further investigation.

Rain pouring through the damaged roof for nearly 20 years has left many sections of the steel-plate flooring, some stairwells, and structural steel rust-encrusted and structurally unsound. Broken glass windowpanes and gaping holes in several sections of the building's edifice added to containment and cleanup problems.

5.3   Aboveground storage tanks (ASTs)

ASTs located between the Nylon and North Buildings were found to be full or partially full and each required sampling and analysis.

5.4   Cooling towers

Cooling towers located adjacent to the Nylon Building presented both environmental and safety hazards. Significant pooling of rainwater housed within these open-air towers was sampled for chromates and other potential contaminants. Wood-constructed stairwells presented worker safety hazards.

5.5   Oil pump and tank

An oil sheen on puddles of rainwater and a strong odor adjacent to an oil pump and storage tank indicated the existence of volatile organic compounds (VOCs). Samples were taken from the rainwater as well as the storage tank to identify specific VOCs.

5.6   Sawtooth Building

The roof at one of the entrances to the Sawtooth Building had caved in, and associated structural steel had either fallen to the floor or become loose and dangerous. As evidenced by charred remains and other forms of debris, this had also been the site of a fire. During inspection of the roof for ACM, several hundred air conditioning units were found attached to the Sawtooth roof sections. Each unit must be examined for Freon prior to disposal.

5.7   Process drainage

All process lines, equipment drains, condensation lines, and process sumps flowed by gravity into sub-grade drains which, in turn, ultimately drained into on-site lagoons, holding ponds, or subterranean cisterns.

5.8   Elevated exterior/interior pipe runs

Elevated pipe runs, many of them bundled, were found inside most of the structures, adjacent to exterior walls, and between major buildings. Many were covered with ACM thermal insulation. Each run had to be inspected for hazardous chemicals and residuals.

5.9   Powerhouse equipment and vent stacks

Oil sheens were observed throughout the Powerhouse that housed several coal-stoked boilers, elevated storage tanks, and vent stacks. One doorway leads to a tunnel that had been used for rail-cart transfer of coal from an on-site railroad terminus.

5.10   Tunnels, cisterns, manholes, and other subterranean storage facilities

Several 1920–1930-vintage process drawings found in the South Building indicated the existence of other sub-grade, on-site tunnels. Ex-Firestone employees confirmed the presence of at least one such tunnel in the Sawtooth Building and a cave-in of a parking area above another suspected tunnel alignment during subsequent interviews.

Due to the obvious potential for contamination and structural instability, the city approved further investigations to determine tunnel alignments, connections to cisterns and other subterranean storage facilities, and the structural stability of concrete platforms above the tunnels.

6   Site preparation

According to the "Agreement and Covenant Not to Sue" with USEPA, the city is required to bring the site to a clean condition, cleared of all environmental liabilities. Soil and groundwater contamination is recognized as a distinct possibility. Investigation of this potential has been postponed until after asbestos removal and demolition activities in order to permit the city to submit an application for Brownfields Assessment Pilot funds. Upon acceptance into the program by USEPA, the city will be able to use some of the federal funds for on-site remedial investigations. Additionally, federal legislation just passed in December of last year opens funding to cleanup activities as well.

The next series of site activities and investigations conducted by Earth Tech targeted confirmation and removal of on-site chemical, physical, and electrical hazards. Elimination of these hazards minimizes the potential for accidents, improves operational logistics, and reduces asbestos removal time requirements and costs.

6.1   Munitions investigation

The munitions casings found in the Nylon Building storage room were individually examined for potential explosive content. Although the contents were determined to be benign, most casings were coated with creosote. All casings were isolated and staged for transportation and disposal (T&D) to an approved and authorized landfill.

6.2   Radiation hazard investigation

Nylon Building boilers marked "Radiation Hazard" were examined, and their atmospheres tested for potential radiation. The boilers, their interiors, and immediate environs were all found to be clear of any radioactive contamination.

6.3   Removal/reinforcement of collapsed roofing

The collapsed and unstable sections of roofing in the North and Sawtooth Buildings were cleared or reinforced to facilitate removal and demolition activities.

6.4   Removal of ground-level debris, PCB ballasts, oil burners, and oil drums

Movement on the ground floors of each building was severely restricted by debris and equipment strewn throughout. All forms of asbestos-free debris were cleared and stockpiled in areas isolated from removal activities. All fluorescent lighting PCB ballasts were removed and stockpiled for T&D.

6.5   Marking/isolating manholes, underground storage tanks (USTs), and tunnels

A comprehensive structural stability evaluation was performed within the areas of known and suspected tunnels to maximize safety for future movement of heavy equipment. Ground-penetrating radar (GPR), test borings, and confined space entry investigations were used to locate tunnel alignments, determine depths, ascertain maximum safe working loads, and confirm connections to manholes, cisterns, and other subterranean storage areas.

All manholes, USTs, open pits and lagoons, and confirmed/suspected tunnels were marked and barricaded to maximize worker safety. Specific tunnel sections were inspected for explosive atmospheres and hazardous materials. The Demolition Plan was revised to accommodate safety hazards and other uncertainties posed by the tunnels.

6.6   Clearing and grubbing

These site preparation activities were significant steps essential to making the site safe for asbestos removal and demolition, but clearing and grubbing of the site revealed other hidden hazards. Vegetation overgrown for 20 years had concealed a pump house, additional tunnels and storage lagoons, process discharge mixing basins, on-site soils contamination, and off-site discharges.

7   Pre-demolition decontamination

The last steps taken to maximize worker safety were to conduct additional product sampling and analysis and drain and dispose of product considered hazardous within designated work zones. Sediment/soils samples were taken from lagoon discharge sluices, a sump/site discharge, and the contaminated sand pit adjacent to the Powerhouse. Product samples were taken from the newly discovered lagoons and an associated concrete vault. Product and residuals were removed from the oil tank, oil pump, several ASTs, process lines, and exterior pipe chases throughout the property.

Those tunnels identified as devoid of safety or environmental hazards and as logistically preferable have been designated for on-site disposal of concrete and masonry debris. Movement of heavy machinery within and across barricaded areas will be prohibited until the debris is sufficiently compacted, rendering the area structurally stable. This will not only maximize safety, but will also reduce T&D costs.

8   Capitalizing on Hopewell's heritage

Hopewell has been America's prototypical industrial community. The city has been the subject of phenomenal economic growth as well as decline during the 20th century. Industrialization in the early part of that century converted this previously small rural community into a robust but rancorous "company" town. DuPont was the first to produce chemicals and pollute the environment in Hopewell, but many others followed. Now most of those firms are gone, but they have left their contaminants behind. Brownfields has become Hopewell's only veritable avenue toward economic revitalization in the 21st century.

The city has timed redevelopment of the Exeter property to coincide with refurbishment of its downtown business district. Since the two are within eyeshot of one another, successful revitalization of one is directly dependent on the success of the other. Receipt of federal Brownfields funds this fall will permit the city to accomplish some of its immediate goals:

- Completion of the Exeter property soil and groundwater contamination investigation;
- Initiation of site remediation required prior to redevelopment;
- Development of a "Greenbelt" pedestrian linkage between Exeter and the downtown area;
- Screening of other idle/abandoned industrial properties for future inclusion into the Brownfields program; and
- Solicitation of additional funds from other public agencies for future remedial investigations and cleanup.

Meanwhile, the city has hired other consultants to formulate and implement a downtown revitalization program. Refurbishment of building facades, signage, and parking will be only a partial solution. The city will also evaluate alternative methods for developing and maintaining economic compatibility between new downtown merchants and the future Exeter mall.

The City of Hopewell's annual budget is still significantly dependent on revenues received from Hercules, Allied Chemical, and other manufacturers. Hopewell's future lies with its ability to not only retain its remaining industrial base, but also successfully redevelop idle properties and refurbish its downtown. Investment of millions of dollars from its own budget, in addition to future utilization of Brownfields funds, will provide the city with a golden opportunity for conversion of its 20th century industrial legacy into economic prosperity during the 21st century.

Successful redevelopment and risk management in Emeryville, California

I. Dayrit[1], R. Arulanantham[2] & L. Feldman[3]
[1]*City of Emeryville, USA*
[2] *California Environmental Protection Agency, Regional Water Quality Control Board, USA*
[3] *Geomatrix Consultants USA*

Abstract

Emeryville, California is a small city in the San Francisco Bay Area that faced the challenge of redeveloping partially abandoned and underutilized industrial properties, while protecting environmental and public health. In the early 1970's, it suffered from high crime and unemployment rates, high vacancies of non-residential properties, and perceived extensive groundwater contamination. Guided by a Community Task Force and Technical Advisory Team, Emeryville implemented a program that significantly reduced the uncertainties to brownfields redevelopment, while enhancing environmental protection. The strategies that the city employed include an area-wide groundwater management program, financial assistance for site assessment and remediation, Internet-based GIS applications for dissemination of environmental, planning, real estate and institutional control information, and adaptive reuse of industrial buildings. In using these strategies in concert, Emeryville redeveloped key projects that served as economic catalysts, including infrastructure and transit-oriented developments, retail and entertainment centers and multi-family and mixed-income housing.

1 Introduction

Emeryville is a 1.2 square mile city located on the eastern shore of the San Francisco Bay. It was home to heavy industry and manufacturing. Beginning in the 1970s, Emeryville's economy struggled amid the exodus of industry and jobs, and the legacy of soil and groundwater contamination. Brownfields were scattered throughout. The cost of cleaning up these sites to pre-industrial conditions was enormous and made redevelopment difficult. With 20 percent of its non-residential property vacant and 40 percent underutilized, the situation imposed economic and social costs on the city and its residents. Property transactions and redevelopment of sites were stymied by uncertainties of environmental regulation, remediation costs, and legal liability. This resulted in the subsequent blighting of the city, and efforts of the city to remove these barriers to redevelopment [1].

While these redevelopment barriers did not deter the redevelopment of several large brownfield properties, these were difficult to overcome in small parcels because potential economic returns were insufficient to offset uncertainties and transaction costs. More important, environmental information on properties that were not remediated to residential standards was not easily accessible, thereby causing concern that risk management measures could be overlooked. Beginning 1996, with a Task Force and Advisory Team composed of stakeholders, the city developed strategies to encourage reuse of industrial buildings and properties to new uses. Four of these strategies are discussed below.

2 Area-wide groundwater management

Many cities have underutilized industrial corridors where perceived area-wide groundwater contamination complicates single parcel remediation, thereby frustrating redevelopment interest. After a decade of some success addressing site specific groundwater contamination in large sites, Emeryville realized that it would be far more difficult to redevelop small sites and that an area-wide approach would facilitate redevelopment of all sites.

Under the area-wide approach, groundwater within the city would be managed to minimize risks to receptors outside the city, and to human health. Risk management measures would be established for the entire city based on proposed land use and regulatory agency requirements. To achieve this, the city assessed the regional groundwater conditions, the quality of groundwater, and current and proposed beneficial uses. Based on those assessments, the city would design the risk management program to ensure protection the groundwater.

2.1 Regional and local groundwater characteristics

Emeryville is located along a shallow portion of the Berkeley Alluvial Plain, which is a subarea of the East Bay Plain. It has low permeability and recharge capability. The two main water-bearing zones include a shallow (less than 60 feet) and a deep (between 200 and 300 feet) water bearing zone. Groundwater generally flows westward toward San Francisco Bay in both water bearing zones. Based on published hydrogeologic parameters, approximately 150 to 300 acre-feet per year of groundwater may flow through the deep, regional water yielding zones underlying the city, representing less than 20 percent of the city's current annual water needs, regardless of land subsidence and salinity intrusion [2].

Based on data from more than 2100 existing sampling locations, including soil lithology, maximum concentrations of chemical constituents in soil, concentrations of chemical constituents in groundwater, and groundwater monitoring well construction information, the city developed regional and area-wide conceptual models describing groundwater sources, destinations and movement.

Approximately 400 locations had groundwater quality data for depths of up to 30 feet below ground surface (bgs). For certain locations, maximum concentrations of benzene, an indicator of a fuel source (e.g., a leaking underground fuel storage tank) and trichloroethylene (TCE, an indicator of a solvent source), for the period 1994 through 1996, were at very high levels [2].

Only 20 wells had groundwater quality data for depths greater than 30 feet bgs. Of these, only two locations had detections of benzene and only six locations had detections of TCE. These data suggest that concentrations decrease significantly with depth, however, there were insufficient data available for the zone greater than 30 feet bgs. Based on this, the city identified the following data gaps: groundwater quality at the San Francisco Bay interface; stratigraphic data deeper than 50 feet bgs; groundwater quality deeper than 50 feet bgs; upgradient groundwater quality at all depth zones; and surface water quality entering San Francisco Bay [2].

Through two sampling events, the city learned the following:

VOCs. A variety of VOC constituents were detected at relatively low concentrations just above chemical-specific maximum levels and were located at upgradient or transgradient boundaries of the city, where the groundwater contamination most likely originates.

Metals and minerals. General mineral parameters were measured primarily to gather information about minerals dissolved in the groundwater. The detected maximum level was exceeded in six constituents, which are common in natural groundwater or were the result of current and historic surface activities, such as application of nitrogen-based fertilizers.

Chemical constituents potentially impacting San Francisco Bay. The chemicals released to shallow groundwater may flow toward San Francisco bay. The chemicals of concern in the bay are some metals (e.g., copper, mercury, selenium), polychlorinated biphenyls (PCBs), some pesticides (e.g., DDT, Dieldrin), and dioxins, some of which were detected in minute concentrations that are unlikely to contribute to the documented impacts to the bay [3].

2.2 Beneficial groundwater use and remedial strategy

In its East Bay Plain Groundwater Basin Beneficial Use Evaluation Report [4], the Regional Water Quality Control Board (Board) determined that groundwater underlying Emeryville is unlikely to be used as a drinking water resource and that risk based corrective action be utilized in establishing groundwater cleanup standards. Passive remediation to restore drinking use as a long-term goal is recommended. Specifically, since groundwater is not currently used for any municipal, domestic, industrial, or agricultural purpose, and no extractive beneficial uses are planned in the future, remedial strategies should focus on protecting potential aquatic receptors and potential future irrigation or industrial uses. Aggressive remediation would be implemented on a case-by-case basis. Ultimately, the remedial options that would be part of a less aggressive strategy are dependent on site specific conditions. Likely options could include restricting groundwater remediation to the source area only, allowing monitored natural attenuation, or implementing pump-and-treat solely to limit plume migration [3].

2.3 Conclusions and lessons on area-wide approach

The city took significant risks in assessing its groundwater, and had assumed that an area-wide groundwater remediation would be necessary. Analysis of area-wide groundwater data dispelled the notion that shallow and deep groundwater, and San Francisco Bay were significantly impacted by more than 50 years of industrial activities, and that area-wide remedial activities were indicated. As the groundwater beneath Emeryville is not being used for any purpose, risk management measures can be designed to protect possible future beneficial use without compromising public and environmental health. Localized shallow groundwater contamination would be managed through targeted cleanup, and the city would implement measures to protect deeper aquifers from shallow contamination, and to track site-specific risk management requirements [5].

3 Financial incentives

It is generally believed that direct financial support to property owners and developers facilitates brownfields redevelopment. The city offered various forms of financial assistance, based on the source of funding.

3.1 *CIERRA* matching grant and loan program

Using federal, state and local funds, the city established a program to provide property owners and developers with matching grants for site assessments and loans for remediation. Called *CIERRA* (Capital Incentives for Emeryville's Redevelopment and Remediation), the city made these funds available to project proponents that were in the redevelopment, site analysis or remediation process. Since the inception of *CIERRA* in 2000, in consideration of the favorable rates and varied requirements of fund sources, the results are mixed.

CIERRA matching grants are available to owners, developers and prospective purchasers of brownfields. While any site in the city is eligible, the guidelines are designed for small properties. Proceeds may be used for any non-civil work activity. The city may grant up to 50 percent of these costs, with a maximum of $25,000. Recently, the city obtained state loan funds that may serve as the grantees' match.

Of approximately 10 parties which expressed interest, two were issued grants, with two pending as of this writing. Most of the interested parties were proponents of larger projects, and believed that the amount of the grant was not commensurate with the effort involved in the application process, despite its simplicity. Those that obtained a grant had most of the application requirements completed in the course of other work related with the project. Proponents of small projects that had an immediate need and had access to matching funds availed of the grant.

Federally funded *CIERRA* cleanup loans have stringent regulations including reporting, disclosure, wage requirements and cross-cutting applications of other federal laws [6]. The city offered interest-free loans for the first year, and 2 to 4 percent thereafter, in the maximum amount of the fund balance. Only one of approximately 20 interested parties availed of a loan.

The concerns of the interested parties were the exclusions on eligible activities, and the requirements on public hearings, reporting and wages. They stated that the attractive terms of the loan were negated by the additional compliance costs, potential for federal scrutiny, and regulatory oversight that is more suited to high risk sites than low risk brownfield sites.

3.2 Other financial assistance

The city provided direct financial assistance on a case-by-case basis for remediation expenses, using state and local sources. The city applied with a property owner to avail of state funds for the remediation of contamination caused by a leaking underground storage tank. In another case, using its redevelopment authority under state law, the city reimbursed remediation costs to a redeveloper. In both cases, the city was participant in the redevelopment project.

3.3 Conclusions and lessons on financial assistance

Cities have limited flexibility in designing loan and grant programs using federal and state funds. Regulations are critical elements to the success of financial assistance programs and must be tailored to potential markets. Key elements to consider in improving the marketability of these programs include:

Technical assistance. In all cases, interested parties need technical assistance to complete application documents and comply with regulations.

Suitability of Requirements. Regulatory requirements should be commensurate to the amount of available funds and risks associated with remediation.

Legal Exposure. Regulations should address borrower and grantees' fear that they may be exposed to additional scrutiny by virtue of having government funds.

4 Internet-based GIS applications

In situations where various stakeholders are involved in or may be affected by a brownfields redevelopment project, access to information is critical to minimize uncertainties and to facilitate decision-making. Much of this data is publicly available from various sources. However, these are not readily accessible. The city's goal was to present disparate data into information meaningful to those that make brownfields redevelopment decisions. The city achieved this by creating an Internet-enabled Geographic Information System (GIS) tool.

4.1 GIS data and software

GIS is a computer-based system designed to capture, store, update, manipulate, analyze, and display all forms of geographically referenced data. The city compiled parcel-specific ownership and land use data for each parcel. Environmental information, including soil borings and groundwater wells, regulatory agency information, and use restrictions, was collected for regulated sites. The city produced background layers for the parcel maps, including aerial photographs, street maps, zoning, building intensity and height maps [7].

The city selected software that was affordable, flexible, expandable, compatible with its current hardware and software, and adaptable to the Internet.

4.2 User interface

With data and map layers compiled, the key task was to design an intuitive user interface to address a set of common inquiries. Called *OSIRIS* (One Stop Interactive Resource Information System), the features of the interface include:

Search function. Query the database by owner name, address, parcel number or by map navigation.

Background function. View the map using different backgrounds, including aerial photograph, zoning, building intensity, height, and street maps.

Navigation. Alternate ways of moving around and zooming in and out.

Output. Results of queries are displayed on the screen, may be printed out or saved to disk.

Legible layout. The screen in laid out to provide the maximum amount of information and function without unduly cluttering the screen.

OSIRIS provides interested parties with access to extensive information about any property in the city, including any environmental activities and institutional controls. It provides custom menus that guide users to select a parcel, choose from several background options, and query the database for specific information, then print out a hardcopy map, or store query results onto a disk for later use. Users can view parcels using an aerial photograph and street map of the city as a base map, and include environmental, planning, zoning, land use and assessor's information. Users can access these details using parcel number or street address, or by clicking on an area of interest on the screen.

4.3 Conclusions and lessons on information systems

OSIRIS and its predecessors were instrumental in facilitating the redevelopment of several sites [7]. Using its information, regulatory agencies were able to identify sources and receptors of off-site groundwater contamination, and allowed risk management measures to be designed accordingly.

Experience in Internet-enabled GIS tools is limited, and the city will continue to polish and expand *OSIRIS* with the following considerations:

Multiple functionality. Single purpose websites are less likely to attract visitors, thus limiting exposure to its content and functions. Integration of other functions within the website will enhance its visibility. In *OSIRIS*, in addition to environmental, land use and ownership functions, the city added real estate locator and public art functions. With the real estate function, users can search for properties for sale or for lease. Searches can be done by use, size, type of use or area. Photographs of properties are also displayed. Additional features are planned.

Data acquisition and conventions.   Data compilation is the most critical and costly component of GIS. Data formats must be consistent and regularly obtained and maintained. Close coordination with all data sources is vital.

Internet integration. *OSIRIS* needs data generated by other institutions. In instances where sources make data available on the Internet, it would be ideal to obtain remote access to metadata, rather than downloading and processing source data. If remote access is not possible, providing links to webpages that display parcel-specific data would be useful.

5 Adaptive reuse of industrial buildings

Emeryville has a long industrial history. Beginning the 1850's, it was one of many industrializing cities in the San Francisco bay area, strategically located across the bay from San Francisco, and at the crossroads of three major freeways. Industrial activities included storage and manufacture of automobiles, pesticides, pigments, petrochemicals and transformers. Companies such as Westinghouse, Judson Steel, del Monte, Shell, Chevron and Sherwin-Williams had operations here.

In the 1970s, industrial activity declined and relocated leaving vacant and underutilized properties with real and perceived contamination. In the mid-1980s, developers built redevelopment projects with adaptive reuse components, involving commercial, retail and *live-work* uses. After 1990, the city set policies to encourage adaptive reuse, which now accounts for a significant portion of redevelopment in the city.

5.1 Factors for successful adaptive reuse

Adaptive reuse opens opportunities to redevelop, to preserve buildings that have significant historic or aesthetic characteristics and identity, and to facilitate projects that are profitable for the developers. The following factors were important to adaptive reuse in Emeryville.

Lower development costs. Adaptive reuse projects offer significant cost savings and higher returns on investment. A developer noted that well-rehabilitated buildings will generate 5 to 10 percent above a comparable new project. Suitable buildings can reap a developer significant cost advantages. For instance, using existing walls and foundation, and avoiding demolition reduces the amount of expensive new construction and can save up to 25 percent of construction costs [8]. Federal tax credits for historic preservation may be obtained to restore a building to a standard higher than normal adaptive reuse would.

Lower remediation costs. In most cases, soil excavation is limited to utility trenches and landscaped areas, thereby saving remediation expenses.

Market conditions. Adaptive reuse projects are more popular with potential residents and homeowners because they have industrial vestiges, such as 10 to 20 foot high ceilings, metal framed-oversized windows that open outward, exposed overhead pipes, wooden and steel beams, exterior walls of concrete, brick, and corrugated metal, original concrete floors, and interior brick walls, that provide aesthetics without sacrificing utility [8].

Live-work and residential use. The *live-work* element appeals to many creative, self-employed people because of the convenience and flexibility it offers.

City policy. The city created an institutional framework for adaptive reuse by prioritizing it in planning documents and making public investments in infrastructure. Adaptive reuse is emphasized in the General Plan and Zoning Ordinance, and in area-specific plans. The *live-work* policy further induced adaptive reuse for residential purposes and provided opportunities to create thriving mixed-use environments.

Community Involvement. The Emeryville community supports adaptive reuse projects because they preserve the urban fabric, attract investment, create neighborhoods, enhance the character of the area, and lead to a greater diversity of retail establishments.

5.2 Factors that inhibit adaptive reuse

Adaptive reuse is not possible in some cases because of issues relating to the following:

Financing. It is more difficult to obtain financing for adaptive reuse because of the additional uncertainty in working with the existing materials. In order to qualify for financing, developers have to submit additional studies verifying structural integrity. The uncertainty endemic in reusing buildings often necessitates higher returns, higher ratios of equity, highly experienced developers and contractors, and more environmental and structural analyses [8].

Structural considerations. Buildings need to comply with more stringent building codes and adapt to anticipated occupant loads. Unreinforced masonry buildings are sometimes unfeasible to retrofit.

Adaptability to disabled access. In constrained sites, retrofitting buildings to provide for disabled access consumes too much of the existing space.

5.3 Conclusion on adaptive reuse

Adaptive reuse is an excellent strategy for brownfields development because it allows communities to reinvent themselves while preserving notable buildings. A successful adaptive reuse program relies on a collaborative, proactive approach coupled with strategic public investments and policy. In Emeryville the community, civic leaders, and developers spent the time and effort to create an environment that allowed adaptive reuse to flourish.

6 Case studies

The following projects illustrate how the city employed these strategies to redevelop brownfields sites.

Warehouse lofts. This project involved the conversion of a 156,000 square foot building into 142 units of live-work lofts, 26 of which are affordable to moderate income households. The site was affected by underground storage tanks, soil and groundwater contamination. The project proponent was absolved of responsibility for remediating groundwater after the regulatory agencies determined that this site was not a source of contamination and that down gradient receptors were not affected by residual contamination. The city assisted in financing the affordable units.

Office lofts. This project involved the reuse of a pipe and fitting manufacturing facility into 230,000 square feet of offices and retail. The site had shallow soil and groundwater contamination. As in the warehouse lofts, analysis of area-wide contamination showed that contamination originated form an upgradient source and there were no down-gradient receptors.

Transit-oriented mixed-use development. This project involved the construction of several office and local serving retail buildings, adjacent to a rail station. The site was a long-abandoned site with historical releases of PCBs. Once again, area-wide analysis helped the regulatory agencies focus remedial strategies. The concerns of the responsible party were eased upon learning that PCB contamination had not migrated off-site. The city provided the project proponent with a matching grant and loan for infrastructure construction.

7 Conclusion

There is no single strategy that ensures brownfields redevelopment success. Aside from the four strategies described here, the city applied many others and is crafting new tools to facilitate redevelopment. These include regulatory assistance to project proponents, city assumption of limited environmental regulation under state law, general and specific plans, and local laws to strengthen institutional control mechanisms. The sum of these technical, financial, regulatory and practical approaches were applied to provide economic growth, environmental protection and equal benefit. For projects constructed after 1996, the city is expected to gain 1000 housing units, 4 million square feet of commercial space, 1 million square feet of retail, 500 hotel rooms and 9000 additional jobs.

Leadership and stakeholder participation were key ingredients in formulating and refining these strategies. The City Council carefully considered contributions from residents, businesses, lenders, developers and city staff. Coordination and technical assistance with the regulatory agencies, including the U.S. Environmental Protection Agency (EPA), California EPA Department of Toxic Substances Control and Regional Water Quality Control Board, were vital in designing the program.

Emeryville's program is considered a model for brownfields redevelopment and has been internationally recognized, including a Bangemann Challenge Award (now the Stockholm Challenge) in 1999, and Phoenix Awards in 1999 and 2001.

References

[1]   City of Emeryville, *Project Status Report*, November 1998
[2]   Geomatrix Consultants, Inc. (Geomatrix), *Conceptual Groundwater Model, Emeryville Brownfields Pilot Project, Emeryville, California*, March 30, 1998.
[3]   Geomatrix, Final Groundwater Report, City of Emeryville, Emeryville, California, July 26, 2001.
[4]   California Regional Water Quality Control Board, San Francisco Bay Region, 2001, *East Bay Plain Groundwater Basin Beneficial Use Evaluation Report, Final Report*, April 21, 2001.
[5]   Dayrit, I. and Geomatrix Consultants, Area-wide Approaches to Groundwater Contamination - Emeryville, California, *Proceedings of Brownfields 2001* in Chicago, Illinois, The Engineers' Society of Western Pennsylvania, 2001
[6]   U. S. Environmental Protection Agency, *The Brownfields Economic Redevelopment Initiative: Brownfields Cleanup Revolving Loan Fund Administrative Manual,* Washington D.C., 1998
[7]   Dayrit, I., Expanding Use of GIS for Planning, Redevelopment and Institutional Controls, *Proceedings of Brownfields 2001* in Chicago, Illinois The Engineers' Society of Western Pennsylvania, 2001
[8]   Dayrit, I. and Charpentier, S, Residential Adaptive Reuse - Case Studies, *Proceedings of Brownfields 2001* in Chicago, Illinois, The Engineers' Society of Western Pennsylvania, 2001

Redevelopment of the former Shell Haven refinery

J. M. Waters[1], C. Lambert[2], D. Reid[1] & R. Shaw[1]
[1]*Environmental Resources Management, Oxford, England.*
[2]*Shell UK Ltd, Stanford-le-Hope, England.*

Abstract

It is proposed to redevelop the former Shell Haven Refinery in Stanford-le-Hope, Essex, England into a world class port and business centre creating up to 16,500 new jobs. The refinery closed at the end of 1999 and a series of site assessments and progressive remediation has been instigated to address areas of historical contamination.

A Quantitative Risk Assessment was undertaken to develop site specific risk based remediation goals which were accepted by the Environment Agency in England and the Local Authority, Thurrock Council. Shell adopted more stringent remediation targets based on addressing potential commercial and reputation concerns.

The result was a pragmatic, sustainable approach to soil and groundwater remediation focusing on: bioremediation of oil impacted soil; removal of free product; monitoring of natural attenuation processes; and off-site disposal of the more untreatable materials.

Phase I remediation activities are currently drawing to a close including the excavation of approximately 115,000m^3 of oil contaminated soil. Almost 80% of this material has been treated on site using licensed mixed biopiles, making this project what is believed to be the UK's largest on-site bioremediation project. The balance of untreatable material has been sent for licensed disposal using a dedicated rail loading facility installed for the purpose. Further works are currently being planned and it is anticipated that a Phase 1a contract will be let in May 2002

1 Introduction

After over 80 years of operation, the Shell Haven Oil Refinery in Essex ceased production at the end of 1999. The 310 hectare site was developed on mud flats and marshes on the north bank of the River Thames and has comprised refinery, chemical works and oil storage operations.

Prior to closure, Shell commissioned a Land Use Study to advise on the pattern of land use likely to be established after refinery use ceases, with priority given to future employment opportunities.

Post closure, Shell initiated a programme of decommissioning, demolition and site remediation works, which in view of their sheer scale will run until 2003.

As part of these works Environmental Resources Management (ERM) was commissioned to undertake a phased site investigation and remedial plan development associated with the closure of the refinery and potential redevelopment as a container port and associated business centre.

2 Pre-Closure Assessment Activities

2.1 Previous investigation works undertaken by ERM

ERM was commissioned by Shell to undertake a Phase II Investigation of the site in September 1999. The purpose of the site investigation was to enable the development of a remedial strategy for the site and to provide appropriate estimation of likely remedial costs. To do this it was necessary to delineate the free phase product and contaminated soil on the site, and to confirm if contaminated water had impacted the shallow aquifer present beneath the site. The investigation built on previous investigative works, in particular, a RBCA Tier 1 risk assessment undertaken by Shell Global Solutions, in order to provide more complete information on the site.

The investigation was undertaken by ERM on-site between October and December 1999 and comprised over 330 intrusive exploratory holes into the shallow made ground and alluvium beneath the site. The holes were advanced as either soil bores (192) or trial pits (140), allowing continuous sample recovery from the shallow soil column to a maximum depth of 6.00m.

Additionally 10 wells were installed into the shallow aquifer in the sand and gravel beneath the alluvium. These boreholes were also utilised for limited Geotechnical assessment.

The exploratory holes and wells facilitated geological logging of the subsurface and the collection of samples for laboratory analysis.

In order to provide adequate data for a risk assessment, the leaching potential of soil contaminants was addressed through collection of site-specific soil leachate data. The volatilisation potential of soil contaminants was addressed through collection of site-specific gas data.

Field geological and laboratory analytical data were recorded onto an Earthsoft EQuIS database.

2.2 Remediation strategy

A risk-based approach was adopted to develop a remediation strategy for the site which addressed human health and groundwater issues within recognised guidelines and applied Shell's stringent commercial and reputational criteria.

2.2.1 Quantitative risk assessment

ERM undertook a Quantitative Risk Assessment based upon the findings of the site investigation undertaken between October and December 1999. The QRA was based on Risk Based Corrective Action (RBCA) Tier 2 risk assessment methodologies. The RBCA methodology for chemical release sites was used in parallel with the Environment Agency methodology for protection of groundwater as a means of integrating the assessment of both human health and groundwater using recognised guidelines.

The results of the assessment for soils were considered consistent with both the nature of the chemical contamination observed to be present, namely low concentrations of BTEX and PAH and relatively higher concentrations of less toxic heavy aliphatic fractions (e.g. TPH), and the nature of the end uses under current and proposed future scenarios. The results indicated that under these conditions risk based concentrations of most of the contaminants in soil would not determine the degree of remediation undertaken.

For groundwater within the first major body, the river terrace deposits, the conceptual site model demonstrated that this aquifer was of low sensitivity. Furthermore, it was concluded that risks posed by it to the River Thames from discharge were low. Improvement in groundwater quality will be achieved by the implementation of source remediation.

2.2.2 Remediation rationale

In addition to the remediation requirement to be protective to human and environmental receptors provided by the quantitative risk assessment, other reputational and legal drivers were considered:

- Surface soil visually impacted with residual product, refinery hydrocarbon wastes in designated waste disposal and treatment areas, and areas of mobile free phase floating product were identified to represent the main areas of concern with respect to reputational liability.
- Remediation work would require licensing either through:
  - a site licence authorising the keeping, treatment or disposal of controlled waste on land; or
  - a mobile plant licence authorising the treatment or disposal of controlled waste.

2.2.3 Remediation strategy

The following overall phased strategy was developed based upon the findings of the quantitative risk assessment and consideration of reputational drivers.

Phase 1 – Works independent of redevelopment

The main liabilities have been identified as:
- Surface residual product (visual surface impact);
- Mobile free phase floating product; and
- 'Waste areas'.

The removal of these areas is a baseline policy requirement of Shell to meet its environmental management targets and to minimise legal and reputational risks and forms the minimum requirement for remediation. These works are considered independent of the current proposed end use for the site.

Phase 2 – Works dependent on the proposed development

The Phase 1 works are intended to remove the significant areas of contamination, but potentially further Phase 2 works may be required, dependant on the planned future use. Such works will be carried out during development and may include capping of selected remaining contaminated areas (residual saturated soils), provision of clean underground service corridors and disposal/treatment of contaminated material from roads, foundations, piles, etc.

Phase 3 – Post remediation monitoring

Continued environmental monitoring of the site is proposed following remediation, chiefly to enable assessment and demonstration of natural attenuation processes in the river terrace deposits.

2.3 Scope of the Phase 1 remediation activities

The Phase 1 remediation strategy included the following:

- The removal of surface residual hydrocarbon product, site wide, to either 0.50m below ground or the top of the water table, whichever is the shallower. Control of the excavation by visual assessment will be backed up by the application of a 10,000 mg/kg total petroleum hydrocarbons (TPH) criteria.
- The removal of on-site 'waste areas'. For the remediation strategy, waste areas have been defined as:
    - Sludge Land Farm – Manor Central North'
    - The identified Acid Tar Pit and Acid Tar Pockets, Industrial Site; and
    - Sludge Lagoons – Curry Marsh West; and

All visually identified waste is to be removed from these identified waste disposal areas. To help control of the excavation, visual identification will be backed up by a target of 10,000 mg/kg TPH between surface and 0.50m below ground, and 50,000 mg/kg TPH below a depth of 0.50m. Both these numbers were lower than the >70,000 mg/kg TPH concentration identified as the risk based criteria.

- The removal of buried mobile free product. For the remediation strategy the following areas have been identified to contain mobile free product:

  - Reedham;
  - Rugward West; and
  - Homelight.

The strategy recommended the adoption of pragmatic criteria for product recovery based upon the practical difficulties associated with recovery.

2.4 Procurement of Phase 1 remediation works

In December 2000, Shell decided to appoint an Environmental Consultant ('Contractor') to design, procure, manage, undertake and report on the remediation works required in the areas defined above. The Contractors role included:

- Detailed delineation of the defined areas;
- Extension of the remediation strategy into detailed remedial design;
- Support of Shell in appointing a Remediation Contractor;
- Management, employment and supervision of the Remediation Contractor;
- Regulatory liaison including assistance in application for site licences;
- Verification of remediation and closure reporting; and
- Provision of warranty covering assessment and remediation activities.

ERM was appointed by Shell to conduct these Phase 1 remediation works.

3 Redevelopment Proposals

3.1 Port and Business Centre

In parallel with the assessment activities, in the year 2000 Shell signed an agreement with P&O related to the development of a container port, and also a business centre comprising state-of-the-art associated storage and distribution facilities to be known as London Gateway.

The proposed port will take up to 75 hectares of the site and an additional 93 hectares of reclaimed land created from dredging the approach channel. This dredged material will raise the elevation of the river frontage section of the site by between 0.5m and 6m. It will provide container and roll-on, roll-off (Ro-Ro) berths plus a relocated oil jetty for existing Shell operations.

The business centre will accommodate up to one million square metres of new buildings that will attract occupiers from the manufacturing, industrial and distribution sectors. There will also be a range of smaller buildings able to accommodate high-tech, light industrial and distribution activities.

3.2 Economic benefits

The benefits to the local economy are considerable including:

- Re-use of a 600 hectare brownfield site which can accommodate a mix of commercial uses whilst retaining green areas;
- The creation of 16,500 new jobs;
- Use of a 3km river frontage which has been used as a port since the 16th century;
- A population catchment of 3.2 million people within a one-hour drive;
- Good road access to central London and the motorway network; and
- An existing and expandable connection to the rail network.

3.3 Environmental benefits

Extensive and detailed environmental standards have been carried out to ensure the scheme is developed in a sustainable manner. These are integral to the planning and statutory applications submitted in early 2002.

The environmental benefits of the scheme include:

- The remediation and clean-up of contaminated industrial land;
- Maximising the use of on-site bioremediation (i.e. waste treatment rather than disposal);
- Minimising both the volume of waste taken off-site and associated traffic movements;
- Widespread landscaping across the site;
- Management of Greenfield land to encourage greater biodiversity;
- The use of energy-efficient construction materials;
- Improvement to the local footpath and bridleway; and
- Enhancement and improved management for nature conservation.

4 Phase 1 Remediation

4.1 Introduction

The Environment Agency and Local Authority accepted the proposed remediation plan in April 2001. A cost benefit study conducted by ERM identified that considerable savings could be achieved if on-site bioremediation could be used. Accordingly a bioremediation trial was undertaken in May/June 2001 on 1,000m^3 of oil contaminated soil accumulated from across the site. This demonstrated that the material was amenable to biodegradation and a tender was prepared for the main works.

In July 2001, Bilfinger and Berger was appointed by ERM to undertake the Phase 1 remediation contract comprising:

- Removal of free product including the bulk excavation of approximately 115,000m^3 of contaminated soils exceeding the site specific target levels;
- Biotreatment of as much of the oil contaminated soil, as was technically possible to minimise the cost of off-site disposal; and
- Installation of monitored natural attenuation wells.

Following a further cost benefit analysis and subsequent tender, Shanks was appointed to conduct the transport and off-site disposal of untreatable materials, such as oil saturated soils, acid tars etc. The transport method for the disposal of this material, approximately 25,000m^3 in volume, was via rail, achieved by constructing a new rail siding. Using this approach, the trains could travel direct to the landfill site in Bedfordshire avoiding the need for unnecessary lorry movements.

4.2 Clean-up goals

As previously described Shell adopted post remediation target criteria that were more stringent than those required through the development of the Quantitative Risk Assessment. For example the criteria for TPH was 5,000 mg/kg, benzene was 1 mg/kg and visual and odour assessment criteria were adopted. Compliance to these criteria was maintained by ERM as the works progressed.

4.3 Progress to date

Work in July 2001 and the Phase 1 remediation works are currently drawing to a close, on the scheduled contract completion day of 31 May 2002. During this time the following has been achieved:

- In seven accessible areas of the site, principally in the former Oil Movements areas over 115,000m^3 of contaminated soils have been excavated for treatment or off-site disposal.
- The excavation works ranged from shallow surface scrapes of near surface oil saturated soils, principally in former tank farm areas, to excavation down to a maximum depth of 4m. This work has been co-ordinated with the progressive demolition of many of these storage facilities.
- Bioremediation has been successfully employed to treat over 91,000m^3 of oil contaminated soils. The bioremediation was conducted in licensed conditions on prepared impermeable concrete bases within three treatment areas, in accordance with the Contractor's Mobile Plant Licence (Waste Management Licensing Regulations 1994).
- The oil contaminated soils were piled in 3m wide and 2m high windrows and processed using two soil turning machines which moved through the piles to promote oxygen transfer and contaminant accessibility. The source of the material in each windrow was logged as was the treatment time and final location of treated material. Treatment times varied between 5 and 9 weeks dependant on the source of the material.

- Where possible recycling measures were adopted. For example, bitsand material was removed from the top of many of the tank bases. This was screened and over 1000 tonnes of suitable material were sent to a local bitumen road base producer. In addition a considerable volume of subsurface concrete, former foundations, were recovered and crushed on-site. Much of this will be used on site during the future redevelopment works.
- Despite the on-site treatment and recycling measures adopted there was still almost 25,000m^3 of untreatable material that required off-site disposal. This included some acid tars and oil saturated/bituminous material that was recovered during the surface scrape activities. The construction of the rail siding was successfully completed. By the end of 2001 31 trains, each carrying approximately 1450 tonnes of waste, were loaded and transported the material to landfill for licensed disposal.
- All these works were conducted under the Construction, Design and Management Regulations, in accordance with Shell's permit-to-work system. To date no injuries or lost time accidents have occurred and works have remained on schedule.

4.4 Cost benefit

The Phase 1 works will be completed for approximately €6.5 million, which through incentive mechanisms to maximise the degree of bioremediation, the use of train transport for off-site disposal of wastes and the recycling measures adopted is well below the target price of €7.5 million set at the time of the Phase 1 contract award in July 2001.

5   Phase 1a remediation

At the time of preparing this paper (April 2002) Shell is finalising proposals for further works to address some further areas requiring remediation before redevelopment can commence. These include areas that have only become available as demolition works have latterly been completed and the works are anticipated to comprise between 64,000 to 85,000m^3 of oil contaminated soil the majority of which will be bioremediated.

At least one further phase of Phase 1 works is anticipated. Once that is completed, probably in 2003, any further remediation works will be dependant on the specific nature of future site redevelopment. For example, measures may be necessary to address any contamination identified during the construction of building foundations or during the installation of service corridors.

6   Conclusions

The Phase 1 remediation works necessary to prepare the site of the former Shell Haven refinery for redevelopment as a container port and business centre is proceeding on schedule.

The remedial targets were developed by the use of site specific risk based goals and supplemented by more stringent commercial and reputational criteria adopted by Shell. Works were discussed with the regulatory authorities, who accepted the remedial strategy proposed by ERM and Shell.

The use of these site specific, risk based targets, the application of on-site bioremediation, recycling of suitable materials and the use of freight train transport for off-site waste disposal all have contributed to the fact that the first stage of the Phase 1 remediation works will be completed on time and well under the target price, despite the fact that additional material was identified.

Remediation and redevelopment of the former highway maintenance yard, Grande Prairie, Alberta, Canada

G. J. Johnson[1], D. Mattinson[2], & D. Friesen[2]
[1] *Komex International Ltd, Calgary, Canada*
[2] *Hazco Environmental Services Ltd, Calgary, Canada*

Abstract

Komex International Ltd., in conjunction with Hazco Environmental Ltd., has implemented a remediation and redevelopment program for a 21.8 hectare industrial property located in Grande Prairie, Alberta, Canada. The property was first developed in the early 1960's as a small petroleum refinery which manufactured heating oil, gasoline and diesel. In the early 1970's, the refinery was decommissioned and the property was given to the Province of Alberta which operated a highway maintenance yard. In 2000, Hazco and Komex entered into an agreement with the Province to remediate and redevelop the site for subsequent industrial/commercial use. This program was completed in 2001.

The property in question was contaminated by crude oil and refined products associated with the original refinery, as well as by salt storage and underground fuel tanks associated with the highway maintenance yard. Organic contaminants were remediated to a level appropriate for industrial/commercial reuse. Salt impacts were managed on-site by installing containment systems. The project is significant because it was completed for the Province of Alberta, who are the responsible party in this case and are also the regulatory body. This shows the Province's support of the Brownfields development concept and the idea of transition of environmental responsibilities. It is also significant because the property is the only remaining industrial land in Grande Prairie, a fast growing northern centre for natural gas exploration. Redevelopment of the property ultimately conserves the natural areas and farmland that surround Grande Prairie.

1   Introduction

1.1   Background

Komex International Ltd. and Hazco Environmental Services Ltd. (the Joint Venture) entered into an agreement with the Province to purchase and remediate the former Alberta Transportation maintenance yard (Site), which comprises a 21.8 hectare subject property that is located in the industrial area of Grande Prairie, Alberta (Figure 1). The Site was used as a small refining facility from 1956 to 1964, producing gasoline, stove oil, diesel fuel and oils. In 1964, the Province purchased the Site for use as a transportation maintenance yard. It operated as such until the 1990's when activities started to wind down. Environmental impacts associated with the refining operations were noted at the time and were further characterised from 1989 to 1997. In 1999, Alberta Infrastructure, a division of the Alberta government and owner of the Site, initiated a process to sell the Site to a remediation/development company with the requirement to complete specific remediation. Komex/Hazco was selected as the successful bidder in 2000 and an Agreement to purchase and remediate was signed in 2001. The Site has been remediated and is in the process of being re-developed for light industrial and commercial uses.

The project is unique in that it involves an agreement between private companies, 2 departments of the Alberta government, and the City of Grande Prairie, to implement a Brownfields remediation program to return the Site to productive industrial use. The project was initiated at a time when environmental guidelines for site clean-ups in Alberta were in a state of change, requiring the Province to 'select' remediation criteria appropriate for subsequent industrial use from an array of federal and provincial guidelines. The project is also unique in

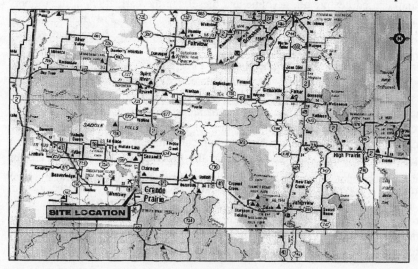

Figure 1: Site location map

that it allowed for on-Site risk management of salt contamination, which is a regulated substance in Alberta.

1.2 Remediation and development agreement

The idea to sell the Site as a potential Brownfields development was initiated by the Province as part of a program to consolidate its maintenance operations and sell lands no longer needed. The process of bidding involved pre-qualification of 3 bidders, followed by selection of the successful bidder through conventional tender.   The process of qualifying required demonstration of financial, environmental and development qualifications.   The bid and selection process involved submission of financial terms and a comprehensive remediation plan consistent with the requirements of the bid form.   The eventual Agreement included the following aspects:

- subsequent land use was restricted to industrial;
- remediation criteria and implementation of the comprehensive remediation plan were specified as part of the contract;
- the Province remained responsible for off-Site impacts;
- the Joint Venture became responsible for on-Site impacts;
- remediation was to occur over a specified period of time; and
- assurances were documented in an Indemnification which was attached to the Agreement as a separate Schedule.

All conditions specified in the Agreement and associated schedules have been fulfilled by the signing parties.

1.3 Remediation criteria

The Site was remediated and re-developed for industrial and related commercial use. To this end, the Province specified remediation criteria for the Site as part of the Agreement. In general, the lands are expected to support light and heavy industrial activities with a few landscaped areas that are expected to support an ecological function. Initially, an agreement with the Province cited the following remediation criteria for these areas of the Site.

- Provincial storage tank standards for bulk and specific hydrocarbon compounds.
- Federal standards for industrial lands for other compounds.

In accordance with the agreement with the Province, the remediation criteria included reducing E.C. < 4 dS/m, within 1.5 m of the ground surface in areas expected to maintain an ecological function.   In other areas, the remedial objective was to manage salt impacts on Site such that the salts did not spread onto adjacent lands or impact areas that are expected to maintain an ecological function.

1.4 Environmental investigations

Investigations and studies of the Site were completed on behalf of the Province in preparation for sale and/or decommissioning of the Site. Knowledge of the Site history and related environmental impacts was developed through these studies as well as additional investigations completed by the Joint Venture in preparation for entering into the Agreement. Together, the studies indicated the presence of salt and hydrocarbon impacts on-Site, as well as the potential for off-Site migration of salts to the south and west at the south end of the Site, and to the west at the northwest corner of the Site.

2 Site description

2.1 General

Grande Prairie is a robust and growing community that serves as the northern Alberta centre for agriculture, petroleum exploration and production, and forestry. Despite its rural location, the resource activity in the area is taxing the available industrial space in the City, to the extent that development of natural lands will be required to meet the ongoing demand. Hence, remediation and redevelopment of the Site was key to putting the lands back into service, improving the City's tax base and reducing the pressure to develop otherwise productive agricultural lands.

The Site is situated in light industrial and commercial lands within the City's original core and adjacent to Wapiti Road, a major highway serving lands to the south of Grande Prairie.

2.2 Site topography and drainage

The topography at the study site is relatively flat to slightly undulating. The highest elevations occur on the west side of the property and the ground surface slopes toward the northeast and southward. Total relief is on the order of 2 metres over the 21.8 hectare area. Surface drainage is controlled by the local topography, with site runoff trending towards the south and northeast. A stormsewer system was originally installed over the southern portion of the Site to service the former highway maintenance yard and discharge to the City's stormsewer system which is located on 84th Avenue and Wapiti Road.

2.3 Site geology

The surface soils are primarily composed of fill material (sand and gravel, or sandy, silty clay) and native soils (sandy clay). The fill material most likely represents foreign material transported to the site during facilities construction and site operations. Previously developed areas comprising approximately 70% of the Site had been covered with up to 2 metres. In addition, lagoons and pits associated with the former refinery had been filled in.

Across the site, glaciolacustrine clay underlies the fill and native soils to a depth of approximately 6 m below ground surface (bgs). The clays are typically varied with alternating light brown silty layers and dark grey clay layers up to 5 mm thick. Occasionally, the clay contains thin sandy layers. The top portion of the clay appears weathered, as evidenced by its brown colour, and contains vertical fractures and visible plant roots.

Beneath the layer of lacustrine clay is a grey, pebbly, clay till. This material is dense, compact, and contains minor sand and gravel lenses. In some locations, the upper portion of the till (2 m or so) is moderately to highly weathered and has a reworked appearance. In these particular areas, the till is fractured and the fracture zones are oxidised to a reddish brown colour. The fracture orientation is predominantly vertical to sub-vertical, with the occasional horizontal fracture. The fracture infilling material consists of light brown rusty material or white gypsum crystals. The till is described as being dry with no visible moisture except along the occasional sand seam or highly weathered fracture. The depth of the base of the till exceeds 15 m bgs.

2.4   Site hydrogeology

There are two main water-bearing intervals in the drift sediments covering this site. The uppermost interval is situated within the fill material and weathered portion of the glaciolacustrine clay. Hydraulic conductivities in this interval are typically on the order of 10^{-7} m/s as measured by single well response tests. The deeper water-bearing interval is confined to the glaciolacustrine clay, and represents a more sluggish interval, with wells requiring up to one year to recover. Hydraulic conductivity values calculated for wells completed in this material ranged from 6.5×10^{-11} to 3.2×10^{-10} m/s. Based on these values, the clay represents an effective containment layer protecting deeper water-bearing intervals from contamination originating from above. In essence, the pits and lagoons excavated into this clay would have their fluids contained within the area, much like a bathtub, due to the very low permeability of the surrounding material. Anticipated lateral groundwater flow velocities in this interval will be negligible compared to the upper fill material.

The Site is assumed to be in an area of groundwater recharge (downward flow direction) because there are no areas of groundwater discharge apparent in the immediate Site vicinity. The containment properties of the low permeability clay layer blanketing the site will impede the downwards component of groundwater flow. There is no groundwater use in the vicinity of the Site.

2.5   Pre-remediation condition

2.5.1 Soils
Contaminated conditions within the soils of the Site were determined by conducting geophysical surveys over the entire Site and by drilling and sampling approximately 100 boreholes and test pits. The following areas of hydrocarbon contamination were identified from these data (Figure 2):

- a former crude oil bunker lagoon at the north end of the property that was associated with the former refinery;
- a former wastewater lagoon at the south end of the property that was associated with the former refinery;
- an underground storage tank associated with the former maintenance yard; and
- a foundation area associated with the former hydrocarbon processing facilities.

The total volume of hydrocarbon contaminated soil was on the order of 25,000 m^3. Contamination in these areas is characterised by the following:

- the hydrocarbon fraction included light and heavy-end fractions up to 8% by weight;
- total BTEX concentrations were relatively low, generally less than 100 mg/kg;
- total phenols concentrations were relatively low, less than 0.5 mg/kg;
- PAH's were present in the former bunker and lagoon areas, at concentrations up to 10 mg/kg;
- elevated nickel and arsenic concentrations were observed, but were determined to be of natural original; and,
- all materials were classified as non-hazardous in accordance with Alberta Regulations and, hence, were suitable for disposal at conventional industrial landfills.

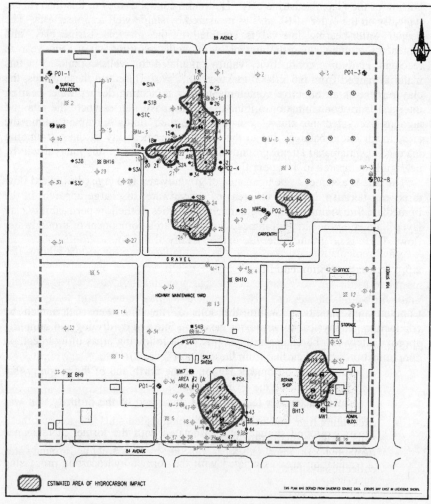

Figure 2: Site plan with all monitoring locations

Salt contaminated zones were present in the northwest corner, where the salt was handled and stockpiled, as well as at the south end of the property, adjacent to the former salt shed. The total area of salt impacts exceeded 2 hectares, and the depth of impact varied from 1 to more than 3 m bgs. The size of the salt impact made removal impractical. Also, the salt impacts affect plant growth, which is not a significant issue for the subsequent industrial land use.

2.5.2 Groundwater

Hydrocarbon impact to groundwater was confined to the immediate areas of past hydrocarbon storage. The only exceedances of groundwater remediation criteria occurred in the monitoring wells installed within the contaminated area associated with the UST in the southeast corner of the Site (Figure 2). Additionally, the only significant zones of hydrocarbon impact to groundwater occurred in zones that were contaminated by hydrocarbons. These impacts were expected to be effectively addressed through the remediation of the hydrocarbon-impacted soils that served as the source of groundwater impacts. No exceedances of hydrocarbon remediation criteria occurred outside of the areas of hydrocarbon contamination that were excavated and either treated or disposed off-Site.

Significant salt impact to the shallow groundwater was noted in the vicinity of the former salt storage shed at the south end of the Site and in the northwest corner of the Site. Salinity impacts occurred within the range of naturally high groundwater mineralization, and are restricted to areas of past salt storage. Because the groundwater seepage velocities are so low, the risk of adverse effect to any ecological receptors is essentially zero. These data supported the position that salt impacts could be managed effectively on-Site.

3 Remediation

3.1 Remediation Program Overview

A formal remediation plan was submitted to the Province for formal review and approval as a condition of the Agreement. A copy of this plan was also submitted to the City of Grande Prairie and area stakeholders, such as adjacent landowners and businesses. The objective of the remediation plan was to establish 'equivalent land capability' in a manner that does not result in 'release of a substance that causes an adverse effect' as is defined by the Environmental Protection and Enhancement Act. These remedial objectives were met by implementing the following:

- excavating heavy-end hydrocarbon contaminated soils that could not be practically treated and disposing of these materials at an approved industrial landfill;
- excavating light-end hydrocarbon contaminated soils and treating these soils in an on-Site treatment area;
- managing salt impacted areas on-Site by constructing groundwater collection systems; and

- dismantling and removing facilities and buildings that were no longer required or useful and disposing of the demolition debris in the Grande Prairie Landfill.

The remediation was completed over 1 year, from approximately April to October, 2001. The work was completed in a co-operative manner, by the Joint Venture partners, Hazco and Komex.

3.2 Verification program

3.2.1 Soils

Excavation limits were verified by collecting representative discrete samples and analysing for purgeable hydrocarbons, extractable hydrocarbons and BTEX. Salinity parameters and metals were also determined for approximately 1 in every 4 verification samples. A minimum of 1 verification sample was collected for every 500 m^2 of excavation area, accounting for both the excavation walls and base, and additional samples were collected whenever results were inconclusive.

Treated soils were verified by collecting representative composite samples and analysing them for purgeable hydrocarbons, extractable hydrocarbons and BTEX. The minimum sampling frequency was one sample per 500 m^3 of treated soils which correlates to approximately 5 representative samples per 0.3 meter lift. Additional samples were collected if results were inconclusive. Salinity and metals analyses were also completed where the treated soils were potentially impacted by salts and metals, such as those located at the north end of the former wastewater lagoon (approximately 1 in every 5 verification samples). The verification analyses were assessed against the remediation criteria specified in Section 1.4.

All samples were sealed in air-tight jars, refrigerated, and delivered directly to the laboratory under standard chain-of-custody protocols. All standard laboratory QA/QC procedures were followed and reported with laboratory data sheets. Based on the consistency of the results that were obtained through the program and the inherent variability of hydrocarbon concentrations in soil samples, duplicate sampling and analyses were completed on only a few samples. Analytical work was completed by Maxxam Analytics and/or ETL Laboratories of Grande Prairie, Alberta.

3.2.2 Groundwater

Additional groundwater investigations and monitoring was completed to confirm the absence of groundwater contamination in areas of possible past impacts and within and/or down-gradient of the areas of remediation, as follows:
- in the northwest corner of the Site;
- along the east side of the Site where previous investigations indicated the potential for impact;
- in the area of the former UST in the southeast corner of the Site;
- in the area of the former wastewater pond and salt storage shed; and
- in the area of the former crude oil storage.

Analyses included full VOC scans, routine potability parameters and dissolved metals. Groundwater monitoring will also be continued in areas where active hydraulic controls are implemented as part of the remediation plan (in the northwest corner and in the area of the former salt storage shed).

4 Development

The land has been re-developed for industrial use in accordance with the layout shown in Figure 3. Based on existing sales agreements, approximately 75% of the lands have been sold or committed. The City of Grande Prairie has welcomed the efficient use of space within the its existing area of industrial development. The majority of purchasers and interested parties are existing businesses that are needing to expand or augment their existing facilities to meet the increasing demand for services in the area. Further, the Province has been able to sell and re-develop an unused facility in an efficient, reliable and responsible manner. The Joint Venture partners (Hazco and Komex) have been

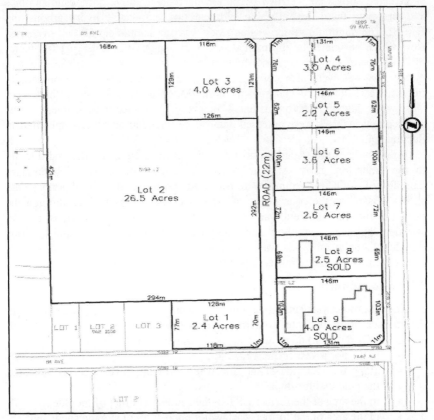

Figure 3: Plan showing proposed concept #12 of part of Plan 5102 LZ

able to work to their respective area of expertise, at the same time as they initiated a new area of business. In essence, all participants in the process were able to realise the beneficial principles of a Brownfields development.

Process of integrating screening and detailed risk-based modeling analyses to ensure consistent and scientifically defensible results

J. W. Buck, J. P. McDonald, & R. Y. Taira
Pacific Northwest National Laboratory, USA

Abstract

The United States Department of Energy manages several installations with waste sites having a potential to impact human health and the environment. The Hanford Site in south-central Washington state is one such installation. The Hanford Site has over 1,000 wastes site with 177 of them being underground storage tanks containing high-level radioactive waste. To support cleanup and closure of these tanks, modeling is performed to understand and predict potential impacts to human health and the environment. Pacific Northwest National Laboratory developed a screening tool for the United States Department of Energy, Office of River Protection that estimates the long-term human health risk, from a strategic planning perspective, posed by potential tank releases to the environment. This tool is being conditioned to more detailed model analyses to ensure consistency between studies and to provide scientific defensibility. Once the conditioning is complete, the system will be used to screen alternative cleanup and closure strategies. The integration of screening and detailed models provides consistent analyses, efficiencies in resources, and positive feedback between the various modeling groups. This approach of conditioning a screening methodology to more detailed analyses provides decision-makers with timely and defensible information and increases confidence in the results on the part of clients, regulators, and stakeholders.

Introduction

The U.S. Department of Energy (U.S. DOE) Hanford Site is a large brownfield site that has radioactive, hazardous, and mixed waste associated with over 40 years of Cold War nuclear material production. Besides having over a thousand waste sites, the Hanford Site has 177 high-level radioactive waste tanks that need to be cleaned up and closed. Many of these tanks are suspected of leaking to the environment and the integrity of many of the tanks is in question. A major focus is the groundwater contamination of past, current, and future activities. The Hanford Site has some complex groundwater problems and issues that are being addressed using environmental and risk assessment models. In many cases, the problems are so complex and large that overall strategies are needed to meet the clean up and closure goals of the site. For many site-specific problems, screening or prioritization level and detailed level models could be used together to understand the environmental and human health risk ramifications of site clean up alternatives.

This paper discusses a process of linking screening level modeling results to more detailed modeling results to allow decision-makers to develop strategies quickly and efficiently while maintaining a certain level of scientific defensibility. In the case of environmental clean up and site closure, the timeframe for developing options and strategies is much shorter than the time it takes to conduct a detailed modeling assessment. Often detailed environmental modeling studies take several months or more to conceptualize, collect appropriate data, and perform the modeling. Also, because of their focused nature, detailed environmental modeling efforts are slow to respond to changing requirements and scenarios.

To provide decision-makers with environmental and risk information in a timely and responsive manner, prioritization-level environmental modeling systems can be used. These models quickly provide decision-makers with results that can be integrated easily into strategic plans. The results of these prioritization-level models can be used to focus the use of the detailed models. These screening level models are generally analytical or semi-analytical in nature and require major simplifications of the environmental problem. Unfortunately, because the screening models are simplified, their scientific defensibility is often questioned.

To meet the strategic planning needs and technical defensibility requirements of these environmental problems and issues, a combination of screening and detailed modeling must be conducted. This paper discusses the process of integrating the screening and detailed modeling to provide decision-makers with understandable information that is timely, flexible, and scientifically defensible. An application of this process will also be presented to provide the reader with an understanding of its applicability.

Benefits of integration

If a waste site has a simple geology, source terms, and/or is a single medium problem, it may not be reasonable to do multiple types of analyses (i.e., screening-level and detailed modeling) or to integrate analyses. But at a large brownfield site like Hanford, the complexity of the source terms and their environmental transport into multiple media (air, surface and subsurface soil, groundwater, and surface water) require different levels of modeling to help identify the key strategies that will effectively reduce human health impacts. In many cases, the clean up of a large brownfield site takes years, involves many specific waste sites, and costs many millions of dollars. Under this scenario, multiple levels of modeling and their integration are not only reasonable, but essential to provide the best overall clean up.

Prioritization-level and detailed modeling should be interconnected to ensure consistency and efficiency of the overall analysis. Unfortunately, it is common to have different groups or contractors performing the different levels of modeling under various programs. In this scenario, there is little, if any, coordination between those conducting the screening and detailed modeling. Neither of the analyses benefits from the other and inconsistencies and inefficiencies routinely occur. Therefore, it is beneficial to coordinate and plan the integration of different levels of modeling to provide decision-makers with information that can be used strategically as well as information that can feed directly into site-specific decisions.

There are obvious cost efficiencies that occur if the different modeling levels are planned and integrated. A consistent conceptual site model is used for all modeling and data can be shared where appropriate (i.e., constituents of concern and their properties, source inventories, exposure scenarios and factors, and receptor locations and characteristics), which ensures a more beneficial result. If the different modeling teams share information and meet regularly on issues, a more cost effective, consistent, and scientifically defensible product can be developed.

Technical approach

Models are integrated through the process of conditioning. To condition a screening-level model to a detailed model, means to identify a set of parameter values and boundary conditions such that the computational results of the screening-level model match that of the detailed model as close as is needed and practical. Conditioning is different from benchmarking, in which the results of two or more models are compared and the differences are documented and explained. This is contrasted to calibration whereby parameter values and boundary conditions are determined such that computational results match field-measured values. Conditioning is similar to calibration in that a set of input parameters is varied until the screening-level model results match a set of target values. The difference is that the target values are obtained from a detailed model.

An exact match of results during conditioning is not expected, because it is very difficult for the results of one model to exactly match another. Various models can be based on different assumptions and use a variety of mathematical methods. These differences need to be understood in order to 1) carefully select input parameters that will compensate, as much as possible, for these differences, and 2) determine which input parameters will be varied to achieve a match. In addition, it is also important to understand how the detailed model was applied. In other words, the same conceptual site model used for the detailed modeling should be employed for the screening-level model.

With knowledge of the assumptions and methods used by the models, as well as the assumptions employed in the application of the detailed model, a base case can be established for the screening-level model. This case will serve as a starting point for the conditioning. Then, the previously identified conditioning parameters are varied to obtain the best match.

The main challenge is the assumptions and mathematical methods of the screening and detailed models can be so different that a good match cannot be obtained. For instance, detailed groundwater flow models can simulate aquifer heterogeneity and transient flow conditions, whereas screening-level models typically simulate only uniform flow. One way of handling this type of situation is to try and capture the important aspects using the screening-level model. Matching a peak concentration and the time to the peak, while ignoring other aspects of a breakthrough curve, may be sufficient to meet the decision-goals of the assessment.

Example application

The approach to integrating screening and detailed level models can vary depending on the type of environmental problem and the maturity of the analysis. In the case of the Hanford Site, groundwater modeling has been conducted at the site for several decades and a specific set of detailed models have been developed and maintained specifically for this site. These models have been peer reviewed extensively and site data have been collected for these specific models. These detailed models were used to develop the baseline clean up strategy. But because of funding changes, politics, and pressure from stakeholders, an accelerated clean up plan has been developed. To meet the changing needs of this new clean up plan, a screening-level modeling system has been developed to rapidly conduct analyses that can be used to determine the best path forward in meeting the new clean up schedule and budget (Lober et al. [1]).

Because a baseline analysis already exists at the Hanford Site using detailed models, the screening level models are being "conditioned" to the detailed modeling results and used to conduct "what if" analyses off the baseline. The conditioning process involves matching the results of the screening-level modeling to the results of the detailed modeling for the same waste site and conceptual site model. This process is covered in more detail in the next section. The conditioning of the screening-level model to detailed model results provides for a more scientifically defensible analysis.

The U.S. DOE's Strategic Analysis Project initiated development of an integrated risk-based strategy for the Office of River Protection (ORP). Simply, this strategy is to reduce the long-term health risk from Hanford's high-level tank waste as quickly and efficiently as possible while achieving interim safe harbor of the tank waste by year 2018 and permanent safe harbor by year 2028 (U.S. DOE [2]). The development of this strategy was driven by the need to protect human health and the environment. It was also driven by the objectives to 1) comply with federal and state regulations, 2) utilize the initial waste treatment and immobilization facilities, 3) maintain annual costs below $1 billion throughout the project life-cycle, and 4) complete waste retrieval, treatment and immobilization of all tank waste within the design life-times of the new facilities.

The Receptor Risk Model (RRM) was developed to assist in evaluating the potential human health risk for different receptor types (e.g., different activity patterns and land use) and locations (e.g., on-site, boundary, off-site, intruder, individual and population) associated with Hanford's high-level waste tanks and related sources. As ORP moves forward, one of the driving issues that needs to be addressed is risk to human health and environment from past, current and future operations. The RRM will address, at a screening level, key questions associated with ORP operation decisions, such as:

- What level of retrieval for each tank and/or tank farm is required to meet performance objectives?
- Considering risk, inventory, and waste volume, which tanks are the highest priority for clean up?
- What is the relative risk-based performance of the different retrieval technologies and options?

The RRM will support decisions that need to be made quickly, at a screening level, which is useful given the changing nature of clean up operations. This work, integrated with other efforts, can also help define and support a consistent risk methodology for activities associated with ORP and the Hanford Site.

Description of the Receptor Risk Model

The RRM is a macro-scale, risk analysis tool that can provide for a comprehensive risk analysis while meeting the schedule and resource requirements for the ORP mission. It is not meant to replace more detailed or site-specific analyses. The RRM produces areal risk contours, as well as constituent concentration contours, for specific tank farms, times, exposure scenarios, and release types. These contours are useful for displaying a large amount of data to decision makers and stakeholders. A key effort of this task is to use information and data from more detailed projects and programs related to the ORP mission to ensure consistency with these other efforts.

The RRM includes the following characteristics:

- It is a macro-level scoping tool for long-term human health risk.
- It is deterministic in nature for this analysis, with an option to be stochastic in the future.
- It builds on information and approaches used by previous and current projects related to the RPP.
- It tiers with more detailed project results as they become available.
- It is coordinated with tank and tank farm risk analyses being performed in the Single-Shell Tank Program.
- It provides an enhanced capability to plan alternative analyses.
- It is linked to schedule and budget models.
- It provides a choice of receptor types, locations, time of interest, and receptor activities.
- It is flexible enough to consider source terms and additional receptor locations beyond those required by the ORP.

The process of integrating a screening-level model with a detailed model is further explained in the following section. To strategically assess remedial alternatives for cleanup and closure of Hanford's high-level waste tanks, the RRM is conditioned to a previous fate and transport assessment.

Conditioning to a detailed model

The Tank Waste Remediation System Environmental Impact Statement (TWRS-EIS) is the generally accepted risk assessment performed on the tank farm complex at the Hanford Site (U.S. DOE and WDOE [3]). For the RRM to provide meaningful information to support strategic decision making regarding tank farm remedial options, it was conditioned to the modeling performed for the TWRS-EIS. The TWRS-EIS modeling was not calibrated, but it was designed to be conservative in nature.

The TWRS-EIS assessment simulated the release of contaminants from the waste tanks, transport through the vadose zone to the aquifer, and computation of concentrations in the aquifer for a human health risk assessment. Subsurface water flow and contaminant transport, in both the vadose and saturated zones, was simulated using the Variably Saturated Analysis Model in Two Dimensions (VAM2D). This is a 2-dimensional, finite element code that simulates both saturated and unsaturated flow and transport.

The 177 high-level waste tanks at Hanford are grouped into 18 separate tank farms. These farms were aggregated into eight source areas for the TWRS-EIS. Modeled constituent concentrations were available for the vadose zone/saturated zone interface beneath each source area and for a 2-dimensional grid in the aquifer. The RRM was conditioned to the vadose zone first, because the results of the vadose zone conditioning will be used as input to the aquifer modeling (at the time of publication, only the vadose zone conditioning was complete).

The first step of the vadose zone conditioning was to research the TWRS-EIS documentation to identify the pertinent differences between VAM2D and the MEPAS vadose zone module. The Multimedia Environmental Pollutant Assessment System (MEPAS) (Whelan et al. [4], McDonald and Gelston [5]) contains separate modules for source release, vadose zone transport, aquifer transport, riverine transport (not used in this assessment), and human health exposure, intake, and impact computations. The task of matching the results from a semi-analytical model (MEPAS vadose zone module) to the results of a 2-dimensional, finite element model (VAM2D) was made easier because VAM2D was applied in a 1-dimensional mode (i.e., 1-dimensional advection and

1-dimensional dispersion) to the vadose zone. The MEPAS vadose zone module is also a 1-dimensional advective, 1-dimensional dispersive model.

The pertinent difference between the two codes is the method used to describe the relationship between unsaturated hydraulic conductivity and moisture content. For the TWRS-EIS, the van Genuchten relations of hydraulic conductivity versus moisture content was used for the vadose zone flow modeling, while the MEPAS vadose zone module makes use of a power curve (Campbell [6]). The van Genuchten relations account for the hysteresis in the drying and wetting curves, whereas the power curve does not. Because of these two different methods, it was expected that moisture content and its effect on pore-water velocity, and thus travel times, would be the major difference in the vadose zone transport between the two models. The power curve makes use of a soil type coefficient, which was used as the conditioning parameter to match the vadose zone travel times computed by VAM2D. Due to the presence of numerical dispersion in VAM2D, which is not present in the MEPAS vadose zone module, longitudinal dispersivity was selected as a conditioning parameter to match the mass flux values entering the aquifer. The conditioning was performed for technetium-99, the constituent believed to be the primary constituent of concern for the tank farms. The TWRS-EIS no action alternative was selected for the initial conditioning.

Once the differences between the two models were understood, the TWRS-EIS documentation was examined to determine the conceptual site model used for the assessment and to obtain input parameter values used by VAM2D. Then, the modeling scenarios were recreated using the MEPAS vadose zone module. Initially, the module was set up with exactly the same input parameter values as used in VAM2D, except that the recommended MEPAS default value was used for the soil type coefficient (after Clapp and Hornberger [7]). A strong emphasis was placed on using TWRS-EIS vadose zone data directly, whenever possible, to ensure that both models were as close to solving the same problem as possible. The initial longitudinal dispersivity was the same as used for the TWRS-EIS.

Both the magnitude and time of the peak mass flux entering the aquifer served as conditioning targets, and the soil type coefficient and the longitudinal dispersivity were varied to achieve a best match. All eight source areas were conditioned simultaneously by minimizing the average error. For each run, the ratios of the time to peak and the peak mass flux to their target values were computed to evaluate the match. Then, the average ratios for all eight sources were computed. The objective was to minimize the error by iterating towards an average of 1.0 for the eight targets collectively. The default values for the soil type coefficients were all uniformly changed by a certain percentage until the average of the time to peak ratios equaled 1.0. To achieve this, the soil type coefficient values were reduced to 44.67% of their initial values. The same methodology was used to set the longitudinal dispersivity values. It was determined that to achieve an average ratio of 1.0 for the peak mass fluxes, the longitudinal dispersivity values needed to be increased to 140% of their initial values. The vadose zone conditioning results for two of the eight sources are given in Figure 1, showing the range in the quality of the matches achieved.

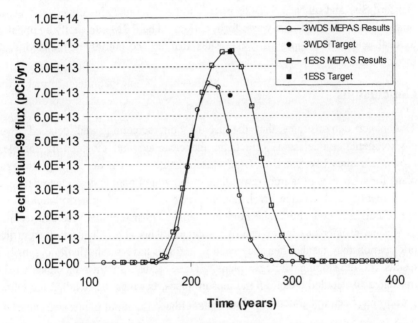

Figure 1: Simulated flux of technetium-99 from the vadose zone to the aquifer compared to the time and magnitude of the target peak flux for both the 3WDS and 1ESS source areas. These two sources demonstrate the range in the quality of the matches achieved.

Conditioning the vadose zone simulations to the TWRS-EIS no action alternative represents only the first step of this process. The vadose zone was further conditioned to a cap alternative, and conditioning of the aquifer simulations is planned. The aquifer conditioning will follow a similar process of identifying conditioning targets as well as the input parameters to be varied.

Screening of tank waste remedial alternatives

The RRM can be used as a stand-alone software package or integrated with other modeling software such as life cycle models that evaluate budgets and schedules associated with clean up operations. The RRM operates on an IBM or compatible personal computer and is very easy to operate. The knowledge required is the actual problem and alternative to be simulated. The RRM can be deployed via the Internet as a download or as a CD that will automatically install the software.

The RRM was originally designed to evaluate any problem or alternative that involves high-level waste tanks but it is planned to be used for other types (i.e., buried waste, landfills, trenches, and liquid disposals) of waste sites at the Hanford Site and other U.S. DOE installations. This includes covers and waste removal, as well as various waste form options. The RRM can easily be updated to deal with other types of waste sites that are commonly associated with brownfield sites.

Conclusions

This paper demonstrates that the use of both screening and detailed level environmental modeling is critical to the clean up and closures of complex brownfield sites like the U.S. DOE Hanford Site. When developing strategic plans for such sites, it is imperative that an integrated plan is developed between the screening and detailed level modeling efforts. The integrated planning and use of data and conceptual site models provides consistent analyses and results, as well as efficiencies in the use of resources. This gives decision-makers the information they need, when they need it, and still have scientifically defensible results for developing strategic plans. These plans can then be more easily translated to detailed clean up and closure plans, because the detailed modeling process has been integrated into the strategic plan.

References

[1] Lober, R. W., Yasek R. M., Morse, J. G. Buck, J. W., Bunn A. L., Henderson, C. C., Sams, T. L., and Vaughn, P. Integrating Risk Analyses and Tools at the U.S. Department of Energy Hanford Site. *Waste Management Symposium 2002*, Tucson Arizona, February 24 –28, 2002, in press.

[2] U.S. Department of Energy (U.S. DOE). *A Review of the Environmental Management Program*, U.S. Department of Energy: Washington D.C., February 4, 2002 (http://www.energyca.org/EMReview.pdf).

[3] U.S. Department of Energy (U.S. DOE) & Washington State Department of Ecology (WDOE). *Tank Waste Remediation System, Hanford Site,*

Richland, Washington, Final Environmental Impact Statement (DOE/EIS-0189), U.S. Department of Energy: Richland, Washington, 1996.

[4]   Whelan, G., Buck, J. W., Strenge, D. L., Droppo, J. G., Jr., & Hoopes, B. L. Overview of the Multimedia Environmental Pollutant Assessment System (MEPAS). *Hazardous Waste & Hazardous Materials*, **9(2)**, pp. 191-208, 1992.

[5]   McDonald, J. P., & Gelston, G. M.   Commentary on fate and exposure models: description of the Multimedia Environmental Pollutant Assessment System (MEPAS) version 3.2, with application to a hypothetical soil contamination scenario. *Journal of Soil Contamination*, **7(3)**, pp. 283-300, 1998.

[6]   Campbell, G. S. A simple method for determining unsaturated conductivity from moisture retention data. *Soil Science*, **117**, pp. 311-314, 1974.

[7]   Clapp, R. B., & Hornberger, G. M.   Empirical equations for some soil hydraulic properties. *Water Resources Research*, **14(4)**, pp. 601-604, 1978.

The contribution of the reuse of historic naval and airforce sites to brownfields regeneration: filling in the 'white holes'

C. Clark
University of Portsmouth, UK

Abstract

Historic naval, military and airforce sites are a specialised subset of brown land, absent from civilian maps. Once the focus of national history, they are of particular importance to the future of local communities, which may have been dominated by the ebb and flow of war and peace for centuries. Regeneration of local economies may depend on new land uses on them. Systems of disposal of government owned assets: at minimum price, low cost, or market value, affect outcomes such as land uses, historic buildings reuse and employment, in positive and negative ways. Local communities may be partners in the reuse process, or excluded from decision-making.

Adaptation proposals also need to be measured against various conservation criteria and degrees of physical intervention. Regulatory regimes for the protection of historic sites and structures which vary from country to country may inhibit or enable their future survival. There are also particular values and meanings associated with historic sites of national defence which need to be respected in beneficial reuse. Twentieth century buildings are regarded as heritage in some countries, but not in others.

This paper draws on experience of reuse of historic naval, military and airforce sites in Europe and the US, examining the issues raised and the contribution of the reuse of large structures such as boatsheds, covered slips, and flying boat sheds to regeneration of local employment and economies deeply affected by defence cuts, and in search of a new future.

The challenge of reuse

'War shapes lives': proclaims the brochure launching the new Imperial War Museum North, in Salford UK. As Greg Ashworth points out in *War and the city* [1], war has also shaped the form and development of towns and cities. Now it's our turn to reshape redundant war dominated sites, many of them in urban areas, into new, hopefully peaceful purposes.

This paper, appropriately enough, given the proximity of San Fernando naval base, focuses on the future of historic military, naval and air force sites. It addresses several aspects of reuse of brown land. These specialised sites' symbolic importance to national identity, their accumulation of important buildings and structures, their protected legal status, degrees of contamination and delapidation, the systems of disposal of state assets to which they are subject, and security measures which mean that they are often unknown to their host communities, make their transition to civilian uses difficult.

Defence rundown and site redundancy

The importance of conventional defence sites, structures and weapons in western Europe has continually diminished since the end of the second world war, arguably because the development of nuclear weapons held the balance of the Cold War; as a result of increasing European integration, and in response to worldwide political change. The disintegration of the Soviet empire led to a breaking up of post-war power structures and the beginning of new alignments. The new military necessity for rapid flexible response requires quite different physical resources from the massive fixed positions of the Cold War, rendering huge tracts of land and buildings, some of them historic, redundant and looking for new uses. According to the United Nations there is an ongoing worldwide reduction in defence spending in relation to gross domestic product; but unfortunately, Isaiah's prophecy has yet to be fulfilled.

The scale of military landholdings is in many countries, huge. In the UK, despite recent selloffs, after the Forestry Commission, the Ministry of Defence has the second largest estate in single ownership (about 1% of the total UK land area), valued at £10.5 bn: over 1% of the UK land area. It owns approximately 226,000 ha and leases a further 15,000 ha of land (MOD 1996a p 56). The majority (66%) is used for training. Other uses include airfields (11.6%), research establishments (9.1%), barracks and camps (4.5%), storage and supply depots (4.5%) and telecommunications stations (2.5%).

Disposal of defence land has rapidly gathered pace: the Bonn International Centre for Conversion estimate that up to 4,733 hectares - more than 8,000 sites - may have been transferred in the period from 1990 to 2000 [2]. The scale of redundancy in the USA and Germany has been far greater, and much academic study has been made of the problem. Another indicator of scale is the money raised from sales. In the US there have been five rounds of base closures - in

1988, 1991, 1993 and 1995 - which even taking into account the costs of environmental cleanups have saved the military about \$16.7 billion already. They are expected to generate more than \$6 billion a year in future savings [3]. These estimates by the General Accounting Office do not take into account \$3.5 billion in anticipated environmental costs or \$1.5 billion still to be spent to help communities.

Redundancy, consolidation and the need to maximise capital receipts from disposals have produced further pressures. Worldwide political change resulting in large reductions in military operations have combined with UK policy initiatives to lead to the closure of many British defence sites. Two factors make this defence reduction unprecedented: its suddenness and the fact that industrial decline is occurring in high technology, high value industries - not those that have traditionally been in decline. Losses of skilled employment are seriously damaging to local economies, and sites long occupied for defence may have accumulated structures which are difficult to adapt.

The historic defence legacy

What makes historic defence sites a special case in brown land redevelopment is their diversity of architecture and engineering. 'The beauty of utility' which historic military and naval buildings exemplify is at once expressive of state power and taste, and a particular challenge to appropriate reuse in physical and economic terms. As Sir Christopher Wren said: "Architecture has its political use: public buildings being the ornament of a country; it establishes a Nation; draws people and commerce; makes people love their native country."

Defence landholding's magnitude, complexity, and the changing requirements of the armed forces for defence estate property are identified as key factors by the British DETR [4]. As far as heritage and ecological value is concerned, the UK defence estate contains over 700 listed buildings and scheduled ancient monuments. In England there are about 160 Sites of Special Scientific Interest and about 40 sites with a significant ecological interest [5]. The MOD's land holdings have been described in the House of Commons Defence Report [6] as "an asset of national significance". Pride in the Swedish navy's history is reflected in the excellent state of repair of the early buildings on Karlskrona naval base.

The closure of military and naval activity and disposal of sites long dedicated - often for centuries - to national defence may be traumatic in physical and social terms for this specialised subset of redundant MOD sites: 'defence heritage'. The problem has an interesting relationship to the socio-political context in which it takes place - as part of the wider challenge to redevelop brown lands - with a particular set of constraints imposed by historic buildings and their legal protection from damaging change.

These sites' redundancy and the search for new uses, raise in acute form questions of value and meaning in the relations between people and place, since they are so often bound up with national identity and an imperial or colonialist past, making the task of finding appropriate new uses - filling in the 'white holes' on maps - both difficult and symbolic of the national search for a post industrial future.

The traditional public exclusion in the name of national security - often behind high walls - is a key barrier to positive local participation in disposals. The special status of military sites from civil law is symbolised by their absence from maps. Paradoxically, although the UK Board of Ordnance set up the Ordnance Survey to produce accurate maps for military purposes, defence sites are blanks on civil maps, containing no detail except for their boundaries. After this long exclusion from local consciousness except for those who worked there, these areas' sudden coming into focus - now that they are deemed surplus to national requirements - can be traumatic, quite unlike and disruptive of the normal property processes of supply and demand and of planning procedures. The closure of military bases clearly has major economic and land use implications, particularly in parts of the country where there has been a historic concentration of military activity. These implications are complicated by the spatial and physical diversity of the estate [5]. These sites offer potential to attempts to bring a peace dividend to once defence-dependent regions; but their release is taking place in the context of complex change in economic, social and planning systems which may not deliver this laudable aim.

Systems of disposal of state assets

Where redundant historic defence sites are sold or disposed of to new users, the terms - at full market value or free gift, time limited, with special restrictions for historic sites - vary enormously from country to country, and so does the weight given to local needs and aspirations in post-defence reconstruction. There are considerable contrasts between different countries' ways of disposing of their redundant defence sites, with vastly different knock-on effects for the value placed on the historic legacy. The commercial sale of the Royal Naval College at Greenwich provoked national controversy, to the extent that the sale was halted and an independent trust set up to find appropriate public uses, particularly education (the University of Greenwich and Trinity College of Music). Much depends on the legal status of state property as far as sale and control of change are concerned. In Norway, the Storting, the legislative body decided to retain certain historic military sites in the hands of the armed forces. In Britain and Norway the Crown exemption
excludes all physical change by the armed forces from control by local authority planning authorities, even after sites have become redundant, although both countries now treat defence heritage as if it were protected in
the same way as civilian monuments. In contrast Danish conservation laws apply to state property, and in Sweden, former naval buildings are protected by the director-general of national antiquities [7]

There may be a cultural difference between northern and southern Europe. Some countries such as Italy take the view that defence heritage sites are national cultural property, which is inalienable under a law of 1909 [8] it may only be leased to new occupants, not sold. The eighteenth century Spanish dockyard in San Fernando near Cádiz is preserved as a 'shrine' to the Spanish navy, but, however, redundant defence property can be sold, and the Spanish Ministry of Defence has a disposal programme under way to try and bolster funding for the defence budget. Defence sites may also be sold in Denmark, the UK and Germany. In Sweden and the UK there was a further disposal round in 1999, including barracks. In Denmark, a prescribed procedure is used for sales, including public advertising and two rounds of public competitive bidding. The German National Estates Department aims to sell its estates on the open market, but for certain social or cultural projects, prices are reduced. The Swedish system allows the local community to gain financially from changes in land use. Local authorities may purchase sites, fix new land uses with public consultation, and sell on to developers, gaining the profit. The British system, dominated by the Treasury, is that land is sold to the highest bidder at maximum planning value within three years, with slight modifications on price and timescale for historic sites. The French system allows for redundant naval sites to be transferred - as in most countries to other government departments. They may also be sold to particular individuals or to 'collectivités territoriales', or they may be given free to 'associations' - partnerships, non-profit associations and, community groups. The timescale under which this is done is not yet established. This system appears to be more locally community friendly than the UK's. The priority in France is local reconstruction, which may be achieved by a state/local government partnership. No European country appears to mirror the American system, where disposals are locally rather than centrally controlled, and sites may be gifted free to local organisations such as hospitals, schools, and the homeless [7].

As these examples indicate, the terms on which state property is disposed of vary considerably, and are bound to have an effect on outcomes, a point expanded later. Which agency is responsible for disposal and redevelopment also affects the outcome. In some countries it is the Ministry of Defence itself. The British Royal Town Planning Institute says that there is no justification for the Defence Estates Organisation remaining responsible for disposal of redundant sites, but despite recent research, the system is unlikely to change. Perhaps in response to criticism of earlier sales, the MOD has retained Devonport South Yard in Plymouth to develop it with DML as an industrial park with maritime and storage uses which take advantage of the high degree
of security. Historic structures may be vulnerable to demolition, since Crown Exemption in the UK allows the state to demolish its own historic buildings. In France the Bureau Affaires Culturelles within the Secrécrétariat Générale de l'Administration of the defence ministry is responsible for the built heritage. La Mission pour la réalisation des Actifs Immobilier de la Défense is responsible for defence estates. Unlike the UK, redundant sites in France are sold by a separate

agency: the Association pour le Développment et la Diffusion de l'Information Militaire (ADDIM).

Despite the British government's efforts to educate naval building budget holders to look after the historic buildings in their care and the annual report to ministers on defence heritage, the British Treasury's strong financial incentive to realise the highest sale price is probably the basis of the MOD's unwillingness to take an interest in military sites' future except to get the highest price within three years, while retaining its top down power over disposals contrasts sharply with the remit of the Office of Economic Adjustment in the United States, which has an orderly, locally dominated disposal procedure, offering sites free to local interests such as the homeless and education, and elsewhere. The Swedish model is a half way house, in that before disposal, plans for the buildings - at regional, outline and detailed level are proposed and confirmed by the local commune which has a monopoly say on future land uses. All sites worth over a certain value must first be offered to them. If they wish to buy, they have priority - at the price the highest bidder has offered. They gain the profit from the new uses which they arrive at via extensive public consultation. This offers local recompense to communities often dominated for centuries by defence [7].

Several countries including Sweden, Malta, and the UK offer redundant state property to other government departments. Examples of the state retaining and converting dockyard buildings are not hard to find. Another national museum in prominently sited dockyard buildings is Amsterdam's Lands Zeemagazijn built in 1613 as a storehouse and arsenal for the Admiralty's ships which has housed the Nederlandsch Historisch Scheepvaart Museum since 1968. Barcelona's grand waterfront storehouses provide accommodation for a social security office, museums, hotel and welcome waterfront bars and restaurants.

Skeppsholmen (ships' islet) in Stockholm close to the centre of the city offers a state controlled and financed model of dockyard regeneration. The ropewalk, arsenal, storehouses, barracks, church, drill halls and workshops have become a lively educational and arts campus [9]. Another such publicly funded conversion is in Turku in Finland, where a twentieth century boatshed has been ̦elegantly transformed into a conservatoire, the ropery partitioned as practice rooms. In Suomenlinna a chandler's building in Helsinki's island fortress has been 'reinterpreted with robust sympathy'.   There are several museums on Suomenlinna, one dedicated to designer, Eherensvård.

State subsidy to former defence-dependent regions

Governments' responses to defence contraction and related state land redundancy vary considerably. Some countries' disposal systems acknowledge the importance to local interests of what happens on these sites post-defence. The scale of redundancy in the USA has been far greater, yet the orderly American system of disposals puts locals in the driving seat: the homeless are identified as a priority group; schools, hospitals and housing are offered free to meet local needs. There are special guidelines for community preservationists on base closures [10].

The International Urban Development Association examined the problem of the Redevelopment of Closed Military Facilities in a seminar in Bremen in 1994 with support from the EU Network Demilitarised. Christopher Thorne of Wiltshire County Council, which had over 50 military sites, said that the scale of the decline in the defence industry was of comparable significance to those areas affected as the earlier structural economic changes of coal and steel, but with significant differences: it is an act, not of market forces, but of public policy; it is a recognition of a safer world;the cuts tackle head on the power of the military/industrial establishment; many of the areas affected are those within the industrial and technological heartlands bringing major redundancies and dereliction to places outside the traditional assisted areas framework; most of those workers are highly skilled, specialist administrators or military personnel very difficult to redeploy, with motivations removed from the commercial world [11].

Another contrast that local authorities find hard to accept is the lack of support their areas receive compared with the comparatively generous state aid given to regenerate areas devastated by the loss of steelmaking, coal or shipbuilding industries. Here it is governments that dealt the blow, but they do not offer compensation. The UK Ministry of Defence is not funded for economic reconstruction and offers no direct subsidies for reconstruction [3] 'defending the country is a big enough role'. As owners of by far the largest public landholding in the UK the MOD's unwillingness to take an interest in military sites' future while retaining its top down power over disposals contrasts sharply with the remit of the Office of Economic Adjustment in the US. The only reference to social objectives in the UK's interim *Guidelines for the Joint Working of Ministry of Defence and Regional Government Offices on the Disposal of Surplus MOD Property* [12] is the recommendation of "a mix of uses.. particularly on larger sites, as this can help to market sites and meet the needs of communities". Timing is critical for local authorities, yet releases do not tie in neatly with the local plan led system [13].

The protests in Britain about operation of the present disposal system by town planners, MPs and other environmental bodies show on the contrary that they expect a greater return to the public interest and greater accountability to local communities devastated by defence closures [14]. The recent transfer of sites such as Royal William Victualling Yard, Plymouth to the South West Regional Development Agency may not deliver this either. Controversy arises in other countries too. In Las Palmas, the local council wanted to take over the whole naval base without payment and evict the navy to a distant part of the port [15]. In Holmen, Copenhagen, the stipulation that the expenses of moving the fleet's facilities should be covered by sale of buildings forced the Defence Building Agency to give the highest bidder priority instead of 'employing a more idealistic consideration to the character and composition of the coming activities and usages, which would give the area the greatest possible value for the city as a whole. Whether this requirement has given an unfavorable result, as opposed to what could have been achieved if sales prices were not a determining factor, is a subject for great discussion' [16].

The UK system forces all parties into a financial and administrative straitjacket [17]. For historic sites, the MOD may accept a slightly lower price [18], but to facilitate a regeneration use much lower or nil value may be needed - which would be cross-subsidy across government, which is prohibited. As a result, formerly defence-dependent local communities in the UK lose out doubly, and local expectations are of necessity focused on new land uses on sites released by the rundown of defence activity. Developers paying a high price clear existing buildings for luxury residential use and leisure/entertainment. Examples of affected communities are the communities of Portsea between Portsmouth Dockyard and Gunwharf, and Chatham and Gillingham after closure.

The US system's strong emphasis on the importance of creating a public/private sector partnership between the key interests involved and in particular local communities [19] is acknowledged in research on the British system, but the disposal of sites in this way is not considered relevant to the UK by the DETR's researchers, 'because it relies on a different set of conditions' [4]. These are not spelt out. Presumably they relate to the contrasted administrative systems: in the UK: top-down, and in the US and Sweden: bottom-up, with local autonomy, as well as to both the past and present governments' high expectations that surplus publicly owned sites will finance other state functions - currently health and education. Since May 1 1997, UK government departments have been given new incentives to recycle receipts from asset sales into new assets, and they have recently adopted a policy of 'priority to the sale of high value assets' [20]. According to Woodward [2] a cultural change is needed within the British MOD, or else the massive investment of public money into defence sites is in danger of being squandered through its reluctance to see beyond its own immediate remit of national defence. Analysis of recent experience in regeneration projects demonstrates that unless public expectations for them

are clearly understood and taken into account by both government planners and private developers, public support will be weak, opposition may arise and projects fail to meet their economic objectives. If a less socially progressive model for the UK were sought, Sweden offers a halfway house: local authorities are offered sites at current value, to which they can then add planning value in full consultation with local communities. They then resell, keeping the profit gained - but, as said, such models were not considered by the researchers for the British Department of the Environment, Transport and the Regions [5].

As said, UK academic analyses of military site redevelopment are rare. In one study [21] sites were selected because they offered three different models of decision-making. Community and conservationist groups had severely limited influence on the redevelopment of Gunwharf, Portsmouth where because of the high sale price, nearly all the buildings were cleared. In Plymouth one community (Mount Wise) was able to influence the redevelopment process and the other (Stonehouse) was unable to do so. The Royal Gunpowder Works, Waltham Abbey, a particularly contaminated and historic site, experienced a 'successful' transition [22]. There, the Ministry of Defence ensured that a wide-ranging group of participants were involved in a collaborative planning process, and the outcomes: enabling housing development, a museum and nature reserve are widely acceptable. These case studies were analysed by relating key factors together: location/commercial value, the range of alternative uses, the level of heritage importance and degrees of public involvement. Location and degrees of heritage value were shown to impact on the potential range of uses. There is also a clear relationship between high degrees of public involvement and low levels of conflict - and its converse, also apparent in the outcomes. Three factors seem to be important to successful transitions: strong proactive local government, which defines what is expected of developers via well drafted planning briefs, which are responsive to local aspirations and concerns about future uses of these specialised sites, matched by similar characteristics in those disposing of state assets at national level, and creative local input.

The British Under Secretary of State for Defence says "not all our cultural assets will be capable of operational defence use. The challenge here is to engage in a pro-active way with our stakeholders to secure the investments and partnerships necessary for important sites" [23]. Further research would offer benefits to policymakers. The need to restore with respect what are architecturally outstanding urban environments poses major challenges of aesthetics, management, funding and values. Strategies need to be developed to ensure that the new uses effectively balance economic rejuvenation and heritage conservation. This transition stage is just one in the organic growth and development in the life of very special places. Despite their shared history as centres of war and defence, these sites have in recent years followed different paths towards different futures, which involve greater or lesser degrees of demolition or building reuse. It is clear that successful regeneration acknowledging their specialised geography and history can only be achieved by longterm effort and purposeful interaction between many local and national agencies, including those controlling degrees of physical intervention, and local people. The key challenge is how the disposal of state assets for urban regeneration is to be managed in a way that ensures a responsive relationship between state asset disposal procedures, acknowledgment of the importance of responsible local government, and active local communities, to ensure the most widely acceptable outcomes.

References

[1]   Ashworth GJ War and the City Routledge London 1991
[2]   Woodward R Military Base Conversion and Rural Development ESRC Application Centre for Rural Economy, University of Newcastle-upon-Tyne 1998

[3]   Skorneck C 'Base closings already saved $16.7B' Arizona Daily Star April 6 2002 p.A7

[4]   Department of the Environment, Transport and the Regions Guidelines for the Joint Working of Ministry of Defence and Regional Government Offices on the Disposal of Surplus MOD Property DETR London p.4 1997

[5]   Doak J 'Planning for the Reuse of Redundant Defence Estate: Disposal Processes, Policy Frameworks and Development Impacts' Planning Practice and Research Vol. 14 No. 2 pp. 211-224 Taylor and Francis Ltd. 1999

[6]   House of Commons Defence Committee (1994) The Defence Estate Vol. I - Main Report London HMSO 1994

[7]   Clark C Vintage Ports or Deserted Dockyards: differing futures for naval heritage across Europe University of the West of England Working Paper 57 116pp July 2000

[8]   Fera C and Arena M 'National Policies towards Heritage - Italy' in Ashworth and Howard 1999 European Heritage Planning and Management Intellect Books Exeter 1999

[9]   Clark C & Pinder D 'Naval heritage and the revitalisation challenge: lessons from the Venetian Arsenale' Ocean and Coastal Zone Management Volume 42 Numbers 10-11 pp.933-956 1999

[10]   Johnson E G and Rhoad D Base Closures and Historic Preservation: a Guide for Community Preservationists National Trust for Historic Preservation, Washington 1997

[11]   Thorne C 'The closure and re-use of military bases' Wiltshire County Council, England Re- Development of Closed Military Facilities INTA Press 1994

[12]   Department of the Environment, Transport and the Regions Guidelines for the Joint Working of Ministry of Defence and Regional Government Offices on the Disposal of Surplus MOD Property HMSO London 1997

[13]   Combes J Head of Defence Lands Service, Ministry of Defence, London 'Re-Development of Closed Military Facilities MOD Disposals Policy and Practice' INTA Seminar, Bremen 1994

[14]   Ingham Councillor J Leader, Plymouth City Council Letter to C Clark 16 September 1994

[15]   Rose D RTPI Letter to Fuller Peiser: Development of the Redundant Defence Estate Research 28 April 1998

[16]   Kvorning J arkitektur DK 4 1998 pp. 177-209 1998

[17]   Clark C 'Disappointing Defence Disposals - the Peace Deficit to continue?' RICS Building Conservation Journal 2000

[18]   HM Treasury Guidance on Property Disposal Procedures DAO(GEN) 13/92 Treasury London

[19]   Department of Defense US DOD Guidance on Establishing Base Realignment   and   Closure   Cleanup   Teams www.dtic.mil/envirodod/brac/bct.html 1999

[20]   Burgess J Royal Gunpowder Mills Waltham Abbey A Report of the Planning Forum University College, London 1992

[21]   Clark C 'White holes: decision-making in disposal of Ministry of Defence heritage sites' PhD Thesis University of Portsmouth 2002

[22]   CIVIX Waltham Abbey Royal Gunpowder Mills From Closure to Disclosure The beneficial re- use of one of Europe's most important monuments London September 1995; Keeping M & Comerford J 'How to dispose of a 'problem' site - conferring allowed' Town and Country Planning May/June ppl148-149 1995

[23]   Moonie Dr. L Annual Report Cultural Heritage Across the Defence Estate Defence Estates Solihull 2001

Unusual aspects of Arsenic distribution, in the area of the industrial district of Scarlino (Tuscany): problems connected with sulfuric acid production from pyrite

M.P. Picchi, L. Fugaro & A. Donati
Department of Chemical and Biosystem Science, University of Siena, Italy

Abstract

The problem of Arsenic pollution, in the area of an industrial plant for the production of sulfuric acid from pyrite combustion, was studied. The plant is located in a former wetland, close to the sea, in the county of Scarlino, southwest of Tuscany.

The main Arsenic source was suspected to be the enormous stocks (1500000 tons) of hematite dust, residue of the pyrite combustion, and the stock of fine particles of pyrite, that were disposed on the soil, close to the plant, without control.

The presence of both continuous phreatic and artesian aquifers in the contaminated area, enhance the health risk for the population resident in the district.

Analytical data obtained at different depth were collected for Arsenic and other metals and compared with the historical records of the area.

Core samples collected in the area, showed a non-regular distribution of Arsenic, regarding both the depth and the distance from the plant.

Concerning the vertical distribution, the clay-rich soil, shows that Arsenic concentration reach the maximum peak at different depth from surface, with variable trends.

Horizontal distribution, shows a quite regular decreasing in the As concentration going far from the plant, but with the presence of hot spot at the distance of 4-6 km from the industry, where the As level were found very high (As>600mg/kg).

In order to understand this behavior, an analysis of chemical data for As and other toxicologically important metals, was proposed.

1 Introduction

Arsenic is a naturally occurring element, with an average abundance that ranges from 1 to 10 mg/kg (dried soil) in the earth crust[1,2] ranking as the 20th abundant element. It is naturally associated with igneous and sedimentary rocks, particularly with sulfidic ores. In a growing number of cases the concentration of this metalloid has been found much higher than average all around the world, and most of the time these anomalies were due to anthropogenic activities. On the other hand, natural phenomena such as weathering, biological activity and volcanic activity, are also responsible for the emission of arsenic into environment, and its redistribution on hearth's surface. The presence of As in the environment at high levels combined with its well-known toxicity has generated a large public concern and a widespread interest in the scientific community and government agencies for health and environment. Beside its acute toxicity, that is observable at relatively high concentration, recent researches have demonstrated that chronic exposure to low concentration of As, increase significantly the risk of skin and internal cancer.

In fact, despite its relative rarity, Arsenic appears in the first position in the USA CERCLA (Comprehensive Environmental Response, Compensation, and Liability Act) National Priorities List (NPL), compiled by ATSDR (Agency for Toxic Substance and Disease Registry) and EPA, in which are posed the most significant potential threat to human health due to their known or suspected toxicity and potential for human exposure.

There is a general agreement that the most anthropogenic input is due to industrial operation (smelting, pesticide production and use, pyrite ores processing, dumping of hazardous substance and fossil fuels combustion) even if still debated is the extent to which man's activities contribute to the overall arsenic cycle. In many cases soil and sediments contamination results the first evidence of the occurred Arsenic pollution.

In order to understand the diffusion routes of Arsenic in the environment and its biogeochemical cycle, a large number publication on its chemistry in soil and ground water have been produced. Recently it became clear that mobility, toxicity and bioavailability of Arsenic is strictly depending on its chemical form or "species" and that the chemistry of this metalloid is strongly affected by the redox condition of the medium in which it is present. In particular it is known that reduced inorganic Arsenic, arsenite (As+3) is more toxic and more mobile than Arsenate species (As+5), and that Arseno-organic compound are the less toxic (usually are produced as detoxifying species by living organisms)[3,4,5].

It is well known that wetlands are now considered fundamental for an incredible number of living organisms, and that these delicate ecosystems need a correct management. On the other side, in the past, all around the world, wetlands were mostly seen as areas sequestered to the human productive use, and have been used as repositories for sewage sludge, industrial wastes and mine wastes[6].

The aim of the present project is to start the study of a site, the Scarlino Plan, located in the south west of Tuscany, in which the soil presents high level of

Arsenic. In this area, a partially filled wetland, close to the sea, in the 1962 was installed a powerful plant for sulfuric acid production from pyrite combustion, using the ores found in the mine-field (Colline Metallifere), far about 15 km west from this site.

This site presents some interesting features that could help to understand the general mechanisms of Arsenic diffusion. Consequently, this could help to discern between different hypothesis about high Arsenic levels in the Scarlino plan, which potentially has been directly and/or indirectly impacted, by pyrite processing wastes and/or by the long distance transportation of As-rich sediments from the minefield, respectively.

The re-use of these polluted area, both for industrial and residential scope is considered as a priority for the local administration. Consequently, understanding and solving the problem of the wide Arsenic dispersion is fundamental for a correct and safe remediation.

2 History of the site

In Figure 1 is reported the map of the area, from which is possible to observe the position of the industry in respect to the other natural and anthropic structure of the Scarlino plan.

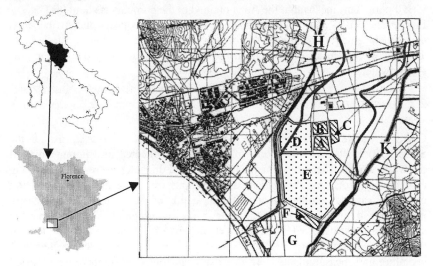

Figure 1: Topographic map of the Scarlino plan and its relation with Tuscany and Italy. Letters show the main, natural and anthropic, structures of the area: A) Pyrite cinders waste stock (hematite); B,C) Fine pyrite particle and Pyrite enrichment sludge stocks; D) Sulfuric Acid plant area; E) Titanium dioxide plant area (Huntsmann-Tioxide); F) Former pyrite cinders waste stock; G) residual wetland. The two watercourse are the Pecora river (H) and the Canale Allacciante (K). The town of Follonica is visible on the left. The medieval town of Scarlino is outside the map. The area comprises between the red curves presents average As concentration > 100 ppm.

It is important to note that the sulfuric acid plant have been located in a former wetland. This wetland was submitted to a gradual and partial filling starting from 1830. A residual pond is still present in the terminal part of the area, close to the sea. The engineering project for filling was completed with the construction of a channel, called "Canale Allacciante", that was collecting the water of creeks coming from south and the partial canalization of the Pecora River coming from north. These two watercourses delimitate the area, which has been found polluted with arsenic. Letters A, B, and C show the storage areas for the sulfuric acid process wastes. Letter E shows a former waste storage that now is remedied. The whole area present a groundwater stream, that flows at variable dept, from 1 m under the surface just in the coincidence of the plant and the waste stocks, to some meters going far from the industrial site. This groundwater is resident in sand lenses that are present in the soil, which is mainly formed by clay mineral.

The main waste stock (A) is formed by about $2x10^6$ metric tons of hematite cinder, coming from pyrite combustion. This enormous amount of material has been disposed on the ground without any protection. Moreover, this mass subsides of at least 6 m under the previous ground level and the groundwater soaks it completely. From analysis performed by current owner of the plant, the Nuova Solmine Company, results that average As concentration in this material is 450 ppm. This means that the As contained in the cumulus is about 900 kg.

Fine Particle of Pyrite (FPP), contained in the milling dusts and in the ores-enrichment sludge, constitutes the other two stocks of residues wastes (B,C). These wastes were also disposed on the ground without any protection for the groundwater.

2 Experimental

Cores of native soil were collected in a 9 km^2 of country area, around the Sulfuric Acid Plant (SAP). Cores of waste were collected in different point of the waste stocks. The cores dimension, length and diameter, were 6 m and 10 cm respectively. Total As, Pb and Zn concentrations were determined on dried soil or waste samples, collected at different depth of the cores. Samples were homogenized and digested with nitric acid, by graphite furnace – atomic absorption spectroscopy (GF-AAS). Technicians of the chemical laboratory of the Nuova Solmine Company, which is the owner of the SAP, conducted both sampling and measurements[7,8]. In this paper, for our convenience, we reported the As concentration in ppm units, where ppm is corresponding to mg/kg of dried material.

3 Results and Discussion

3.1 Horizontal profile for As, Pb, and Zn

In Figures 2A and 2B are reported the minimum and maximum concentration of Arsenic, registered in the soil cores respectively, as a function of the distance from the plant. It is possible to observe that in both cases, on the average, As concentration was higher in the neighbourhood of the plant and lower far from the plant. The only exception regards a point located at about 3.5 km from the plant.

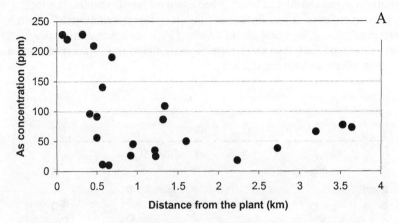

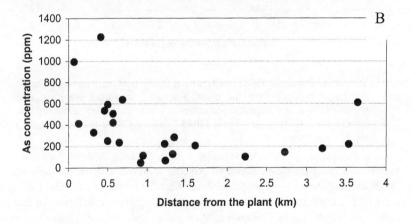

Figure 2: Two-dimensional plots obtained reporting the As concentration in respect to the actual distance from the plant. The abscissa values were reported regardless the geographical coordinates. A: maximum As level in soil cores; B) minimum As level in soil cores

This kind of behavior is compatible with a diffusion of the element from the Arsenic source that can be identified with the three production waste stocks located in close proximity of the plant. Three different diffusion routes can be hypothesized: 1) waterborne transportation of ionic forms of As by the groundwater; 2) airborne transportation of As_2O_3 containing cinders; 3) mechanical dispersion of hematite cinders for country roads sanding.

The first route is probably the most important. In fact, the freatic groundwater directly absorbs both the Hematite and FPP stocks at about 1.5 m of depth. Under anaerobic conditions, the As could become more soluble as arsenite and migrate. The migration stops when different redox and/or mineralogical conditions occur and As can be adsorbed again on the clay mineral surface.

This hypothesis is confirmed by the fact that groundwater, sampled in correspondence of the waste stocks contains high As concentration (up to 1200 ppb). Moreover, high As values were recently found in the water of wells that are about 400m far from the stocks.

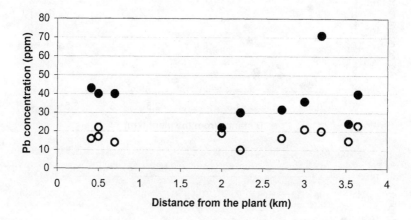

Figure 3: Two-dimensional plots obtained reporting the Pb concentration in respect to the actual distance from the plant. The abscissa values were reported regardless the geographical coordinates. Empty circles indicates the minimum Pb level in soil cores; Filled circles indicates the maximum Pb level in soil cores.

In order to compare the behavior of Arsenic with other inorganic pollutants, for the present discussion we will take in account the levels of Zinc and Lead. In the case of these last metals, the behavior appears completely different. In fact as it is shown in Figure 3, the Pb concentration in the soil, remains almost constant with the increasing distance from the plant. Also the Zinc concentration does not vary significantly, between points located in the proximity of the plant and points located far from the plant (Figure 4). This strongly suggests that the mobility of Arsenic is much higher in respect to other metals. These data confirm the findings of other authors in analogous cases.

As reported elsewhere by Marchettini et al.[9], these data also confirm the fact that the Arsenic anomaly was probably due only to the dispersion of the metalloid in the environmental matrix from the pollution sources, which are represented by the piles of waste coming from sulfuric acid production by pyrite ores.

Unfortunately from available data, the actual mechanisms for As pollution in the Scarlino plan are not completely understood, and a deeper investigation about different aspects of its diffusion is needed. In particular, a more accurate sampling of the soil should be conducted, in order to improve the statistical significance of the data.

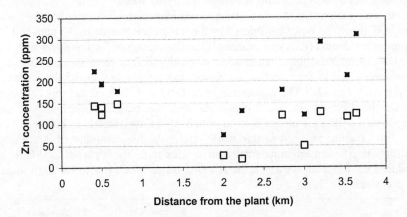

Figure 4: Two-dimensional plots obtained reporting the Zn concentration in respect to the actual distance from the plant. The abscissa values were reported regardless the geographical coordinates. Empty squares indicates the minimum Zn level in soil cores; Filled squares indicates the maximum Zn level in soil cores

3.2 Vertical profile of Arsenic, Zinc and Lead

The analysis of the vertical profile of concentrations, also, reveals a different trend for Arsenic in respect to Zinc and Lead.

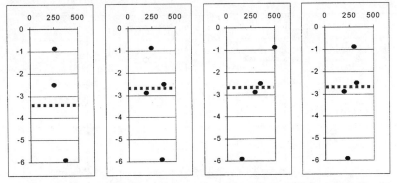

Figure 5: Depth profile for Arsenic. Cores were collected in four different points of the waste stock of fine pyrite particle. This kind of waste was derived from the milling plant. The red horizontal bar indicate the thickness of the waste stock in the sampling point. Below this line native soil was collected.

In Figure 5 the concentration of Arsenic at different depth is reported. It is possible to observe that the level of this metalloid does not decrease significantly below the limit of the waste pile. It is evident from the figure that high As values can be found also in the native soil under the pyrite waste.

It is also important to note that the values found for Arsenic in the native soil are well above the limits considered by the Italian law, which are 10 ppm and 50 ppm for residential (RD) and industrial (ID) destination of the site respectively.

On the contrary, the concentration of Lead and Zinc drop suddenly to very low values just at the interface waste/soil and remain low in the deeper layer (Figure 6A and 6B). The concentration of Lead and Zinc, in the soil collected under the FPP waste, was comparable with the natural background of the unpolluted soil. This means that migration of the two metals is blocked. The vertical profiles, reported in this work, reflect the higher ability of Arsenic to diffuse in different environmental matrix. This mobility is particularly high in reduced environment, in which the Arsenite form, which is also the most toxic, is prevalent.

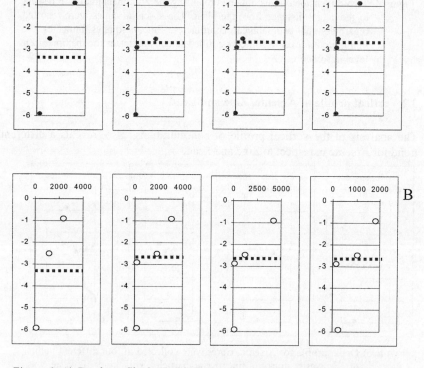

Figure 6: A) Depth profile for Lead. B) Depth profile for Zinc. Analysis was conducted on the same samples which were used for As measurements. The red horizontal bar indicate the thickness of the waste stock in the sampling point. Below this line native soil was collected.

In the case of Lead and Zinc the threshold concentration in the soil considered by the Italian law are the following: Pb, 100 ppm (RD) and 1000 ppm (ID); Zn, 150 ppm (RD) and 1500 (ID). In both cases, the actual concentration found in the soil under the waste area is well below the two limits.

These, analytical evidences, suggest that the remediation project of the site and the subsequent use that is considered for the site, must accurately take in account the different behavior of the various pollutant. For example, in the Scarlino case, the sole removal of the pyrite processing waste, could give a safe remediation for Zinc and Lead, but not for Arsenic, which migrated across wider areas.

Conclusion

Industrial area presenting high Arsenic and heavy metals concentration, presents some peculiar features that have to be accurately taken in account, for the remediation projects and for the subsequent re-use of the sites. In particular, the analysis of the data suggests that Arsenic migrates across wider areas with respect to the area of the waste site, and this could lead to only partial remediation of the sites. In these cases a risk analysis must be conducted, in order to evaluate the impact of residual pollutant concentration on the resident population and/or the environment.

From available data, the actual mechanisms for As pollution in the Scarlino plan cannot completely be understood, and a deeper investigation about different aspects of its diffusion is needed. In particular, a more accurate sampling of the soil should be conducted, in order to improve the statistical significance of the data. At the same time a higher number of measures for each core sample, together with As speciation and an accurate mineralogical study of the soil extracted, will better account for the basis of the vertical distribution of Arsenic. Moreover a microbiological study, in order to understand how the mobilization occurs, should be conducted.

References

[1] Cullen W.R. & Reimer K. J. Arsenic speciation in the environment. *Chem. Rev.* **89**, pp..713-764, 1989.
[2] Yan-Chu, H *Arsenic in the environment, Part I. Cycling and characterization*; ed. Nriagu, J., John Wiley & Sons: New York; pp. 17-49. 1994
[3] Ahmann, D.; Krumholz, L.R.; Lovley, D.R. & Morel, F.M.M. Microbial mobilization of Arsenic from sediments of the Aberjona watershed. *Environ. Sci. Technol.*, **31(10)**, pp. 2923-2930, 1997.
[4] Spliethoff H.M. & Hemond H.F., History of toxic metal discharge to surface waters of the Aberjona Watersheed. *Environ. Sci. Technol.* **30(1)**, pp. 121-128, 1996.
[5] Aurilio, A.C., Mason, R.P. & Hemond, H.F. Speciation and fate of Arsenic in three lakes of the Aberjona watershed. *Environ. Sci. Technol.* **28(4)**, pp. 577-585, 1994.

130   *Brownfield Sites: Assessment, Rehabilitation and Development*

[6] La Force, M.J., Hansel, C.M. & Fendorf S. Arsenic speciation, seasonal transformation and co-distribution with iron in a mine waste-influenced palustrine emergent wetland. *Environ. Sci. Technol.*, **34(18)**, pp. 1217-1226, 2000.
[7] Nuova Solmine s.p.a. report – *Environmental characterization and preliminary project for the remediation of the polluted site GR72 in the Scarlino plan.* 1997.
[8] Nuova Solmine s.p.a. report – *Environmental characterization and preliminary project for the remediation of the polluted site GR72 in the Scarlino plan.* 1999.
[9] Marchettini N., Donati A., Pulselli F., Bastianoni S. and Tiezzi E. High soil As level in the area of a plant for pyrite processing: natural anomaly vs pollution hypothesis *Proc. of the 17th World Congress on Soil Science.* In press.

Section 2:
Community and public involvement

Eco-industrial redevelopment of LA brownfields

C. S. Armstrong[1] & C. E. Tranby[2]
[1]*Policy, Planning & Development, University of Southern California, U.S.A.*
[2]*Environmental Affairs Department, City of Los Angeles, California, U.S.A.*

Abstract

This paper focuses on the use of Geographic Information Systems (GIS) within the City of Los Angles (LA) municipal government in the context of eco-industrial redevelopment of brownfield sites. LA, like many cities around the world, has implemented a Brownfields Program to address the problem of urban blight arising from environmental impairment. The program has now expanded to consider eco-industrial concepts. The increasing and dual interests of quality job generation through industrial development and community-focused environmental protection have initiated investigative efforts on eco-industrial redevelopment by the City's Brownfields Team. Although the definition and understanding of eco-industrial development varies, the City's efforts revolve around a broadly interpreted encouragement of networks among businesses to achieve maximum economic and environmental performance. This may include waste exchange, sustainable design, pollution prevention, service sharing, and life cycle analysis.

The "Industrial Symbiosis" park in Kalundborg, Denmark, while exemplary, does not provide a model directly relevant to LA, with its different mix of industries and scarcity of undeveloped large tracts of land. Instead, the Brownfields Program may adapt the traditional model to its dense urban fabric by facilitating waste and materials exchanges among existing businesses and using redeveloping brownfields to fill-in the gaps in potential eco-industrial relationships. This approach extends the eco-industrial concept to redevelopment by involving brownfields themselves as an integral component of the reclamation, recycling and reuse process—a scenario that is highly relevant to densifying urban settings with intermittent patterns of decline. The paper illustrates how demonstration GIS analyses on target areas will assess existing industrial conditions, facilitate optimal matching of businesses, and inform policy making for future sites.

Introduction

The use of GIS has become widespread within the LA municipal government and proves particularly helpful in facilitating early action on contaminated properties or brownfield sites because it enables many environmental and socioeconomic factors to be considered simultaneously. Thus, the City's dual interests in encouraging quality job generation through industrial development and community-focused environmental improvement may be addressed in concert using GIS-based analysis of brownfield sites. This strategy takes the form of an eco-industrial approach to redevelopment, which is consistent with efforts to encourage the growth of recycling markets—as the City has been designated a Recycling Market Development Zone by the California Integrated Waste Management Board (RMDZ 2002). The City's Brownfields Team, comprised of experts from a variety departments and agencies, has expressed enthusiasm for this approach: the Mayor's Office of Economic Development is evaluating candidate businesses for certain brownfields, the Community Redevelopment Agency (Rodino Associates 2002) has examined the potential for attracting environmental technology industries to LA, and the Environmental Affairs Department has encouraged sustainable infrastructure and amenities on brownfields developments.

LA, like many cities around the world, implemented its Brownfields Program in order to address the problem of urban blight arising from environmental impairment. Specifically, brownfields are "abandoned, idled, or under-used industrial and commercial facilities where expansion or redevelopment is complicated by real or perceived environmental contamination." (USEPA 2001) Brownfields often exist because landowners and potential developers lack sufficient information to quell their fears of stigma and liability associated with federal mandates pertaining to past examples of more severely contaminated sites.

This situation has resulted in governments taking action to return brownfields to productive use by providing early decision-making information on sites, incentives for cleanup, and liability protection. The eco-industrial approach transcends traditional redevelopment by allowing governments and industry to collaboratively dialogue about optimal strategies for promoting industrial growth that is compatible with more systematic environmental stewardship. Although the definition of eco-industrial development varies, the key principle behind it is: "an emphasis on fostering networks among businesses and communities to optimize resource use and reduce economic and environmental costs…including pollution prevention, byproduct exchange, green design, life cycle analysis, joint training programs, and public participation." (Schlarb 2001, iv) Eco-industrial development is derived from the field of industrial ecology (e.g., Ausubel 1997) which emphasizes a symbiotic understanding of industry and the natural environment; both of these are closely linked with the wider concept of sustainable development (Allenby 1992)—emphasizing a simultaneous focus on the environment, economy, and social equity in planning and policymaking.

Eco-industrial approaches benefit the redevelopment of brownfield sites because they not only encourage site cleanup, but a more sustainable stewardship of problem properties into the future—whereby tenants engage in more environmentally-sensitive business practices that minimize their impacts on the local environment. This can be done through waste reduction via reuse or recycling among businesses (e.g., one business' wastewater output is another business' input) or by pledging to use renewable energy or less-toxic production inputs, packaging, etc. Ultimately, eco-industrial development not only benefits the local environment, but it saves money in the long-run, while also providing jobs, training, and tax-base revenue—the typical benefits of traditional economic development. (Schlarb 2001) Necessarily, eco-industrial development includes a partnering of businesses and heightened social vigilance on the part of each participant business—since they are investing in the success of each other. The National Center for Eco-Industrial Development lists a variety of implemented eco-industrial park cases on its website.

The "Industrial Symbiosis" park in Kalundborg, Denmark, provides the classic example of a successful eco-industrial development. The park uses bilateral agreements among businesses that include a power plant that produces steam and heated water, which is used by a fish farm and pharmaceutical company within a web of similar exchanges. (Ibid., 19) In LA, the scenario is different because eco-industrial relationships must be forged among already-operating firms. Assessing existing industrial conditions surrounding brownfield sites is thus the first step in determining the feasibility of eco-industrial redevelopment. GIS facilitates the streamlining of necessary data and helps make it palatable to a wide variety of decision makers. Although the City has potentially thousands of brownfield properties, the Brownfields Program focuses on a limited number of these, largely due to a combination of community priorities, non-systematic reporting practices, and limited fiscal resources. Each of these factors directly influences the creation of an adequate GIS model to assess eco-industrial redevelopment potential.

Overview of GIS

As mentioned previously, the Brownfields Program is not a stand-alone agency within the City government; it is implemented by a Resource Team of staff from various agencies, including the Community Redevelopment Agency, the Mayor's Office of Economic Development, the Community Development Department, and the Environmental Affairs Department. While no single department has jurisdiction over the Brownfields Program, the Environmental Affairs Department (EAD) provides most of the Program's GIS capability. Brownfield sites are reported to the EAD primarily from City Council offices, the Mayor's office and various City departments that deal with problem properties. EAD staff records newly reported sites in a spreadsheet database linked to a GIS maintained in ArcView 3.2 software. Approximately fifty sites now receive some level of assistance in the City of LA Program, which has been in existence for about five years. (EAD 2001)

The City offers five categories of assistance to the selected properties, which range from the least intensive "on-call technical assistance" to "major demonstration" status involving the use of funds for issues such as cleanup and site purchase. Program funds come primarily from federal sources—including the U.S. Department of Housing and Urban Development and the U.S. Environmental Protection Agency (USEPA) Showcase Community designation. A portion of these funds has been allocated to information technology strategies including the development of a more efficient GIS to identify and monitor brownfield properties. The City's Bureau of Engineering (BOE) and Information Technology Agency (ITA) combine to provide parcel polygon shape-files (BOE) and other useful layers served through internal City networks. Though the contamination itself may deviate from parcel boundaries, the City relies on such boundaries to spatially define its brownfield sites due to the underpinning philosophy that brownfields solutions often take the form of complex real estate transactions.

Model Creation Process

Brownfield sites with the best eco-industrial potential consist of either a larger mostly vacant tract of land or an area with a number of blighted parcels that may be assembled into a coherent exchange network. The challenge in LA usually stems from the latter because of the City's dense and well-entrenched development patterns. Exceptions include the large vacant sites of the Cornfields and Mission Road River District, which currently are not available for eco-industrial development. Most brownfields, however, are small stand-alone parcels, such as abandoned gas stations, or vacant/abandoned sites interspersed among operating enterprises. Examples of the latter are Beverly-Virgil Gully near Hollywood, and the Wilmington Industrial Complex near the Port of LA, each with many parcels and zoning for manufacturing usage. The GIS model will focus upon these two sites.

In order to understand the eco-industrial potential of Beverly-Virgil and Wilmington, it is necessary to acquire data on the businesses within each area. A promising strategy for assessing eco-industrial potential is to develop a proxy on (1) the kinds of wastes generated by each business and (2) the kinds of wastes that may be a potential production input for each business. These data may be used to estimate potential waste exchange among businesses or to recruit new businesses that may be in the market for use of the aggregate waste generated by the businesses within the target area. The California Integrated Waste Management Board recently initiated waste exchange modeling which relies upon the input of locally-relevant data. (CIWMB 2002) The CIWMB presents a waste characterization model that allows localities to assess local wasteshed parameters by using proxy disposal rate codes correlated with business type and size (by number of employees). This proxy assumes the form of a waste-disposal-per-business-per-year statistic that is widely understood in waste management policy circles. (Tseng 2001) The waste disposal per business may be aggregated for wider areas and, subsequently, the data may be used to either (1) recruit new businesses that may use the waste in their production processes as "feedstock" or (2) develop new waste exchange relationships among existing businesses.

The CIWMB system is based upon survey work done to assess the waste inputs and outputs of a sample of California businesses. (Ibid.) This has enabled the development of a model that may be linked to business type as per the U.S. Department of Commerce's Standard Industrial Classification (SIC) code system. Local business data containing this SIC coding system are available from the American Business Institute (ABI) as well as the U.S. Small Business Administration. The proposed GIS model incorporates ABI data (as it is readily available to the City through contract) for all of LA, with the ultimate goal of implementing a waste characterization analysis for each of the Brownfield Program sites. Here, the Beverly-Virgil and Wilmington Industrial Park sites comprise the initial attempt at adapting the CIWMB model to a community scale within LA.

The ABI data were transmitted from the vendor in electronic spreadsheet format and then converted to spatial data in the form of a point-based shape-file compatible with the EAD's ArcView 3.2 software. The network of points enables consideration of key factors regarding potential eco-industrial relations among businesses: *proximity* (because of transport relations among inputs and outputs); *adjacency* (possibly more conducive for service-sharing); *containment* (whether any processes may be completed within the eco-park area in closed-loop use fashion); and *direction* (to determine compatibility with the wider ecosystem and transportation infrastructure).

Echoing Pequet's (1990) discussion of the stages in data abstraction, the target entity in this case is the piece of land, but it is abstracted through several levels—from the parcel—to the business contained on the parcel—to the SIC code business definition—to the waste behavior classification of that business based upon the SIC code. Depending upon the variation among the types and sizes of businesses and their waste characterization profiles within each of the two target areas, different themes and layers will need to be created to represent them most realistically.

Research Issue and Question: The targeted research question that model will address is: What feedstock exists in urban waste streams within and around the two selected brownfields? The motivation for this question stems from the City's desire to redevelop large brownfield sites where abandoned or underutilized parcels are interspersed with operational businesses; the intention is to redevelop certain parcels in an environmentally-sensitive manner—via reuse and recycling of local waste in their production processes—without disrupting the surrounding businesses. Characterizing waste types and amounts in a certain area provides policymakers with data-ammunition to target the appropriate industries for recruitment.

GIS—Limitations and Future Needs: Drawing a boundary around any area is problematic because functioning relationships could exist across those boundaries that are not captured. Recognizing this limitation, a number of obstacles exist in implementing the kind of GIS model proposed here. First, it is obviously a rough approximation of actual business production and consumption exchanges. Second, this kind of approach is relatively new and lacks a precedent of widespread success; it is therefore extremely exploratory in nature. Third, even given that the waste exchange could be approximated and target types of businesses could be identified, there is another element of complication regarding owners' willingness to either sell their properties, relocate to a remediated brownfield, change their practices, and/or engage in dependency relationships with other businesses.

Anticipating this, however, invites a further challenge to somehow incorporate the feedback of such stakeholders into a GIS model that might more closely approximate real-world conditions, such as which neighborhoods and types of industries might be more amenable to implementing eco-industrial redevelopment strategies. While such a public outreach endeavor is beyond the scope of the subject addressed here, it exists as an important opportunity for future research. However, even if the model lacks in technical prescription, it does provide value in advancing inquiry in these new directions—in a manner described by Haggett and Chorley (1967): "Successful application of models in geography ensures no teleological progress toward full understanding, for scientific effort does not reduce the sum total of problems to be solved—it rather increases them." Thus, the truest value of the model discussed here is likely found in its function as a link in a chain—moving toward a better understanding of how cities develop and change. The next sections will address key implementation issues.

Business Data: The ABI data are presented in tabular spreadsheet format, containing fields on business name, address (with street number), city, state, zip code (zone), employee class, and industry code. SIC codes were used to classify waste by industry type—using an approximation that has been calculated by the CIWMB that divides these into 39 industry groups. The industry codes were matched with employee class (size) codes to determine waste-generated per employee—a measure for approximating waste volume. These rates have been predetermined by the CIWMB also, yielding a measure of "tons of waste disposed per employee per year". The table below indicates the number of businesses in all of LA by industry group type.

Ind Grp	Group Name	# Bus.	Ind Grp	Group Name	# Bus.
1	Agriculture and Fisheries	1,123	20	Communications	1,013
2	Forestry	2	21	Utilities	193
...	...	...	...	...	...
18	Trucking and Warehousing	1,395	37	Services (Other Professional)	13,539
19	Transportation (Air)	366	38	Services (Other Miscellaneous)	27,541
Source: CIWMB 2002			39	Public Administration	1,200

One data limitation of the employee class sizes is that the employee class is represented with a letter corresponding to a *range*, e.g., "A" equals businesses with 1-4 employees. In order to convert the letters to usable numbers, the midpoint range is determined (Tseng 2002), e.g., in this case, 2.5; however, this becomes problematic with larger businesses when the class size range is much larger, e.g., 1000-4,999. Other limitations concern those businesses that were not assigned an employee class code, which made them unusable for the model, as well as those labeled with a 9999 SIC code, which were not classified under the CIWMB system. A breakdown of the dataset includes 180,422 businesses in the full usable set, with the unusable portion of businesses at approximately 5 percent (These include businesses with no employee codes assigned or "other" classifications not incorporated in the CIWMB's 39 groups.).

Site Selection: The brownfield sites selected for this study are located in the City of LA—Beverly-Virgil in the north and Wilmington to the south. The Beverly-Virgil site is in the Wilshire Center/Korea Town redevelopment area, covering approximately 57-acres and 209 separate parcels. The Wilmington site is in the LA Harbor/Wilmington redevelopment area and covers hundreds of parcels within approximately 232-acres.

Beverly-Virgil was selected for this analysis because of its socioeconomic conditions, strategic location, and local political activism. It is one of the few areas near Hollywood that is zoned manufacturing, has nearby freeway and subway access, and offers an opportunity to better match jobs to local skills—via the vacant or underutilized parcels within the manufacturing zone. Also, there has been considerable attention paid to the sustainable development potential of the region—instigated locally by the Eco-Village project and spearheaded within the Council Office and City Planning Department in the Station Neighborhood Area Plan (SNAP). The SNAP was adopted by the City Council and is a first-of-its-kind policy proposal to take a systematic approach to environmental improvement in an

LA neighborhood via integrated open space creation, reduced auto-dependency, industrial ecology, and worker training and education (among other prescriptions). The somewhat larger Wilmington Industrial Park area was selected for many of the same reasons. It has been a targeted redevelopment area for over twenty years, but remains an underutilized manufacturing area with significant access to transportation and material flows. The Mayor's Office of Economic Development has collaborated with the NCEID at the University of Southern California to investigate eco-industrial redevelopment options at this site. Also, a recent Economic Development Administration-funded development/market study will evaluate eco-industrial potential.

GIS Format: The GIS fundamentals for the model should be vector-based given that topology is important for the analysis—e.g., containment, adjacency and connectivity will be central to the area-wide waste characterization. For instance, once businesses are geocoded by their physical location and a point file is created, the nodes (and their data attributes, including industry type and employee class size) will then need to be aggregated using buffers around the site areas. Also, the nodes representing businesses' waste disposal behaviors will form the basis of network analysis—to include calculations of shortest transportation routes for the waste exchange relationships and/or determining which parcels are most eligible for construction of new facilities to increase the area-wide capacity for absorbing the extant waste. In this way, pooling waste collection, transport services, and/or treatment becomes a manageable strategy.

Identification of Themes: A number of different themes are useful to development of the GIS model; however, they may be subsumed within three groups: parcels, firms, and waste. Parcel data are necessary because they include information about site size, ownership, tax status, shape, and, of course, location. Given these different attributes, sites can be converted to other shape-file groups—such as by parcels with the same owner and/or those that are vacant. Tax status may be used to locate parcels that are tax delinquent and therefore (potentially) more easily acquired by the relevant government. Parcel shape is also useful in making decisions, such as where to site recycling facilities, e.g., in the case of siting materials recycling activities in back-lot areas not in full view of a street. Firm-level data are derived from the ABI spreadsheet as a point file geocoded to a street network file. The street network file is necessary in order to geocode the business addresses, but also to evaluate waste transport scenarios. New fields can be added to the business point file in order to calculate waste-by-employment size category (referenced earlier). Also, other themes can be created based upon dominant industry- or waste-type categories in each area.

The waste category involves the use of themes to demonstrate both waste type and amount. These are designed to adhere to the CIWMB categories that calculates the percentage of waste stream for each of the 39 business groups according to the following waste categories: PAPER (uncoated corrugated cardboard, paper bags, newspaper, white ledger, color ledger, computer paper, and other office paper, magazines and catalogs, phone books and directory, other miscellaneous paper, remainder/composite paper); GLASS (clear glass bottles and containers, green glass bottles and containers, brown glass bottles and containers, other colored glass bottles and containers, flat glass, remainder/composite glass); METAL (tin/steel cans, major appliances, other ferrous, aluminum cans, other non-ferrous, remainder/composite metal);   PLASTIC (HDPE containers, PETE containers, miscellaneous plastic containers, film plastic, durable plastic items, remainder/composite plastic); OTHER ORGANIC (food, leaves and grass, prunings and trimmings, branches and stumps, agricultural crop residues, manures, textiles, remainder/composite organic); CONSTRUCTION AND DEMOLITION (concrete, asphalt paving, asphalt roofing, lumber, gypsum board, rock, soil and fines, remainder/composite construction and demolition); HOUSEHOLD HAZARDOUS WASTE (paint, vehicle and equipment fluids, used oil, batteries, remainder/composite household hazardous); SPECIAL WASTE (ash, sewage solids, industrial sludge, treated medical waste, bulky items, tires, remainder/composite special waste); and MIXED RESIDUE.

The CIWMB model, like the one proposed here, is based upon the premise that "Knowing what materials different types of businesses typically dispose can help them reduce waste, recycle more, and save money...[and]...This information is usually collected by examining waste discarded by individual businesses (a.k.a. "dumpster sorting")." (CIWMB 2002) To calculate the end statistic of tons of waste disposed of per business per year, the following calculations are used:

Example of ABI data with CIWMB codes:

BUS_NAME	SIC_CODE	INDGRP	DISPRATE	DISPEMP	EMPGRP	EMPSIZENUM
Business X	019101	1	0.9	2.25	A	2.5

Example of CIWMB waste type codes:

Waste Type		Business Group		
Code	Description	1	2	3
1	PAPER	0.134	0.360	0.309
1A1	Uncoated Corrugated Cardboard	0.044	0.016	0.073
1A2	Paper Bags	0.003	0.003	0.018
1B1	Newspaper	0.030	0.080	0.023
Source: California Integrated Waste Management Board 2002				

1. *Business X, which is classified with a SIC code of 019101 is assigned to CIWMB Industry Group 1 based upon the SIC code groupings assigned in the CIWMB model (each is assigned to one of 39 total groups).*
2. *Business X, which is classified in Employee Group A (1-4 employees) is assigned an employee number (which is the midpoint value of this employee range in all cases except where the lowest value is used for the category of 10,000 employees or more); in this case, 2.5.*
3. *The employee number of Business X is then multiplied by the Disposal Rate 0.9 corresponding to Business X's Industry Group (determined in Step 1 above), which yields the total disposal for Business X in tons per year; in this case, 2.25.*
4. *To determine the composition of the waste amount derived in Step 4, the waste type factor (e.g., for PAPER), which is unique to each of the 39 Industry Groups, is then multiplied by the total disposal amount from Step 4 to yield the percentage of the total waste that is comprised of each type; in this case 0.134. The resulting amount of paper disposed of per year at Business X is then: 2.25 * 0.134 = .3015 tons.*

Since the data collection method of CIWMB is sample-based on "dumpster sorting", it is important to note that all calculations derived from it are rough approximations. Therefore, certain reservations should be considered in at least three points within the analysis: (1) in using the midpoint range for employee class size; (2) in extrapolating from business type to waste stream type and size; and (3) in inferring that the random sample collected statewide applies to the two LA sites.

Once the basic themes were included in the GIS model, then new buffer themes were created to demonstrate how much waste of each kind is generated within a certain distance of each site. Determining the buffer distance settings is a relatively exploratory procedure given that it is unclear how far potential waste exchange participants might be willing to travel/operate to engage in the transaction. For each of the two sites, buffers were created at three radii: 1 mile, 2.5 miles and 5 miles (See example map of the Beverly-Virgil site.) The buffers were used to create new shape-files

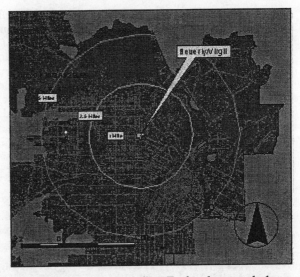

containing only the businesses within the specific radius. Each subsequently larger radius incorporates the businesses within the smaller radii.

Below is a table detailing the results of the buffer analysis for the two sites (Note that only the major categories are presented here; however, the analysis also provides results for the subcategories listed in the *Identification of Themes* section.):

		Waste in Tons Per Year					
Code	Description	Beverly-Virgil Gully			Wilmington Industrial Park		
		1 Mile	2.5 Miles	5 Miles	1 Mile	2.5 Miles	5 Miles
1	PAPER	18,336	135,291	348,508	4,958	13,416	42,262
2	GLASS	1,442	10,050	25,241	410	1,078	3,466
3	METAL	3,250	18,746	54,418	1,159	2,622	8,191
4	PLASTIC	4,976	32,926	110,716	1,777	4,247	12,848
5	OTHER ORGANIC	18,568	118,571	301,044	4,524	11,689	38,363
6	CONSTRUCTION AND DEMOLITION	4,265	23,094	72,914	2,152	5,039	13,824
7	HOUSEHOLD HAZARDOUS WASTE	145	969	2,345	33	73	276
8	SPECIAL WASTE	1,261	10,602	25,829	510	1,589	3,922
9	MIXED RESIDUE	251	1,720	4,874	119	332	870
Total No. Businesses		4,251	22,404	58,070	1,015	2,100	6,960
Total Usable for Analysis		3,991	20,714	54,202	993	2,040	6,742
Total Percent Usable		94%	92%	93%	98%	97%	97%

Next Steps

The GIS model uncovers important information about waste types and amounts by location, but to provide value in an eco-industrial exchange context a corresponding reuse scenario must also be designed. The City has embarked on this effort by creating a database of environmentally preferable manufacturers and prioritizing it by the ability to match conditions in market and infrastructure. Since the intention of the model is to look at innovative ways to use waste and divert it from the landfill-bound stream, this kind of analysis will be critical to the model's success.

Therefore, once waste compositions are profiled for each of the two brownfield sites, then assessments need to be made regarding (1) characterization of parcels available for development within each site and (2) types of businesses that would be able to make use of the identified waste stream as feedstock in their operations. The latter involves identifying operational businesses in other locations that make use of similar feedstock inputs and determining what might influence establishing a similar operation within the two target locations, i.e., interviews might uncover that certain government incentives would be necessary or that crime is a major deterrent, etc. Seen this way, the GIS model is important as a lens in looking at old, used places in a new way. Notably, it serves as a catalyst for reaching a better understanding of the nexus between small-scale entrepreneurship and civic environmentalism.

References

Allenby, Braden R. (1992) "Achieving Sustainable Development Through Industrial Ecology," International Environmental Affairs, Vol. 4, No. 1, pp. 56-68.

Ausubel, Jesse H. (1997) "The Virtual Ecology of Industry," Journal of Industrial Ecology, Vol. 1, No. 1, pp. 10-11.

California Integrated Waste Management Board. Waste Characterization Model. Sacramento, CA: http://www.ciwmb.ca.gov/WasteChar/.

Environmental Affairs Department. (2001) City of Los Angeles Brownfields Program. Electronic web site: http://www.lacity.org/ead/labf.

Haggett, P. and R.J. Chorley. (1967) "Models, Paradigms and the New Geography," in Chorley, R.J. and Haggett, P. (eds.) Models in Geography. Methuen, London, pp. 19-42.

National Center for Eco-Industrial Development. (2001) University of Southern California/Cornell University, Los Angeles, CA: http://www.usc.edu/schools/sppd/research/NCEID/index.html.

Recycling Market Development Zone (RMDZ). (2002) California Integrated Waste Management Board, Sacramento, CA: http://www.ciwmb.ca.gov/RMDZ.

Rodino Associates. (2002) Project to Develop Integrated Economic Development and Community Redevelopment Strategies Through Industry Clusters, Task IV Report: Recommendations and Implementation. Submitted to the Community Redevelopment Agency, City of Los Angeles, CA, March 8, 2002.

Schlarb, Mary. (2001) Eco-Industrial Development: A Strategy for Building Sustainable Communities. U.S. Economic Development Administration, Award No. 99-06-07462. Washington, DC.

Tseng, Eugene. (2001) Personal communication. Consultant to the City of LA Environmental Affairs Department Materials and Waste Resources Division, February 14, 2001, Los Angeles, CA.

U.S. Environmental Protection Agency. (2001) Office of Solid Waste and Emergency Response, Washington, DC, http://www.epa.gov/swerosps/bf/index.html.

Strategic communications can enhance brownfields public participation

J. G. Duffy[1] & R.M.Omwenga[2]
[1]Conover & Company Communications, Inc.
Mass, USA
[2] School of Community Economic Development
Southern New Hampshire University, USA

Abstract

Public participation is generally defined as the process of engaging stakeholders such that those most impacted by a particular activity can influence outcomes of that activity. Although not a recent practice, the importance of public participation has gained prominence in recent years where projects impact natural resources and communities. The United States Environmental Protection Agency (EPA) has and continues to revise its policies with respect to public participation in several agency programs. The purpose of this article is to evaluate the efficacy of public participation within the context of EPA's Brownfields program, and to make recommendations about how the use of strategic communications can enhance the overall success of the revitalization of the project.

1 Introduction

The definition of public participation has evolved over the past several decades. Accordingly, we turn to a variety of sources to provide a comprehensive definition of public participation. It is interesting to observe how its definition is modified to meet unique institutional objectives. In 1981, the U.S. Environmental Protection Agency (EPA) [1] articulated its public participation policy as:

"To ensure that managers plan in advance needed public involvement in their programs, that they consult with the public on issues where public comment can be truly helpful, that they use methods of consultation that will be effective both for program purposes and for members of the public who take part, and finally that they are able to apply what they have learned from the public in their final program decisions."

Presently, EPA is revising its 1981 public participation policy. A recent Draft Public Involvement Policy [2] published in December 2000, revises EPA earlier policy and includes

"The full range of actions and processes that EPA uses to engage the public in Agency's work, and means that the Agency considers public concerns, values, and preferences when making decisions."

EPA's RCRA Program originates from the Resource Conservation and Recovery Act of 1976, the law that sets forth the intent, according to Cox [3], to promote conservation of resources through reduced reliance on land filling. In this program, EPA uses the term public participation to denote:

The activities where permitting agencies and permittees encourage public input and feedback, conduct a dialogue with the public, provide access to decision-makers, assimilate public viewpoints and preferences, and demonstrate that those viewpoints and preferences have been considered by the decision-maker.

Public participation has also received considerable attention by international organizations, including numerous non-governmental and international organizations, including the United Nations. In 1975, the United Nations [4] defined "popular participation" in relation to development as:

"Active and meaningful involvement of the masses of people at different levels (a) in the decision-making process, for the determination of societal goals and the allocation of resources to achieve them and (b) in the voluntary execution of resulting programmes and projects."

Academicians also have an interest in public participation. For example, in 1969, Arnstein likens the process to eating spinach: no one is against it in principle because it is good for you. Arnstein defines citizen participation as:

"Participation of the governed in their government is, in theory, the cornerstone of democracy - a revered idea that is vigorously applauded by everyone... It is the redistribution of power that enables the have-not citizens, presently excluded from the political and economic processes, to be deliberately included in the future. It is the strategy by which the have-nots join in determining how information is shared, goals and policies are set, tax resources are allocated, and programs are operated out. In short, it is the means by which they can induce significant social reform, which enables them to share in the benefits of the affluent society." Arnstein [5].

More than 30 years since Arnstein wrote this article, public participation has received little, if any, resistance. In fact, the opposite is true. Public participation has gained widespread endorsement. Scholars including Kasemir, *et al* [6]; others support this trend. Indeed, Kasemir, *et al* [6] notes that public participation is becoming increasingly essential in assessing environmental problems.

Gibbons, Attoh-Okine and Laha have echoed this sentiment, in their examination of Brownfields (discussed below). Specifically, these researchers assert that stakeholder involvement is important because several different interests are directly impacted in Brownfields projects (Gibbons [7], pp 151-162).

2 Public participation in the context of brownfields

One of the initiatives EPA has developed with a strong public participation component is the Brownfields Economic Redevelopment Initiative (US Environmental Protection Agency [8] p.2). Brownfields, as defined by EPA, are "abandoned, idled, or under-used industrial and commercial facilities where expansion or redevelopment is complicated by real or perceived environmental contamination."

In its *Brownfields Assessment Demonstration Pilot* program, EPA awards grants that typically do not exceed $200,000 dollars to cities and municipalities to retain environmental consultants to conduct assessments to determine the nature and extent of environmental contamination. EPA stipulates that Demonstration Pilot Grants are to be used to conduct environmental site assessments; clarify liability and cleanup issues; build partnerships and outreach among federal agencies, states, tribes, municipalities, communities; and other entities; and foster local job redevelopment and training initiatives.

The Brownfields Demonstration Pilot Program places strong emphasis on public participation. In fact, EPA has set public participation guidelines for applicants while completing grant applications. These guidelines require that Brownfields pilot funds may be used for outreach activities that educate the public about assessment, identification, characterization, or remedial planning activities at a site or set of sites. EPA recommends that outreach should be directed toward obtaining more effective public involvement and/or environmental assessment and cleanup of hazardous substances, pollutants, or contaminants at affected sites (US Environmental Protection Agency [8], pp 5-6).

Demonstration Pilots require grant recipients to develop and implement public participation programs during the environmental assessment process. Indeed, EPA's RCRA public participation guidelines state that:

A vital and successful public participation program requires a dialogue, not a monologue Without an active two-way communication process, no party will benefit from the "feedback loop" that public participation should provide Public participation cannot be successful if the permitting agency or the facility is reluctant or unable to consider changes to a proposed activity (US Environmental Protection Agency [8], pp.4-5).

EPA's public participation programs are intended to include effective two-way communication strategies and tactics to ensure that a dialogue, or a "feedback loop," is created such that stakeholders learn about and can influence decisions as they relate to specific projects.

3 Strategic communications

Achieving a "feedback loop" is the result of a carefully planned communications process, not isolated communications events. A comprehensive communications program should address questions such as these:

- Which audiences, or stakeholders, should be included in a "feedback loop"?
- How should audiences, or stakeholders, be identified?
- What messages should be delivered?
- How should the messages be delivered, and when?
- How shall complex scientific and engineering principles, including risk, be communicated, and who shall do it?
- Can strategic communications establish credibility and build trust?

The authors recommend that to achieve EPA's objectives, public participation programs should be comprised of integrated, proactive, and strategic communications activities. A program of such dimension would ensure that stakeholders [1] have adequate opportunities to fully understand environmental, legal, and engineering issues relative to Brownfields, and [2] that stakeholders have sufficient opportunities to meaningfully contribute to project outcomes.

EPA's public participation objectives resemble a communications model developed by Grunig, *et al.* According to these scholars, two-way symmetric communications occurs when public relations is practiced as a management function where corporations, government agencies, associations and non-profit organizations identify the stakeholders they affect and those affected by them. Once identified, public relations managers develop ongoing programs of communication with these publics (Grunig [9]). Organizations practicing the two-way symmetric model use bargaining, negotiating and strategies of conflict resolution to bring about symbiotic changes in the ideas, attitudes, and behaviors of both the organization and its publics (Grunig [10]).

It is evident that EPA recognizes the importance of two-way symmetric communications, a process that evokes dialogue and creates a "feedback loop." Although the terms used by Grunig and EPA differ, the communications process espoused by each is the same. It is appropriate, then, to examine public participation programs as proposed by various EPA Brownfields contractors to determine whether two-way communications could be achieved

4 Public participation at specific brownfields pilots

While researching this topic, the authors sought information from EPA Region I (which includes the New England states of Maine, Vermont, New Hampshire, Massachusetts, Connecticut and Rhode Island) to determine whether a two-way symmetric model of communications was in fact implemented in the Brownfields public participation context. We submitted a Freedom of Information Act (FOIA) letter in which we requested information specific to public participation activities with respect to Brownfields Demonstration Pilots. In response to our letter, we received public participation plans, among other documents, relating to approximately one dozen individual projects.

We analyzed the public participation plans submitted to EPA by engineering firms, municipalities and community development corporations to identify proposed public participation activities for specific Brownfields projects. Based upon our review of these plans, we observed that events identified as constituting public participation activities included only:

- Public meetings
- Fact sheets
- Media coverage, and
- Document repositories.

These public participation plans did not suggest the consideration of any two-way symmetric communications based on dialogue which effectively creates a successful "feedback loop."

Although the submitted plans identify a common objective of proactively engaging stakeholders, the proposed activities (public meetings, fact sheets, media coverage and the creation of document repositories) do not, according to the authors, constitute two-way communications. In fact, these activities suggest the opposite, in that they appear to be better suited for a monologue where information is merely disseminated without a meaningful or definitive opportunity for stakeholders to influence decisions, which, in turn, affect project outcomes. This is noteworthy, for the public participation plans submitted by grant recipients, as they were provided to us in response to our FOIA request, did not appear to propose activities that would afford stakeholders the opportunities to engage in public participation consistent with agency policy objectives.

5 Recommendations

The authors recommend specific communications activities that will enable EPA contractors to achieve the desired feedback loop. Communications activities should include (but are not limited to) community affairs, media relations, government relations, collateral material development, and outreach programs. These activities will enable stakeholders to understand important issues, provide stakeholders opportunities to contribute to project outcomes, and build relationships in the process.

Community Relations. Brownfields often are located in residential and commercial areas. Accordingly, it is important to reach out to abutters and inform them of the Brownfields redevelopment plans and objectives *prior to* projects starting. Project abutters, (including residents and businesses) are likely to be impacted by project activities (e.g., equipment traffic and noise) and have a right to timely information so they can take steps to prepare. In addition, reaching out to abutters has other benefits including the possibility of learning of historical activities at the site, as well as collectively exploring reuse opportunities.

Media Relations. The authors recommend that project officials meet with local journalists and reporters to inform them of the Brownfields project, educate them about important environmental, legal, and engineering issues, and offer to serve as a resource for them as they report on project-related activities in the future. Informed journalists can report more accurately about issues and it makes good sense to keep them apprised of project issues and milestones. There is more to media relations than merely media coverage, and we advocate that quality relationships with media professionals can benefit both the community and the project.

Government Relations. Local elected and appointed officials do not necessarily understand the complex issues surrounding the redevelopment of Brownfields properties. Accordingly, the authors also recommend meeting with community leaders to inform them of project issues and plans. Their awareness and involvement will enable them to serve their respective constituents better. For example, by understanding important project-related issues, elected and appointed officials can better understand redevelopment constraints and opportunities, thereby serving their constituents more effectively when advocating for the most appropriate reuse(s).

Collateral Material Development. We recommend creating educational materials including project fact sheets, brochures, backgrounders, diagrams, and other educational information. These materials can help explain the issues and objectives of the project to all stakeholders, as well as identifying points of contact in case of emergency. In addition, these materials can serve multiple purposes, including the basis for web site content and documents for a public library repository.

Outreach Programs. There are numerous creative ways to bring the project to the community and the community to the project. For example, site tours are beneficial in that they offer stakeholders a first-hand perspective of the project. Special events at particular project milestones can also help stakeholders keep informed of progress. Public and other meetings are also helpful, particularly if they are part of an ongoing and integrated communications program. Rather than serving as the *only* opportunity to learn about the Brownfields project, public meetings will be only one part of a multi-dimensional program and will reinforce relationships and prior communications. In addition, an important benefit from an integrated and proactive communications program is that members of the project team will build trusting relationships with stakeholders. Relationships built on trust are critical when discussing site-specific risk issues and when building consensus about reuse alternatives, among other important topics.

6 Conclusion

The authors have determined that EPA's public participation policies, both agency-wide and within certain specific programs (*i.e.,* RCRA), embody a constructive, effective, two-way communications model. However, it seems unlikely that such a model could be implemented based solely upon the proposed communications activities, namely fact sheets, public meetings, media relations and document repository, as identified by the Brownfields public participation plans submitted to EPA by grant recipients as obtained by the authors. Such activities embody one-way communications and fall short of that which would establish meaningful dialogues and feedback loops.

Accordingly, the authors recommend that Brownfields public participation programs be required to include more proactive, integrated and strategic communications activities such as community affairs, media relations, government relations, collateral material development, and outreach programs. Public participation programs that embody such commitments are more likely to achieve the two-way communications "feedback loop" EPA has defined as an important objective.

References

[1] U.S. Environmental Protection Agency (46 Federal Register 5736, 1981)

[2] U.S. Environmental Protection Agency (Federal Register 6923-9)

[3] Cox, D. B. *Hazardous Materials Management Desk Reference*, McGraw-Hill Inc:, New York, p.11, 1998.

[4] United Nations, Department of Economics and Social Affairs. *Popular Participation in Decision Making for Development*. United Nations: New York, p. 4, 1975.

[5] Arnstein, S.R. A Ladder of Citizen Participation. *Journal of the American Institute of Planners*, 35 (4), pp. 216-224, 1969.

[6] Kasemir, B., Schur, D., Stoll, S., & Jaeger, C.C. *Involving the Public in Climate and Energy Decisions*. Environment, 42 (3), pp. 32-42, 2000.

[7] Gibbons, J.S., Attoh-Okine, N.O. & Laha, S. *Brownfields Redevelopment Issues Revisited*. International Journal of Environment and Pollution, 10 (1), pp. 151-162, 1998.

[8] U.S. Environmental Protection Agency. *The Brownfields Economic Redevelopment Initiative*. EPA document 500-F-97-156, 1998.

[9] Grunig, J.E. & Grunig, L.A. *Models of public relations and communication. In James E Grunig* (Ed.), Excellence in public relations and communication management. Mahwah, New Jersey: Lawrence Erlbaum Associates, Publishers. 1992

[10] Grunig, J.E., Grunig, L.A., Sriramesh, K., Huang, & Yi-Hui, Lyra, Anastasia. *Models of public relations in an international setting*. Journal of Public Relations Research, 7(3), pp. 163-186, 1995.

Involving stakeholders to achieve successful development of brownfield sites

R. D. Stenner[1], R. N. Hull[2] & R. F. Willes[3]
[1]*Battelle Northwest/Pacific Northwest National Laboratory, USA.*
[2]*Cantox Environmental Inc., Canada.*
[3]*Reon Development Corporation, Canada.*

Abstract

Our overall quality of life depends on balancing the interrelationship between human and ecological health, socio-cultural values, and economic well-being. Achieving appropriate balance of these components is critical to modern environmental (e.g., Brownfield site) decision-making. The American Society of Testing and Materials (ASTM) is involved in developing a standard guide to facilitate the analysis and management of Quality of Life decision making. This guide will provide a process to help identify, analyze, and resolve stakeholders' issues associated with environmental problems. A key component to the Quality of Life process is to empower the affected stakeholders to enable genuine participation in the decision making and management process. The basic components of the Quality of Life process will be presented along with an example case where the methods have been applied successfully to the development of a Brownfield property in urban Toronto, Canada.

The application of the Quality of Life process enabled participation of all the affected stakeholders (people in the community, the developer, local government and regulators) from the very beginning. The stakeholders participated in all decision-making of the redevelopment process; from planning the types and locations of buildings through landscaping/community art for the site, traffic flow/public transportation, day-care requirements and a variety of specific community amenities (up-grading lake access portals, various water recreation facilities, community playground equipment). Application of the Quality of Life process resulted in a win-win situation for all stakeholders (i.e., people in the community, regulators and the developer). The derelict industrial property is being replaced by a residential development that will improve the overall quality of life of the community. The developer is completing a profitable, successful real estate project, without the excessive delays and resulting expense typically associated with such a project. The historical environmental issues are being successfully resolved with the necessary due diligence and care, but without the fears and apprehensions that arise from uninformed perceptions that the general public often associate with such issues.

1 Introduction

Industry and government alike are faced with complex environmental decisions that affect a variety of affected sub-populations with very different values and issues, all caring very deeply about their quality of life, as it will be affected by these decisions. The expressed quality of life issues often get very complex. What one group thinks is a bad thing another group is likely to think is a good thing. As a simplistic example, consider the morning commute:

- A Bad Thing: Surely commuting an hour to work, morning and night, would decrease the quality of the commuter's life. Time is wasted, fuel is spent, tires are worn out, and money is put into restoring the car. The probability of a scene like that presented in Figure 1 is greatly increased.

Figure 1: Common scene associated with the morning commute

- A Good Thing: Tire company workers, fuel providers, and mechanics have their quality of life increased as a direct result of commuters funding their paychecks.

Can we then say people's lives are better or worse because of a longer commute? This good news/bad news scenario is a simple example of the type of challenge facing decision-makers. Now, increase the complexity of the issues to consider balancing the quality of life for a diverse community facing a problem of how to balance the need for growth and economic stability with strong deep-seated religious and cultural values, such as those of First Nations. How does one get both sides to meet somewhere in the middle, so to speak, to move ahead with solving specific environmental problems? Conflicting issues such as these need to be resolved through "informed consensus building" and the direct hands-on involvement of "affected stakeholders."

Figure 2 diagrams the complexity and interrelationship of the components of a Quality of Life assessment and management effort.

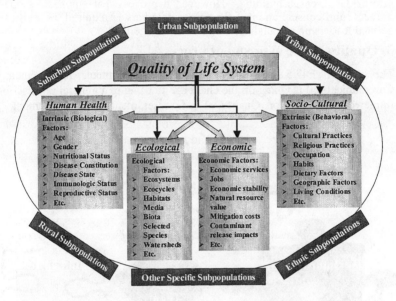

Figure 2: Quality of Life System

2 Quality of Life defined

The World Health Organization has defined Quality of Life as:
"The individuals' perceptions of their position in life, in the context of cultural and value systems in which they live, and in relation to their goals, expectations, standards and concerns" [1]

ASTM is taking on the challenge of seeking to improve Quality of Life by enabling and placing effective, science-based tools in the hands of key stakeholders. The intent is for the stakeholders to decide what areas and issues are most important (human health, ecology, economics, socio-culture) and use the tools best suited to assess and decide on the best course of action for resolving issues and making sound environmental decisions.

2.1 Stakeholder involvement goals

Stakeholder is a term applied to a mix of affected peoples associated with a particular environmental decision. They are made up of individuals whose lives are directly or indirectly affected as a result of the decision. The types of stakeholders will be discussed later under the Affected Stakeholder section, but it is important to note here that the goals of these stakeholders can be quite diverse and focused. One of the major challenges in environmental decision-making is to determine which of these goals are essential to a fair and successful decision.

Many decisions have impacts that affect various stakeholders in completely different ways. For example, an increase in commute time to and from work burdens drivers, yet benefits mechanics and petroleum companies. Balancing the needs and desires of multiple stakeholders can be accomplished through informed consensus building. However, a process that emphasizes openness, fairness, and consideration of the values of others is required, as well as well-defined leadership and understood rules of engagement.

3 Quality of life assessment process

The ASTM E47.5 Sub-committee (Risk Assessment, Management, and Communication) is taking on the challenge to develop a general process-focused framework standard on Quality of Life assessment and management. This

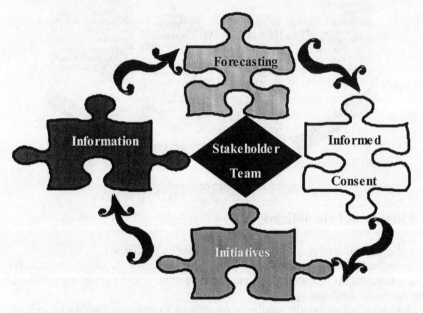

Figure 3: Main components of Quality of Life framework

quality of life process standard will then be supported by existing and new tailored specific analysis and management standards designed to address very focused issues. Figure 3 shows the main functional components of the framework.

3.1 Affected stakeholder identification

The identification of affected stakeholders and the assembly of the Stakeholder Committee is probably the most important step in the whole process, as they will be empowered and responsible for the facilitation of the complete Quality of Life process. Stakeholders can often be grouped into the following three categories: 1) affected stakeholders, 2) interested party stakeholders, and 3) regulatory or oversight stakeholders. The affected stakeholders are those people whose lives will be directly impacted by decisions with respect to their health, economic condition, personal environment, and social-cultural-religious life style. The business or "responsible party" owners are certainly part of the affected stakeholders. The interested party stakeholders are those people who have a vested interest, but who do not personally live and work in the impacted area. The regulatory/oversight stakeholders are usually the local, state/provincial, and federal regulatory agencies charged with legal responsibility for controlling the effect of the environmental decision. Often, one or more of these regulatory agencies are part of, or choose to be part of, the affected stakeholder group.

The Stakeholder Committee is established drawing primarily from the affected stakeholder group and is charged with the responsibility for managing the assessment process and making the decisions. Once the Stakeholder Committee is established, it is essential that the "rules of engagement" for all stakeholders be established and communicated. All stakeholders need to be encouraged and shown how to get involved upfront in the process. The Stakeholder Committee will be empowered and responsible for the issues and information gathering, analysis and forecasting activities, establishing "informed consent," and managing the initiatives and actions resulting from the decision(s).

3.2 Information/issue establishment

In today's complex society, fair and equitable environmental decisions require the balancing of many issues and concerns expressed by the stakeholders impacted by the decision. These issues and concerns generally will fit into these four categories:
- Human health issues
- Ecological issues
- Economic issues
- Socio-cultural issues

Often environmental decisions are based on the in-depth analysis of the issues and concerns of just one, or possibly two, of these categories, with human health usually being the most favored. This practice has often left many stakeholders frustrated and upset that their "real concerns were never addressed," and it is not because they did not want the in-depth analysis regarding human health. They expect that to occur, but they had important issues from the other categories that they felt were equally important and not adequately addressed. The framework standard being developed is aimed at establishing a process for a balanced approach to assessing the issues associated with all four of these categories.

3.3 Analysis/forecasting

After the initial issues and information have been gathered from the stakeholders and it is known where the priorities and values of the stakeholders rest, then the necessary analyses associated with the environmental decision at hand can take place. At this point, the Stakeholder Committee will bring in the technical experts necessary to adequately assess the agreed upon issues and their associated impacts. It is expected that technical impact modeling and analyses will need to be conducted for all four of the issue categories (i.e., human health, ecology, economics, and socio-cultural issues). Once the results from the modeling and analysis activities are available, the Stakeholder Committee will be responsible for establishing agreed-upon weighting and valuations of the forecast range of possible outcomes.

Each of these four areas of analysis potentially encapsulates hundreds of possible forecasting methods and approaches. Also, the analyses performed in each of these four areas can be intimately related with one another. The analyses should not be conducted in isolation. For example, an overall increase in peoples' incomes often results in increased use of natural resources with greater environmental degradation, an increase in human health due to the fact that health care is more affordable, and can result in more money being spent on cultural preservation. In essence, measuring one of these four variables will require that the other three variables be taken into account.

There is no one specific set of analysis methods that will work for all situations. Instead the Stakeholder Committee will need to consider a variety of models and methods in the "tool box" associated with the Quality of Life process to address the specific issues and questions raised regarding the decision at hand. A host of risk analysis tools are currently available from a broad range of sources (e.g., through ASTM, the Environmental Protection Agency, and many others).

To adequately implement the forecasting/analysis stage, a great amount of communication between the stakeholders and the expert advisors will need to take place. Oftentimes stakeholders are turned off immediately when their needs and values are thrown into a "black box" and an answer suddenly appears. Although this cannot be avoided completely, stakeholder facilitation throughout the forecasting/analysis stage can help alleviate much of this skepticism. Also, it is essential that the four forecasting/analysis areas (i.e., human health, ecological, economic, and socio-cultural) be able to "speak" with one another with respect to their results. If all of the economics results are in money terms while all of the socio-cultural measures are in qualitative form, then there will be no real way to analyze these measures together, which is the opposite result of that intended for the Quality of Life Process.

3.4 Informed consent establishment

Once the analyses have been completed, keeping in mind the whole process will likely be quite iterative as the analyses can uncover new issues, it will be necessary to reach (an) agreed upon solution(s). In order to do this, criteria need to be created to decide which solution(s) is(are) preferred. The stakeholders have to agree upon what is most important to them in balancing the human health, ecological, economic, and socio-cultural impacts to establish criteria that cater to what they value most. This structured area of "solution selection criteria" is essential to guarantee that all of the needs of stakeholders are accounted for during the selection process. Without this structure, certain needs could easily go unaccounted. The stakeholders, through the leadership of the Stakeholder Committee, will have to begin making trade-offs among the different forecasting results. Not every forecast will be positive, so the stakeholders must decide what is most important (from the information stage and their solution selection criteria) among all of their options. Decision assessment tools can be used at this point to prioritize the stakeholders' decisions and to help analyze the trade-offs that will be made depending on the solution(s) chosen.

3.5 Initiatives/actions

This step of the process involves the implementation of the selected solution(s). Impact and benefit analyses must be run throughout this stage to assess the actual realized impacts of the decision and any associated changes that need to be made to the original decision.

At any point throughout the Quality of Life assessment process, the participants can go back through previous stages to reassess the progress. If certain stakeholder values were not fully accounted for, then it will be necessary to gather more information before making and implementing a decision. If the expert advisors cannot produce accurate forecasts with the information provided, it will be necessary to go back and obtain the necessary information. At any point in the Quality of Life process, there are opportunities to renegotiate and reassess the stakeholders' needs and additional issues.

4 Case example

The basic components of the Quality of Life process as outlined in the ASTM standard guide have been applied successfully to the development of a Brownfield property in urban Toronto, Canada. This industrial site operated from 1885 to 1991, and was left as an unused contaminated property within a residential community on the shoreline of Lake Ontario. A photograph of the unused contaminated site is presented in Figure 4. The process usually followed in the development of such a site would have been to plan the chosen use of the property, evaluate site contamination issues, complete a human health and ecological risk assessment, then proceed with remediation, as required. Stakeholder approval would be solicited when the final plan for remediation and redevelopment was near completion. Such a process does not allow for participation and input by a large segment of the affected stakeholders. Consequently, "combative" behaviours between the community and real estate developer are commonplace.

As with most Brownfield properties, the Toronto site is an integral part of a well-established community. As with most communities, issues related to the physical aspects of the development of the property and its integration into the surrounding community are most readily understood. In this case, the prime physical concerns focused on the possible interference with lake-views from existing homes, community access to the lakefront, public transportation, traffic congestion, availability of parkland, and impacts on already deficient day-care facilities and schools. Aside from these concerns, the community was generally happy with the idea of removing the derelict industrial buildings from their community. The impacts of these physical concerns, and their alternate approaches, were well understood by community members. There was a general awareness of the environmental contamination issues of the property, but the details of these issues were not well understood resulting in misperceptions and apprehension by community members.

Figure 4: Unused Contaminated Site

Applying the Quality of Life process enabled participation of all the affected stakeholders (people in the community, the developer, local government and regulators) in the redevelopment plans from the very beginning. The process involved more than open communication between the affected stakeholders. The stakeholders actually participated in all decision-making of the redevelopment process, from planning the types and locations of buildings through landscaping/community art for the site, traffic flow/public transportation, day-care requirements and a variety of specific community amenities (up-grading lake access portals, various water recreation facilities, community playground equipment). The stakeholder communication, cooperation and mutual trust established by working together during the planning of the redevelopment of the site facilitated stakeholder understanding and participation in the planning of the environmental remediation of the site. An artist's depiction of the site after redevelopment is shown in Figure 5.

The application of the Quality of Life process to this Brownfield redevelopment project resulted in a win-win situation for all stakeholders; people in the community, regulators and the developer. The derelict industrial property is being replaced by a residential development that will improve the overall quality of life of the community. The developer is completing a profitable, successful real estate project, without the excessive delays and resulting expense typically associated with obtaining the various approvals required to proceed with such projects. In fact, the developer was awarded one of the first "Brownie Awards", in recognition of "outstanding achievement in building and maintaining effective working partnerships with professionals, the local community and others involved in Brownfield redevelopment." The historical environmental issues are being successfully resolved with the necessary due diligence and care, but without the fears and apprehensions that arise from uninformed perceptions that the general public often associate with such issues.

Figure 5: Artist's depiction of the site after redevelopment

References

[1] Hanna, K. and C. Coussens. 2001. "Rebuilding the Unity of Health and the Environment - A New Vision of Environmental Health for the 21st Century." Workshop Summary for the Roundtable on Environmental Health Sciences, Research, and Medicine, Division of Health Sciences Policy, Institute of Medicine, National Academy Press, Washington, D.C.

The brownfields cookbook: a redevelopment guide

S. Villavaso[1], P. Sinel[1] & L. Dauterive[2]
[1]Villavaso & Associates, LLC, United States of America.
[2]Regional Planning Commission for Jefferson, Orleans, Plaquemines, St. Bernard, and St. Tammany Parishes, United States of America.

Abstract

The Brownfields Cookbook: a redevelopment guide is meant to serve as a toolbox for communities looking for assistance and ideas for redeveloping brownfields in their localities. This guide is not only meant to serve the needs of those communities located in Southeastern Louisiana, but nationwide as well for brownfields can exist in any community, in any state. The redevelopment of such sites into productive and safe uses is a major issue for Louisiana communities, as well as communities nationwide, where urban sprawl and environmental issues are major factors affecting land use and development today.

By focusing on lessons learned and the examples put forward by the Southeast Louisiana Brownfields Economic Redevelopment Assessment Demonstration Pilot Program, the members of the Regional Planning Commission and participating parishes (i.e. counties) provide this toolbox example of redevelopment practices and principles. To demonstrate these practices and principles, this cookbook closely examines the unique area of Southeast Louisiana and offers outstanding examples of community efforts to encourage brownfield redevelopment to meet the needs of the citizens, the economy and the environment of the region.

In addition, the cookbook provides a partial list of resources available to communities looking for funding, or just assistance with getting a brownfields redevelopment program off the ground. These resources range from voluntary cleanup programs, to liability assurances for the protection of the community, and to new federal brownfields legislation.

Introduction

Communities need the tools to assess, cleanup, and reuse contaminated properties through partnerships, streamlining, research, and community based-projects.   The Regional Planning Commission for Jefferson, Orleans, Plaquemines, St. Bernard, and St. Tammany Parishes (RPC) in Southeast Louisiana (www.norpc.org) undertook the regional brownfields effort to provide these tools.   The Southeast Louisiana Regional Brownfields Redevelopment Model Program (www.epa.gov.swerosps/bf.html-doc/selouis.htm) was funded in March 1999 by a $200,000 grant from the Environmental Protection Agency (EPA). The regional brownfields effort is structured to provide the resources necessary for Southeast Louisiana's parishes and municipalities—rural as well as urban—to develop local solutions to local brownfields problems.

The RPC created the regional policy group, the Southeast Louisiana Regional Brownfields Consortium, to assist the communities and the parishes of Jefferson, Orleans, Plaquemines, St. Bernard, St. Tammany, St. Charles, St. John, St. James, and Tangipahoa to achieve a common goal—sustainable development.   The Southeast Louisiana Regional Brownfields Redevelopment Model Program is the first and only regional brownfields assessment pilot in EPA Region 6.

Southeast Louisiana is one of North America's most distinctive and culturally diverse areas, covering 3,399 square miles, and includes the parishes (i.e., counties) of Orleans, Jefferson, St. Bernard, St. Charles, St. John the Baptist, St. Tammany, St. James, and Plaquemines.   The region was settled in 1718 and has been a leading commercial center ever since.   After World War II, the area's rich cultural heritage contributed to its emergence as a major international tourist center.   According to the 2000 Census, the parishes of Orleans, Jefferson, St. Bernard, St. Charles, St. John the Baptist, St. Tammany, St. James, and Plaquemines have a combined population of 1,338,000—a rise in population of 53,000 since 1990.   The descendents of New Orleans original settlers and newcomers to the area continue to move into the surrounding parishes, resulting in explosive growth.   However, the economic development that accompanies this progress brings with it a threat to the environment and the continued economic vitality of the area in the form of contaminated, or potentially contaminated, industrial and commercial facilities that are abandoned, idled, or underutilized.

Brownfields redevelopment projects attempt primarily to encourage the recycling of these sites and promote redevelopment through encouraging public/private investments. Communities that allow brownfield sites to remain inactive lose the tax revenue and employment opportunities generated by thriving operations—for some cities, this can total hundreds of jobs, millions of tax dollars, and hundreds of thousands of dollars in wages that might circulate through the area, bringing still more economic benefits. Existing streets and roads, water lines, rail spurs, and other infrastructure systems go unused. In jurisdictions with numerous brownfield sites, this means that billions of dollars in prior public and private investment are essentially wasted. Given land-use patterns prevalent earlier in the last century, many brownfields sites are well located, often along waterfronts or adjacent to downtown centers. Their decaying presence can drag down efforts to revitalize nearby sites, stalling a community's revitalization efforts and undermining its tax base.

The *Brownfields Cookbook* has been very effective in offering communities a "taste" of what brownfields development is like, in different parishes, and with different outcomes. With this combination of results, *The Brownfields Cookbook* provides an adaptable toolbox for communities looking to redevelop brownfields in Southeast Louisiana and nationwide.

Brownfields redevelopment: the recipe

Ingredients needed: One Concerned Community, One Brownfields Team, Several Potential Brownfields Sites, Several Interested Developers, One Community-Supported Cleanup Plan, Voluntary Cleanup Plan (season to taste), and One Reuse Plan

Take the community and thoroughly inspect for potential and public involvement. Look for possible federal, state, and local funding sources. Separate the brownfields team from the community, but keep the team in close proximity to the community. Blend the team into one cohesive unit. Use the team to assess and improve on public involvement. Extract sources of valuable information on traffic patterns, local work forces, and public desires. Conduct public meetings in affected neighborhoods to gather community desires and needs. Take this public outreach effort and set aside for continuous use. Add to it as needed.

Have the team collect all potential brownfields sites and place them in a row. Select the best site for redevelopment using appropriate selection criteria for community tastes. Perform a site ranking. Take the best site and inspect it thoroughly. Place remaining sites in a GIS database to simmer, for later use. Determine if any contamination may be present that could threaten or hinder redevelopment of selected site. Replace if necessary. Assess the history of the site thoroughly. Environmental risks and potential cleanup costs must be thoroughly investigated.

Prepare a community-supported cleanup and reuse plan. Season to taste with an appropriate voluntary cleanup plan. Mix reuse plan with one or more developers. Stir often. Wash brownfields sites as necessary to achieve cleanup goal. Look for other funding ingredients to add to the pot. Bring to a boil and serve immediately. Season to taste. Enjoy!

Getting started: gathering the ingredients

Brownfields redevelopment is a voluntary undertaking by members of both the public and private sectors, focusing on converting land from idled, polluted states into a cleaner, useful, and potentially profitable condition. Initiating a brownfields redevelopment plan means that one or more entities has recognized the benefits associated with this redevelopment. It is intended for communities to modify the "recipe for redevelopment" to fit the needs and desires of their community.

Form a brownfields team: a fine blend

Strategies for successful redevelopment are invariably community based. Dealing with brownfields sites is a multi-disciplinary challenge. No one person has all of the skills needed. Those communities and developers who have been successful have assembled teams of people to address the issues.

The role of the brownfields team is to provide a forum and an organized series of public meetings to address the issues and concerns associated with the redevelopment of brownfields. The brownfields team serves to advise and assist the municipality or governing agency in the identification of potential public and private brownfields sites and development initiatives desired in each locality. The brownfields team should be formed as soon as a community has noted that brownfields might exist, and there is a desire to redevelop these sites. Ideally, the brownfields team should operate as long as questions remain as to the identification and redevelopment of brownfields sites.

To guide the RBP, the RPC established a technical steering committee, named the Regional Brownfields Consortium, consisting of representatives from each parish government. Representatives of the Louisiana Department of Environmental Quality (LDEQ), the real estate community, and private citizens are also participating in meetings to provide advice to the Consortium. The Consortium initially served to identify potential public and private brownfields site opportunities in each parish.

Public involvement: the main ingredient

Public involvement allows citizens to become partners in the brownfields redevelopment process. The time and effort invested up front to involve the community will eliminate costly delays and plan revisions. Many citizen groups are inviting redevelopment into their neighborhoods. The community can be a source of valuable information on traffic patterns, local work force, and other qualitative factors.

Stakeholders should be involved from the outset of a brownfields redevelopment project. Regional, parish, and city Planning Commissions can be excellent resources to help identify active community groups in a project area. Non-traditional partnerships, including those consisting of churches, chambers of commerce, and schools, are excellent cooperatives when working with neighborhood revitalization. Education is crucial to the stakeholders' understanding of information relating to a brownfields redevelopment project. By educating the neighbors, valuable time and money can be saved as the project proceeds. Information should be widely shared with the community to encourage support and open communication.

The formation of the Southeast Louisiana Regional Brownfields Consortium presented the best opportunity for the RBP to disseminate information to the public and gather information from the public. The Consortium held a series of public meetings, in several different locations throughout the region, so no one area would be over represented. By holding these public meetings in a range of locations, the best opportunity to reach the widest audience of residents was achieved.

Site identification: filling the pantry

Site identification is undertaken to diminish the costs of site assessments, meaning the actual analysis of contaminants. Site identification is used as the initial qualifier to determine if a site is indeed a brownfield. Identification can also provide the community with the beginnings of a database of all brownfields sites in the area. All communities can contain sites that are potential brownfields. There is no one-way to determine whether a site is a brownfield or not. More often than not, research into the past history of a site will yield the operations on site and the types of potential contamination that might exist on site.

After identifying potential brownfields sites through historical records and visual surveys, the members of the RBP Consortium recognized that a site inventory process, including both short- and long-term considerations, was necessary. As a result, they formulated a comprehensive process to list, record, and publicly disseminate information about properties that are or could be considered brownfields.

At times, there may be little information remaining about the operation of a site due to the loss of records to natural hazards. Some property owners and adjacent property owners are reluctant to allow access to sites and records.

Site ranking: select the ripest

The purpose for ranking a site is to determine which sites have the best chance for an economically successful redevelopment based on the needs of the community. This means developing a set of criteria unique to your individual community that meets its needs and goals. This system also gives the community an opportunity to create a developer's package for that site. Ranking scenarios can include criteria for the potential for redevelopment, inclusion of community priorities, and the targeting of local efforts.

The ranking of sites can be an ongoing process as new sites are identified and the needs of the community change. Typically, however, site ranking will provide community leaders with a definitive list of sites prime for redevelopment. This list can also increase public awareness of those areas with environmental challenges, as well as providing an opening for the provision of incentives for redevelopment in these areas.

Numerous "breakout" meetings supplemented the Regional Consortium meetings. The meetings were designed to help each participating parish develop the parish's brownfields inventory, delineate ranking criteria for their brownfields sites, and to select a candidate site for site assessment and site evaluation.

The final step for the parish members was the challenging step of ranking their parish's sites. While each of the parish members felt that all of their sites were "top" sites in terms of characteristics and needs, each site had unique issues, which distinguished it from the other sites. Ownership, size, level of suspected contamination, and land supply are but a few of the differences.

These sites will be the first in each parish to be eligible for further funding and development incentives under state and federal brownfields programs. The candidate sites for the member parishes selected for the assessment and evaluation process are as follows:

Jefferson Parish: Jefferson Parish Incinerator site

This site is located on David Drive near Airline Drive (also known as U. S. Highway 61). The site is the location of the former Jefferson Parish incinerator. Most recently the site has been used as the parish recycling center and parking area for the parish maintenance facility. For over twenty years the site was used as one of the public facilities that burned solid waste collected from the residents of Jefferson Parish. The concept for the redevelopment of this site is based on a dual approach that incorporates the planning concepts of public-private partnerships and mixed-use development. The anticipated use of this property is a combination of a public recycling facility and a privately operated sports complex for in-line skate hockey.

Plaquemines Parish: Jefferson Lake Canal site

This site is located on Louisiana Highway 23 approximately 25 miles south of the urbanized area at the Jefferson Lake Canal. The site is parish-owned land, which dead ends where the Jefferson Lake Canal intersects Louisiana Highway 23. The former site of a public boat launch and other parish marine activities, it now lies abandoned with sunken vessels and is listed as the top brownfields redevelopment potential for Plaquemines Parish.

St. Bernard Parish (1): St. Bernard Parish Main Road Yard site

This site is located at 3940 Paris Road in Chalmette. The site itself is a vacant, rectangular parcel of flat land that covers approximately two acres. Concrete slabs indicate the location of previous structures. Behind the site is the abandoned parish incinerator, partially converted to office space currently being used by parish personnel. End use of the redeveloped site is projected to be an extension of parish-based services.

St. Bernard Parish (2): St. Bernard Parish Old Paris Road Landfill site

This site is located at 5039 Paris Road in Chalmette. The site was the location of the now closed St. Bernard Parish landfill that once operated as an open dump. There are no existing buildings on the site; however, there is a waste transfer station in the front of the property.

St. Tammany Parish: Camp Villere Landfill site

The site is adjacent to the Camp Villere National Guard Training Facility off Airport Road in Slidell. The site was the location of the now closed landfill.

Creating a database of sites: simmer for future use

A Geographic Information System (GIS) is a computerized map, which is connected to a database. This combination allows the user to juxtapose, or layer, a wide variety of information about a given geographic area, from lot lines and sewer locations to census population and income data. In addition, GIS can be a useful tool in site identification and ranking efforts for a brownfields team.

Working with the resources of the RPC and aided by a grant from the EPA, the New Orleans region has been selected as a part of a national pilot grant to map brownfields sites using the latest GIS software. By using GIS to map sites, the RPC can and has placed these maps on the Internet for use by the general public. In addition, this database provides a valuable tool for tracking the progress of sites targeted for redevelopment, as well as documenting successful redevelopment stories.

How to assess site conditions: adjusting the recipe

After you have your ingredients for brownfields redevelopment together, now is the time to further tailor the recipe. Once you have developed a list of potential brownfields sites in your community, or you have selected one that is of particular interest, you will want to determine if any contamination may be present that could threaten your neighborhood or hinder redevelopment. Not all old industrial or commercial properties will be contaminated— this will depend on past land uses and housekeeping practices. A brownfields site may also be contaminated as a result of illegal dumping after operations ceased.

Sources of public information that can help assess the possibility of property contamination include newspapers, historical documents, local or county government records, and environmental permits. For example, details about site ownership are available from land titles and property tax files. Past property uses also can be documented from historical fire insurance maps (also known as Sanborn Maps), which may be located on microfilm at libraries, colleges, and local historical societies.

Once a brownfields site has been identified, the environmental risks and potential cleanup costs must be thoroughly investigated. This is called an Environmental Site Assessment (ESA). An ESA is generally conducted by a qualified specialist in conjunction with the sale, purchase or refinancing of real estate. The investigatory phase of an ESA has two parts. Phase I is basically a site history. Phase II involves on-site sampling and testing. The actual cleanup plan of the site is sometimes referred to as the Phase III. Sometimes the results of Phase I and II assessments show that the costs of cleanup (Phase III) are too great to support redevelopment and reuse.

Attract developers and establish a redevelopment plan: season to taste

When considering a site for brownfields redevelopment, communities should consider sustainable developments. Communities should ask themselves questions in order to realize if a development is sustainable. For example, can individual sites be combined into larger tracts to facilitate greater redevelopment opportunities, or will the project build on public or private redevelopment efforts already underway? Working with local economic development agencies and organizations, as well as local residents, will ensure that the brownfields redevelopment is compatible with surrounding neighborhoods and fills a need in the local market. The more diverse the economy is, the more sustainable it is.

Brownfields liability: the splatter shield

However, most brownfields sites are not federal Superfund sites. Brownfields cleanup and redevelopment is occurring largely under a rubric of state laws that reduce barriers erected by the Superfund liability approach, and provide numerous incentives to spur brownfields cleanup and redevelopment. Voluntary Cleanup Programs (VCPs) encourage voluntary brownfields cleanup, as well as making cleanup a faster, more efficient process by assuring property owners that the EPA will honor remediation results that have complied with the state's VCP requirements. This assurance is achieved through a Memorandum of Agreement (MOA) that the state signs with the EPA, making the state and the EPA partners in the redevelopment process. Some brownfields developers will utilize new liability protection tools now available from private insurance markets.

Completed activities and quantitative measures of success

As a result of RPC efforts to date, the parish members of the Consortium have identified over 300 brownfields sites where immediate development is warranted and desired. Candidate sites in each parish have been selected to demonstrate Phase I and Phase II environmental site assessment procedures to the public and potential real estate developers. The environmental site assessments underway are scheduled for completion in 2001 with the aim of developing full site development plans by early 2002, attracting developers and financing of the cleanup and improvements to these sites in 2003.

Brownfields resources

Voluntary cleanup programs

Voluntary Cleanup Programs (VCPs) are state-sponsored programs that encourage private parties to conduct cleanups of contaminated properties in the absence of state enforcement measures. VCPs establish clear cleanup goals that when met, absolve new site owners and operators from any state liability stemming from the original contamination. There are no federal laws or standards that regulate state VCPs. VCPs can vary widely in eligibility, cleanup standards, and liability provisions.

More than 40 states now have VCPs under which private parties that voluntarily agree to clean up a contaminated site are offered some protection from future state enforcement action at the site, often in the form of a "No Further Action" letter or "Certificate of Completion" from the state. Such state commitments do not affect the EPA's authority to respond to actual or threatened releases of hazardous substances under CERCLA.

The EPA does, however, encourage its regions to use the negotiation of voluntary cleanup program Superfund Memoranda of Agreement (SMOA) as an opportunity to define a division of labor between the region and the state by defining what kinds of sites fall within the SMOA. The EPA has developed a framework for these negotiations. This framework provides suggested language for stating a region's intended treatment of sites participating in a VCP program covered by a SMOA.

In 1995, the Louisiana Legislature passed Act 1092, known as the Voluntary Investigation and Remedial Action law, which allows property owners and other persons who clean up properties to risk-based standards to obtain a Certificate of Completion from the Louisiana Department of Environmental Quality (LDEQ). With this certificate, the property owner and any subsequent owners of the property are released from further liability under state law for past contamination at the site. In effect, the certificate allows potential buyers to acquire and remediate brownfields properties without fear of state Superfund liability.

The Louisiana Legislature established the framework for its VCP by passing the Voluntary Investigation and Remedial Action Act (LA R.S. 30:2272.1 and LA R.S. 30:2285) in July 1996. The act has as its primary goal, the redevelopment of former industrial and commercial properties. LDEQ published the minimum remediation standards entitled, "Risk Evaluation/Corrective Action Program" (RECAP) on December 20, 1998.

On April 20, 2001, LDEQ promulgated the Louisiana Voluntary Remediation Regulations to implement this statute and formalize its VCP, the Voluntary Remediation Program (VRP), within the department. LDEQ will use the statute and these new regulations to facilitate voluntary cleanups.

Liability assurances

The EPA's Brownfields Economic Redevelopment Initiative is designed to empower states, communities, and other stakeholders in economic redevelopment to work together in a timely manner to prevent, assess, safely clean up, and sustainably reuse brownfields. The EPA's Brownfields Initiative strategies include funding pilot programs and other research efforts, clarifying liability issues, entering into partnerships, conducting outreach activities, developing job training programs, and addressing environmental justice concerns.

The EPA is working with states and localities to develop and issue guidances that will clarify the liability of prospective purchasers, lenders, property owners, and others regarding their association with activities at a site. These guidances will clearly state the EPA's decision to use its enforcement discretion in specific situations not to pursue such parties. The EPA anticipates that these clear statements will alleviate concerns these parties may have and will facilitate their involvement in cleanup and redevelopment.

Federal brownfields legislation

President George walker Bush signed the Small Business Liability Relief and Brownfields Revitalization Act (P.L. 107-118, United States of America) on January 11, 2002. This bill if the first major environmental law of the mew millennium. Its provisions clarify and expand many of the concepts, funding opportunities and "tools in the toolbox" of brownfields redevelopment.

Beginning with a legal and concise definition of brownfields: `*brownfield site' means real property, the expansion, redevelopment, or reuse of which may be complicated by the presence or potential presence of a hazardous substance, pollutant, or contaminant.* This bill continues in several key areas including: increased funding for site assessments; streamlined grant procedures; direct grants for cleanup; new provisions for petroleum-contaminated sites; Superfund liability relief for small business, innocent property owners, purchasers and nearby property owners; and increased funding for the new roles for tribal and state brownfields programs.

The specific provisions of the "Brownfields Act" of 2002 enhance and expand advantages and opportunities such as those detailed in *The Brownfields Cookbook*. In terms of the "Cookbook" approach; look for specific provisions in applications in the Congressional Budget for Fiscal year 2003 (beginning October 1, 2002) such as:

- increased funding for Phase I and Phase II site assessments,
- direct grants (instead of revolving loans) for site cleanup,
- new funding for petroleum contaminated sites,
- new opportunities for direct grants to localities and non-profit groups,
- clear liability protection for stakeholders groups,
- new funding for sites and tribal responsibilities in brownfield redevelopment programs, and
- clear delineation of federal and state roles on site cleanup completions issues.

The overall effect of this new bill will be seen from two different perspectives. The first is the detailed and specific application of the above listed provisions and their enhanced impact on the brownfields toolbox. The second and more important effect will be new support, a positive attitude, and increased acceptance of the brownfields philosophy of cleaning up contaminated sites. Moreover, re-entering these properties into active use through national environmental legislation and new funding will buttress this powerful land use redevelopment solution for the next millennium.

White book of urban Spanish brownfields

P. F-Canteli[1], A. Callaba[1], C. Alonso[1], H. Palacios[1] & I. Iribarren[1]
[1] *Department of Mineral Resources and Environment, Geological Survey of Spain (IGME).*

Abstract

The Geological Survey of Spain (IGME) in a co-operative effort with the Spanish Ministry of Environment, is currently setting up the foundations for a white book on urban brownfields. This document aims to fill the knowledge gap on the extent and nature of this emerging issue. In order to acquire an overall and precise view of the Spanish situation, information is being collected from the different Regional Authorities involved in this question (Environment, Land Planning, Economy and Social Services departments). Diverse projects carried out on important industrial areas affected by this situation are being consulted to compile the measures adopted to make a better neighbourhood, a better community and a better quality of life around them. Among the primary goals of the above mentioned paper is the introduction of enough information on environmental problems of these properties and their redevelopment possibilities in order to design fitted policies, and also the increase of the awareness on this topic between policy makers and land planners.

1 Introduction

For the last 30 years, Spanish economy has experimented several changes. At the end of the fifties, Spanish economic system depended on some sectors, which were hardly influenced from international Market such as mining, iron and steel, metallurgic, or naval industries. A new concept of international market policy appeared in the early 1960s, brought a deregulated and more competitive Market. Industries, mines, and factories were obligated to improve their technology and processes. However, the need of considerable amounts of investment to modernise the obsolete equipment and plants made a lot of them (mainly medium and small sized) go bankrupt.

Some of significant but non-competitive public companies, in which the social dependence was absolute in the area, were saved thanks to a protectionist policy of the government. However, that situation could not be indefinitely maintained. At the beginning of the eighties, an industrial rationalisation was initiated, and most of these companies and other satellite ones had to close -total or partially-. Hence, facilities, some of them with several associated contamination problems, were abandoned giving rise to emerging marginal zones inside the cities.

It is obvious that not all Spanish brownfield sites have appeared because of industrial rationalisation. However, the number and extension of those permit us to use them as the starting point to acquire a precise view of the Spanish problems.

2 Research Group of Geological Survey of Spain

The Geological Survey of Spain (Instituto Geológico y Minero de España, IGME) is a public organisation which is addressed to natural resources and land investigation, and is traditionally linked to mines and industries. It can be underlined from its activities "to research, develop and apply analysis methods, characterisation, assessment and protection against contamination; and to the remediation of polluted soils and aquifers; and to land used as waste deposit". These aspects explain the arisen interest to establish a brownfield sites research group.

On the other hand, it is a mission of IGME to provide knowledge and information to local, regional or National Authorities and to the society in general, about land resources and technologies. In this way, the Geological Survey of Spain in co-operative effort with the Spanish Ministry of Environment is currently setting up the foundations for a white book on brownfield sites.

3 Urban brownfields *versus* rural brownfields.

As it was mentioned above, the idle or underused soil in which social and pollute effects are worst, is situated inside or next cities. That circumstance, similar to other industrial countries, has particular features here.

In spite of the fact that the total need for land in Spain is less important than other countries, Spain shows a strong polarity on density of population and production. That means that whereas next to the coast and metropolitan cities are between 500 and 800 habitants per square kilometre, less that 25 habitants per square kilometre are in the rest. Consequently, land use and land planning is unbalanced since both industry and urban growth is concentrated around certain zones. This fact allows us to focus on urban brownfields and to consider rural brownfields as a less priority problem.

4 Objectives

The main goals that the Geological Survey of Spain aims with this white book are: i) To avoid the knowledge gap on urban brownfield sites. ii) To compilate information about current situation on Spanish brownfields. iii) To identify the problems related to the reuse these properties and to suggest measures to make a better neighbourhood, a better community and a better quality of live around them.

4.1 Avoiding the knowledge gap on urban brownfield sites.

The first condition necessary to correctly reuse brownfield sites is knowing the real problems associated with them. Nevertheless, early interviews have made clear that there is a wide ignorance about them. For instance, it is observable that the most of the projects carried out have been developed as independence project of remediation and new construction, with the consequent unnecessary waste of time and money. However, these gaps can be avoided by releasing the concept of brownfield site and the environmental, social, economic and land planning issues involved in its management.

On the other hand, it can be considered that there are generally different authorities in charge of the environmental, social, economic and land planning aspects, and sometime there is apparent confrontation between their proceedings. Therefore, as a previous step to define our framework, the legislation, programs and actions that can be applied to brownfield areas and developed by every one of them has been codified and classify.

4.2 Compilating information about current situation on Spanish brownfields.

A questionnaire has been designed with the purpose to define the magnitude of the urban Spanish brownfields and the needs to revitalise these properties. Two different sections have been considered. The first of them deals with brownfield sites no involved in a reuse project, and the other one is focused on brownfields sites in which reuse project has been yet developed. Starting from these data, it will be possible to define why these areas are not attractive to new investment

Both sections have different groups of questions which have been defined based on all the possible aspects involved in urban brownfield sites, that is, environmental, social, economic and land planning issues.

Diverse sources have been selected to answer this questionnaire. First, people in charge from local, regional and National governments. Secondly, groups or foundations set up to promote a reuse project. And finally, communities currently affected by abandoned facilities. Some European Programs have been also consulted since there are some revitalisation projects which have been financed by them, just as RESIDER, RECHAR, URBAN, RETEX, KONVER, and by other European financial aids as CECA.

4.3 Identifying the problems to reuse these properties and suggesting measures to improve the community and the quality of live.

Once the different features associated with brownfield sites are collected, their assessment will be carried out.

When the main needs are identified, a set of actions to promote and stimulate the redevelopment of abandoned sites and against the use of greenfield sites will be able to propose.

5 Questionnaire

In order to complete the paragraph about data collection procedures, a brief description of the questionnaire including both sections and their corresponding features is presented.

5.1 Brownfield sites no involve in a reused project

This section is designed to the abandoned or underused sites in which real or perceived contamination problems can be present, and in which intervention before new use is required.

The main source of information is local, regional and national authorities in charge of Environment, Social Services, Economy and Land Planning Departments. So this form is been e-mailing and posting to the Autonomous Regions, to the chief town of every province, and to some city in which it be suspected by their industrial history that brownfield sites are presented. The questionnaire goes with an explication text about how to fill it in order to obtain homogeneous information.

The collected features are described next.

5.1.1 General description of the site
Total affected area; abandoned or underused; time without use; former land use; information source.

5.1.2 Environment
Real or perceived contamination problems; extent of contamination for soil and groundwater; pollutants; migration paths of pollutants; how much people can be affected?

5.1.3 Social features and economy
People dependence on the former activity; unemployed rate; economic level; estimation of land value.

5.1.4 Land Planning
Nearness to urban centre; surrounding area type; singular buildings and structures; future land use planning.

5.2 Brownfields sites in which reused project has been developed

This section aims to record material about land recycling. The main goal are identified the possibilities to new projects and the measures to eliminated the environmental, social, financial and planning barriers.

The consulted sources, in addiction to above mentioned, will be group of stakeholders involved in the redeveloping program.

The aspects involve in this section are described next.

5.2.1 General description of the project

Total affected area; sites involved in redevelopment project and former land use of every one; project status; information source.

5.2.2 Design of the project

Starting and finishing date; objective of redevelopment; future land use type; surrounding area land use; stakeholders; problems associated to negotiation process.

5.2.3 Environment

Type and extent of contamination problems for soil and groundwater; migration paths of pollutants; pollution assessment; type of remedial planning; remedial measures; aftercare monitoring.

5.2.4 Economic features

Total expenses; property value before and after project; financial model.

5.2.6 Social features

Type and measures for public information; new public buildings.

5.2.6 Land Planning

Nearness to urban centre; singular buildings and structures reused and future land use planning.

6 Current Status

Currently the questionnaires have been sent to selected organisms and groups and their answers are awaited.

First interviews have pointed out the following main barriers to redevelop of these properties: a) Uncertainty about environmental liability since gaps are still present; b) Ignorance of general associated problems to brownfield sites; c) Insufficient programs and activities to lead private investment to reuse brownfields.

It is expected that the white book of urban Spanish brownfields will be ready by the end of 2002.

Section 3:
Environmental assessment

Public health role in redevelopment efforts

J. J. Reyes, R. C. Williams & P. McCumiskey
Agency for Toxic Substances and Disaease Registry, Department of Health and Human Services, United States of America

Abstract

Brownfields should be remediated and redeveloped, but what should replace them? What kind of assessment and cleanup is warranted? What are the public health implications of the proposed reuse? All of these are questions that public health practitioners and investigators can and do respond to when needed. As professionals sensitive to the complex needs of low-income and minority communities, public health officials can facilitate closer integration of environmental protection, economic sustainability, social justice, and health promotion. The U.S. environmental public health community has played an important role in redevelopment initiatives and continues to make important contributions in this field.

1 Introduction

In the United States, there is a renewed interest in urbanism driven by individuals seeking to live closer to work and shorten their daily commutes. As a result, federal and state governments, and local communities are providing incentives to restore to productive use abandoned or contaminated lands. Such revitalization helps to create jobs, remove physical decay and contamination, improve public safety, and eliminate social disparities. The resulting benefits, and improvements greatly contribute to the establishment of a sustainable and healthy community. In fact, redevelopment is typically and correctly promoted as a means to economic revitalization and better health through improved quality of life. Stated simply, redevelopment, the resulting community revitalization, and public health are inseparable. The quality of housing, unemployment, green and recreational spaces, transportation, social services, and economic prosperity all influence human health outcomes and help define the health status of a community. An effective public health system must participate in decisions about housing, public safety, land use, and other vital realms of social life that affect the creation and sustainability of a healthy community. [1]

What has not always been evident in revitalization initiatives in the United States is how federal, state, and local public health departments can be involved in the redevelopment and reuse of contaminated properties. Indeed, there is no legislative mandate that requires involvement of public health entities. Despite the recent (January 11, 2002) passage of Public Law 107-118, the Small Business Liability Relief and Brownfields Revitalization Act (which under Title II, Subtitle A, allows for up to 10% of the financial assistance funding for revitalization to be used for "monitoring the health of populations...at Brownfield sites"), the participation of public health entities in the United States is still a voluntary proposition. This situation poses various important questions. Is the public health contribution justified given the complexities inherent in redevelopment efforts? Does the participation of public health entities further complicate the relationship-building process that must be established for a successful project? We believe that public health can contribute appreciably and facilitate the revitalization effort by preventing community health concerns from reaching a crisis stage that may affect project progress and by ensuring the future development of healthy and sustainable communities. [2]

2 Case studies

Across the U.S., the redevelopment of an estimated 500,000 abandoned, or underutilized properties is complicated by real or perceived environmental contamination. As a result, the question of public health involvement becomes one not only of how but when. However, the need for public health involvement might not always be evident and this misunderstanding might sometimes have serious implications. Such is the situation in these case studies. Although they predate more recent redevelopment initiatives under the U.S. Environmental Protection Agency (EPA) Brownfields Economic Redevelopment Program, they illustrate why public health considerations are important when determining future property uses.[2]

2.1 East 10th Street Site, Delaware County, Pennsylvania

The East 10th Street Site is a 36-acre property located in Delaware County, Pennsylvania. Beginning in 1910, the site was used for the manufacture of rayon and cellophane. In 1977, the property was divided into 23 lots owned by six different entities. These lots contain nine buildings that formerly housed the rayon/cellophane production or storage facilities. In the late 1980s, considerable on-site demolition and building renovation converted the lower floors of two buildings (Nos.1 and 2) into individual offices housing commercial and retail establishments including a day care center, candy manufacturer, restaurant, dental office, a Boy Scout meeting room, and a senior citizen center.

In 1990, EPA Region III Emergency Response staffs in Philadelphia were called to the site because of concerns about improper storage of poly-chlorinated biphenyl (PCBs) containers and of the existence of free asbestos in on-site buildings. EPA's inspection revealed asbestos-contaminated bulk material and asbestos fibers in the air on the upper floors of Building 1. In addition, isolated locations of bulk asbestos and numerous physical hazards were noted outside the building. Concerned about the public health implications of existing conditions, EPA requested that the Agency for Toxic Substances and Disease Registry (ATSDR) evaluate the situation and make appropriate recommendations for follow-up actions. ATSDR and EPA recommended that the day care and senior citizen center operations be relocated until the problem could be addressed. EPA then arranged for cleanup of the building to the extent permissible under authorizing legislation (case law limits EPA's ability to address asbestos when it is considered a structural component of a building). Additionally, ATSDR has provided guidance for protecting public health during the cleanup.

Because of community concerns about the effectiveness of the asbestos cleanup and subsequent maintenance efforts, in 1993 ATSDR and the Pennsylvania Department of Health arranged for indoor air sampling in Building 1 that revealed elevated asbestos levels. Additional consultations recommended implementation of an effective operation and maintenance program for asbestos in the buildings. Under governing asbestos law, the organization responsible for implementing this recommendation is the building owner and operator with oversight by the Pennsylvania Department of Environmental Protection under the state's Voluntary Cleanup Program.

2.2 Hoboken, New Jersey

In 1993, a condominium association in Hoboken, New Jersey, purchased a 5-story building that once housed various manufacturing operations, including production of mercury vapor lamps. After undergoing an environmental audit and obtaining clearance, the building was converted into apartments and studios. As renovations advanced, members of the association began moving into the building. Soon after, residents observed drops of elemental mercury in their apartments and hired a private contractor to remediate the contamination. Subsequent sampling revealed that mercury still existed at levels of health concern. Faced with this disturbing information, the residents notified the Hoboken Board of Health, who in turn requested assistance from the New Jersey Department of Health and Senior Services (NJDHSS) and ATSDR.

Additional sampling results showed mercury vapor levels higher than that recommended for residential properties. Indoor air mercury concentrations in the breathing zone were of public health concern especially for children. ATSDR and NJDHSS arranged for medical testing of residents, which revealed elevated mercury urine levels. As a result, ATSDR issued a public health advisory and the Hoboken Board of Health declared the building to be unfit for human habitation and ordered the premises vacated. EPA arranged for temporary relocations and ATSDR referred residents for medical follow-up. [3] The contaminated building and surrounding area were subsequently remediated. The public health response prevented the further exposure of the residents. Subsequent clinical evaluations supported the relocation decision and because of the rapid response, only one long-term, medically significant outcome was identified. Children underwent followup to evaluate the effects of on their growth and development.

3 Federal, state, and local public health action in redevelopment

EPA defines brownfield properties as "...a site, or portion thereof, that has actual or perceived contamination and an active potential for redevelopment or reuse." [4] Clearly, environmental issues and propensity for economic gain drive redevelopment activities. Without the brownfields program, there would be few incentives for lenders, investors, and developers to clean up and redevelop areas where they could be held liable for contamination they did not create. To promote these projects, EPA and state programs provide grants for site assessment and clean-up planning; clarify and provide liability relief; promote partnerships for the redevelopment effort among various government agencies and communities; and foster local job development and training initiatives. These programs remove barriers to cleanup and redevelopment and return abandoned and underutilized sites to productive functions. Most declare that minimizing or eliminating human exposure to pollutants is a key benefit. In this context, a critical partnership is the participation of public health.

Public health practice assures conditions where people can live in a healthy environment. Science and social norms and values influence and shape the practice of public health. Reliance on science to provide a basis for remediation and clean-up levels is expected, but public sentiment about government action and personal beliefs on what constitutes a health hazard might dictate the acceptance of redevelopment projects in communities. Particularly for brownfield sites, organized community efforts are essential to addressing the public's interest in health and safety and the success of the project. Public health practice ensures that urban redevelopment and reuse result in environments in which people can live healthy lives. Here as in other areas where public health operates, its role is one of problem definition, hazard assessment (on the basis of the problem), intervention (on the basis of the problem and the hazard it poses), and ultimately assurance to make sure the intervention was appropriate or to identify further action needed. Federal, state, and local health entities have successfully responded to the public health needs of the brownfield initiatives, and their involvement continues to grow as public health practitioners work to protect the health and quality of life of persons living and working near redevelopment properties.[2]

3.1 Federal public health efforts

ATSDR analyses of brownfield sites points out that nearly 30% contain chemical hazards, 50% contain physical hazards, and approximately 50% require further public health analysis.[5] This issue becomes more complicated in that many redevelopment sites are located in or near racially and ethnically diverse communities that may already be affected by economic and environmental inequities. The public health role is to ensure that protection of human health is considered in all redevelopment and sustainability activities. To that end, ATSDR has worked with other health partners to develop methods and tools for addressing public health concerns.

In collaboration with EPA, ATSDR has funded a cooperative agreement program for enhancing local health department capacity to effect public health interventions at brownfield and redevelopment properties. ATSDR's program assists local health departments to develop and implement strategies in support of property remediation and redevelopment, and to ensure that reuse does not present environmental public health hazards to current and future community residents. A critical part of this program is the collaboration of local health departments with local governing officials, community-based organizations, and state governments to ensure that public health is an integral part of the redevelopment team. These partnerships are crucial to developing public health practice related to reuse issues, building environmental health capacity locally, assuring the principles of environmental justice, and implementing communication and empowerment strategies to enhance community support and participation in brownfield and redevelopment initiatives.

The program stresses the need for the consideration of public health issues in the earliest phases of remediation and redevelopment so that interventions can be timely. Ten (six in 1998 and four in 2001) local health departments have received assistance under this program and have produced notable advancements in public health practice that demonstrate the value of this partnership. ATSDR is currently working with the National Association of City, County Health Officials (NACCHO) to develop a synthesis of effective methods used and lessons learned in these pilots. This synthesis can serve as a guide to other public health officials engaged in similar activities.

In collaboration and with assistance from ATSDR, NACCHO has also developed guidance entitled *Public Health Principles and Guidance for Brownfields Policies and Practices* [6], and *Community Revitalization and Public Health* [7] to assist local health departments in understanding and carrying out their role. This guidance stresses the importance of working closely with the community and the value of public health participation in the economic redevelopment processes. It also encourages an enhanced vigilance to health consequences during redevelopment. This guidance stresses that protection of public health is the highest priority, that the health of the community is inextricably linked to economic prosperity, and that economic redevelopment is vital to creating healthy and sustainable communities. The advice presented is strengthened by the input and direction gained from four public forums on community revitalization and public health. The forums provided information on the role of local health departments, the resource needs to sustain public health involvement, how health entities and communities can work more collaboratively, and what strategies help to ensure success in economic redevelopment.

After more than 5 years of work evaluating several hundred redevelopment projects, ATSDR and its state and local public health partners have gained considerable experience on the appropriate public health response to redevelopment. This work indicates that this kind of response can follow a similar framework as that used for abandoned hazardous waste sites. For these sites, ATSDR and state health partners use the methods in the *ATSDR Public Health Assessment Guidance Manual*[8] and information in *ATSDR's Toxicological Profiles*[9] series to determine who is exposed, what are the health implications of such exposures, and what additional actions and interventions are needed. The public health assessment process is a weight of evidence evaluation that draws upon the expertise of a multidisciplinary team with experience in epidemiology, toxicology, medicine, risk assessment, risk communication, engineering, and other sciences. This framework includes (1) involving and seeking input from stakeholders, (2) evaluating contamination, (3) assessing exposures, (4) assessing toxicity, (5) conducting epidemiological activities, as warranted, (6) providing health professional and community-based education, and (7) designing and implementing public health interventions.

Public health issues at redevelopment properties might not be as complex or extensive as those found at hazardous waste sites. However, key investigation methods of the public health assessment process (e.g., contaminant identification, exposure analysis, population characteristics and susceptibility, health impact evaluation, and baseline health outcome assessment) are useful resources and have served environmental health professionals in redevelopment initiatives. Realizing that an abbreviated approach could provide more clear guidance to local health officials, ATSDR has developed and is currently testing an exposure assessment algorithm for brownfield properties. The algorithm contains the major elements of the public health assessment framework that can be successfully applied to most redevelopment efforts, including the critical components of identifying contamination, assessing hazards, involving the community, providing education, and designing appropriate interventions. The algorithm is a scoping guide to evaluate exposure and address community health concerns either directly or through referral to other agencies.[2]

As part of the Brownfields National Partnership, ATSDR is also working with other federal agencies to address clean-up and reuse issues in a coordinated manner. This multiagency partnership lead by EPA offers technical, financial, and other assistance to brownfields communities and works to demonstrate the benefits of collaborative activities on brownfield properties. Because of this effort, ATSDR and the U.S. Department of Housing and Urban Development (HUD) are currently working to establish a partnership agreement to support redevelopment needs and spur economic growth in distressed communities that is protective of public health. This agreement will focus on fostering an environmental public health evaluation and perspective at HUD properties designated for redevelopment/reuse.

3.2 State public health efforts

State public health participation in ATSDR's cooperative agreement program ensure that environmental investigations and remedial actions consider public health hazards. This state support can produce cleanups that are more efficient and focused on the most relevant health concerns, and early involvement by the state agencies greatly improves the scope of the public health issues evaluated. Through a range of activities (e.g., training of local public health staff, site visits, input into sampling strategies, review of sampling data, community outreach and education, and clearly delineating the public health implications of these properties), state public health entities have assured that the protection of public health is foremost in redevelopment programs. However, our collective experience shows that in order for these outcomes to be realized, public health activities must be initiated at the time of property selection, and maintained through the assessment, cleanup, and redevelopment phases.

Connecticut's Department of Public Health (CTDPH) brownfields and redevelopment assistance starts with developing linkages between federal, state, and local governmental entities and ensuring the identification of public health issues at the earliest stages of redevelopment. As evaluations progress, CTDOH and local health departments take an active role in evaluating site information, addressing community concerns, and identifying and resolving public health issues. During site remediation and redevelopment, their focus turns to exposure prevention and ensuring that reuse of these properties does not present environmental health hazards to current and future community residents. On the basis on their experience, the CTDPH has developed the *Public Health Protocol for Brownfields* [10], which assists local public health agencies in identifying and evaluating environmental exposures, interpreting their relation to potential adverse health outcomes, providing recommendations regarding actions to protect public health, communicating their findings and providing information to the community, and evaluating future-use impacts on public health.

The Wisconsin Department of Health and Family Services (WDHFS) has an active role in administering the Wisconsin Land Recycling Act and Brownfields Environmental Assessment Program (BEAP) programs. Early participation and assessment of properties by WDHFS helps to address concerns from nearby residents about the risks to health posed by environmental contamination in a property. Similar to the Connecticut protocol, Wisconsin staff trains, collaborates, supports, and provides assistance to local public health departments in evaluating contaminated properties, determining if future property uses might pose a public health hazard, and communicating the health hazards of these properties to communities. [11]

3.3 Local public health example

The essential element of redevelopment is local participation, which is necessary to increase community involvement, promote the principle of sustainable redevelopment, and provide for site assessment and cleanup. Redevelopment success greatly depends on community collaboration from initial assessment to decisions on future uses of the property. Through local health department participation, community involvement on redevelopment properties can initiate and facilitate information exchange among the stakeholders. Local public health entities also can work with community groups to identify health issues and implement actions to address these concerns. During property assessment, local health departments can assist in the examination of historical uses from a public health perspective, evaluate sampling plans to focus on human exposures, and interpret sampling data for health significance. During remediation or cleanup, local health departments, in concert with state and federal health entities, can provide site-specific advice. When continued involvement has occurred, public health officials can address and often assuage community health concerns based on their detailed and intimate knowledge of the project plans.[2],[7]

3.4 Weld County, Colorado

The redevelopment of a waste disposal site in Weld County, Colorado, illustrates the benefits of local public health participation. The site proposed for reuse was a former liquid waste disposal facility that operated from 1971 to 1986. The facility was abandoned in 1986, and came under investigation by EPA. During its 15 years of operation, the facility discharged liquid waste through a subsurface infiltration system, which was later upgraded to a liquid evaporation system. Aware that the site was being unused and of community concerns that its present state might present a hazard for local residents, the county governing commission asked the Weld County Health Department (WCHD) to investigate the situation and determine what, if any, contamination remained from the operation of the facility. The WCHD solicited the assistance of the Weld County Public Works Department, which excavated several observation pits and performed environmental sampling in the area. The assessment revealed the soils beneath the property were contaminated. The WCHD began to work with a prospective local developer to establish a clean-up plan involving soil treatment. The developer implemented and paid for the agreed-upon remediation and the site is now under development for residential use. The county believes that future public health impacts from past contamination have been addressed and is pleased to see that the site is back on the county tax rolls as productive property. [12]

3.5 Woonasquatucket River Greenway, Providence, Rhode Island

In Rhode Island, the smallest state in the nation, the state health department has jurisdiction over all public health functions including those normally carried out by local and municipal health departments. In this situation, the Rhode Island Department of Health (RIDOH) performed the roles of both municipal and a state public health department, but largely functioned as a local health entity. The Woonasquatucket River Greenway project contains two brownfield sites, the Riverside Mills and the Lincoln Lace and Braid properties. The 6-

acre Riverside Mills site, in Providence's Olneyville neighborhood, is a former mill that has been used for waste disposal and as waste-oil storage facility. The Lincoln Lace and Braid site is a vacant 9-acre parcel in the city limits of Hartford. With resources from an ATSDR cooperative agreement, RIDOH began to work with local planning and developing entities, state environmental agencies, and nonprofit community organizations to address public health concerns.

In this role, RDOH provided guidance and advice to ensure that the future reuse of these properties would be protective of public health. RIDOH also worked with community organizations to organize a public health team, which helped to set up various meetings and workshops (in both English and Spanish) and responded to residents' concerns about remediation and reuse proposals. Because of this engagement, two nearby communities, where redevelopment projects on previously contaminated properties were being considered, requested and became part of the team and made use of the services provided. The RIDOH, with input from the public health team, worked with the University of Rhode Island (URI) to produce a chapter, entitled *Home*A*Syst*, in URI's cooperative extension educational system for communities. The chapter (also published as a stand-alone guide), is bilingual and addresses brownfields and redevelopment issues and is part of a handbook that focuses on pollution identification, risk assessment, and source reduction for communities.

4 When to involve public health

From the case studies described above, public health resources would appear to be readily available for involvement in redevelopment projects. However, the scope and complexity of environmental health problems faced by local, state, and federal public health agencies continue to increase, requiring new dimensions in community participation, dispute resolution, applied science, and public health practice. Environmental health challenges will increase in the future, but because of competing priorities, the number of public health professionals and the services they provide is not likely to increase. New challenges, such as bio-terrorism, continue to emerge. Meanwhile, economic globalization will raise new public health and quality of life issues not previously considered. In this setting, public health entities must now begin their work in property redevelopment and reuse initiatives. Faced with a myriad of exposure scenarios at thousands of properties, public health professionals must rely on a triaging process to determine where their involvement is needed most.

To aid communities, developers, and environmental entities in determining when to engage health officials, properties with priority characteristics have been suggested as triggers for public health involvement. Such priority properties include those whose future development includes residential land use, child-care centers, schools, other sensitive population facilities, or environmental justice concerns. Other situations include properties in or near residential areas, where current levels of contamination is of health concern, and where future development includes industrial or commercial use and significant contamination or suspected contamination exists. These are properties where early and consistent involvement of public health entities can not only assure a project's success, but also result in better communication and involvement with the community and more timely and efficient cleanup and development. Those properties where future use includes intensive human contact with soils, water, or other media, such as in residential, school facilities, require greater characterization of contamination and often a more rigorous cleanup than traditional industrial property. Significantly contaminated areas might continue to pose public health hazards even when redeveloped as industrial or commercial facilities. When these properties are in or near a residential area, the remedial methods chosen must adequately protect those currently living and working near the site as well as those who might use the property in the future. Because of their location, brownfield property development might disproportionately impact low-income and/or minority communities. Hence, environmental justice issues must be fully considered. For each of these priority properties, the level of effort required and the extent of public health involvement will vary depending on site-specific conditions.[2]

5 Future challenges and opportunities

Community revitalization projects provide incentives for various stakeholders to embrace these initiatives. Restoration of abandoned and unsightly areas, more jobs, crime reduction, and improvements to the local economy are some of the value-added factors that drive redevelopment and are vital components to creating healthy and sustainable communities. Because of these projects, communities will be placed in situations where decisions and compromises must be made. Brownfield properties are promoted as attractive lures to developers with subsequent benefits to communities. However, these same communities remain attentive to possible public health issues arising from redevelopment and reuse of contaminated properties.

This is where public health involvement becomes inextricably linked to redevelopment. The strengths of public health entities in resolving health problems such as infectious and communicable disease control are well understood and as a result community trust often resides with public health officials. Their ability to provide timely assessments of environmental contamination and to closely interact with the communities they serve enables public health entities to respond effectively to revitalization efforts. Serving as collaborators, communicators, facilitators, translators, and trusted resources, local and state health officials can assure property redevelopment and reuse results in conditions where people can be healthy.

Integrating public health and economic revitalization efforts can bring success to brownfield and redevelopment initiatives. The involvement of public health in resolving health issues associated with the reuse of properties will require the availability of technical and fiscal resources. The success of this involvement greatly depends on factors like early and timely involvement, use of the public health framework for assessment and intervention, and triaging through priority properties. This involvement facilitates efforts to communicate and share information with communities that in turn help to gain community trust and cooperation.

Public health entities also need to have a positive economic outlook to achieve broader trust in redevelopment and reuse actions, while maintaining sustainable public health vigilance. This perspective will help to ensure that a community's economic revitalization will outlast its redevelopment project.

References

[1] Greenberg, M., Lee, C., Powers, C. Public health and brownfields. *American Journal of Public Health*, **88(12)**, pp.1759-60, 1998.

[2] Williams, R.C., Skowronski, E., Williams-Fleetwood, S. Brownfields: focus on public health. *Brownfields and Redevelopment Workshop Proceedings*, Water and Environment Federation: Washington, DC, October 1998.

[3] Orloff, K, et.al. Human exposure to elemental mercury in a residential building. *Archives of Environmental Health*, **52(3)**, pp. 169-173, 1997.

[4] U.S. Environmental Protection Agency. *Brownfields economic redevelopment initiative.* EPA 500-F-98-001, U.S. Environmental Protection Agency: Washington, DC, March 1998.

[5] Agency for Toxic Substances and Disease Registry. *Brownfields Fact Sheet*, U.S. Department of Health and Human Services: Atlanta, GA, 2000.

[6] National Association of City and County Health Officials. *Public Health Policies and Guidance for Brownfields Policies and Practices*, NACCHO: Washington, DC, August 1998.

[7] National Association of City and County Health Officials. *Community Revitalization and Public Health*, NACCHO: Washington, DC, June 2000.

[8] ATSDR. *Revised Public Health Assessment Guidance Manual (Draft),* 2002. Available ATSDR's Web site (http://www.atsdr.cdc.gov/publiccomment.html).

[9] ATSDR. *Toxicological Profiles*. Available at ATSDR's Web site (http://www.atsdr.cdc.gov/toxpro2.html).

[10] Connecticut Department of Health. *Public Health Role in Brownfields Initiative*, internal guidance, CDH: Hartford, CT, 1998.

[11] Wisconsin Department of Health and Family Services. *Public Health Participation in Wisconsin Land Recycling Act and Brownfield Environmental Assessment Pilot Properties*, WDHFS: Madison, WI, August 1998.

[12] Milne, T.L. *Testimony of the National Association of County and City Health Officials before the Subcommittee on Water Resources and Environment*, Committee on Transportation and Infrastructure, U.S. House of Representatives: Washington, DC, May 12, 1999.

Environmental balancing of brownfield redevelopment

V. Schrenk
Research Facility for Subsurface Remediation (VEGAS), Institute of Hydraulic Engineering, Universität Stuttgart, Germany

Abstract

Within the context of sustainable development, it is important to always consider the impact of activities, such as the preparation of brownfield construction sites (subsurface remediation/demolition of buildings), on the environment. In the Federal Republic of Germany, the consideration of environmental impacts, including those resulting from the remediation of contaminated sites, is defined in the German Federal Soil Protection Act.

To date, first "isolated" approaches for the environmental balancing of brownfield redevelopment measures exist which are comparable to the life-cycle assessment of industrial products. The Environmental Protection Agency of the Federal State of Baden-Württemberg (LfU) has developed a software tool for the environmental balancing of remediation techniques [1]; in Switzerland, investigations concerning the ecological impacts of building rubble disposal exist [2]. An examination or environmental balancing of complete brownfield redevelopment projects (subsurface remediation of the property, demolition of buildings, preparation of the site for new residence) has not been published so far. This may be due to the difficulties in obtaining the extensive data necessary for such a procedure. However, an extensive environmental balancing of brownfield redevelopment projects could be the basis for both an ecological and an economic optimization of brownfield redevelopment.

At the research facility for subsurface remediation (*VEGAS*) at the Universität Stuttgart, a research project regarding environmental balancing has been started with the aim of extensively balancing brownfield redevelopment projects under ecological criteria.

The main objective of the project is the development of a readily applicable instrument for the estimation of the ecological impacts during the redevelopment of brownfield sites for construction purposes. This tool should help to determine the ecological consequences of civil works quickly and easily and serve as a basis for an ecological optimization of the projects.

In order to reach this aim, numerous project examples are being examined and transferred into available environmental balancing tools. First results of these investigations show that the transport of contaminated soil and building rubble for off-site treatment or disposal has a high impact on the environment, which is reflected by the high total energy consumption and emissions resulting, amongst other things, from the distances to soil washing plants or to plants which recycle building debris. Furthermore, natural area is consumed if the excavated area is filled with clean gravel from a gravel pit or if the contaminated soil is deposited in a landfill. Under ecological criteria, in-situ and on-site remediation techniques have thus far proven to be better methods than off-site techniques. During further investigations, the ecological impacts of new innovative technologies for subsurface remediation will be determined.

1 Introduction

1.1 Legal conditions in Germany

The remediation of contaminated sites or the preparation of construction sites on derelict land (digging, transportation, operation of pumps and units) causes impacts on the environment (e.g. emissions, energy consumption, waste).

The consideration of these so-called secondary impacts on the environment has been receiving increased attention. This is reflected in the German Federal Soil Protection Act (BBodSchG) and the German Federal Soil Protection and Contaminated Sites Ordinance. In the context of investigations of contaminated sites in accordance with BBodSchG, measures (according to §6 BBodSchV (2)) are to be pointed out that are suitable to "permanently avoid dangers, considerable disadvantages or considerable nuisances for the individual or the general public". In particular the effects of the measures on the environment, the estimated costs and the necessary permissions are to be considered. In Annex 3 of the BBodSchV "requirements in respect of investigations for remediation and the remediation plan", the analysis of the "impacts on the parties concerned within the meaning of § 12 sentence 1 of the Federal Soil Protection Act and on the environment" is mentioned.

The consideration of the costs and the benefits of remediation and therefore indirectly the secondary impact on the environment is shown in the discussion of Natural Attenuation as a subsurface remediation technique.

1.2 Brownfield redevelopment and subsurface remediation

Brownfield redevelopment projects differ from remediation projects in that in the first case several methods/technologies that go beyond mere subsurface remediation are used on the site in question. These steps are e. g. the demolition of buildings, the disposal of waste, the remediation of subsurface contaminations (soil/groundwater), and the modernization or rebuilding of feeder and sewer pipes. A derelict site is prepared to the extent that it can be used for construction again. In some cases it is possible to reuse the former buildings, making numerous building measures unnecessary. In many cases brownfield redevelopment projects can be characterized as complex and consisting of numerous steps with impacts on the environment. A concerted project management can use the synergisms between different steps of the procedure to reduce the impacts on the environment.

Brownfield redevelopment can reduce the preparation of agricultural land for building whereas subsurface remediation can eliminate negative impacts on soil and groundwater. To date it has been difficult to balance and assess these primary impacts on the environment. First examinations on this topic were conducted by Doetsch and Rüpke [3].

2 Eco balancing

Eco balancing or life cycle assessment plays an important role in conjunction with production processes. Numerous investigations and publications exist on the life cycle assessment of different products. The requirements for life cycle assessment are defined in the ISO regulations 14040 to 14043. Formally, these standards are universally applicable. However, during the elaboration of the regulations, the focus was on the life cycle assessment of technical procedures. In Germany, several software tools for the life cycle assessment of products exist, e.g. UMBERTO, GABI, GEMIS. An eco balancing of subsurface remediation projects and other remediation procedures is possible with these programs; however, extensive data must be collected, so that an application of these tools is very time-consuming.

During the last years, investigations of the impacts of subsurface remediation on the environment were done in different countries ([4], [5]). A software tool for environmental balancing of soil remediation measure was developed in the Federal State of Baden-Württemberg [1]. With this tool, secondary impacts on the environment can be determined. A description of the tool is available in Volkwein et al. [6].

Further investigations concerning the impact of building processes on the environment are in progress. Doka [2] shows that the demolition and disposal of certain building materials have substantially larger negative effects on the environment than the production of these same materials.

So far the research project shows that an eco balancing of brownfield redevelopment projects does not exist to date.

3 Objectives of the project

The objective of the presented research project is the analysis of the environmental impacts of brownfield redevelopment projects, including those impacts stemming from raw material and energy requirements, which in turn induce land consumption. In particular those process steps with the largest secondary impact on the environment are to be identified. Recommendations for optimizing future projects under environmental criteria are to be formulated based on the results. An additional objective is the derivation of simple evaluation criteria, so that the impact of an action on the environment can be determined simply and quickly.

4 Methodology

4.1 Description

To determine the extent of the secondary impacts on the environment, suitable brownfield redevelopment projects located in southern Germany were examined and balanced. Advertisement data, accounts and disposal proofs were investigated and analyzed in co-operation with contractors and contract awarders. Appropriate assumptions were made for sizes and values, which could not be determined. Then the data was transferred into the software-tool "environmental balancing of soil remediation measures" [1]. Alternative scenarios were developed, also taking alternative remediation ideas discussed during the planning phase of the projects into consideration.

4.2 Program "Environmental balancing of remediation techniques"

In the program "environmental balancing", remediation techniques and to some extent building measures can be reproduced with 60 predefined modules. These modules are partly comparable to the individual positions in advertisement texts. The consumption data, special balances and impact balances of individual procedures and measures are calculated by internal data bases implemented in the program. The results are summarized in ten impact categories, i.e. cumulative energy demand, waste accumulation, fossil resources consumption, utilization of land, greenhouse effect, acidification.

A comparison between two procedures (remediation techniques) takes place in the same impact category.

4.3 Projects

Until April 2002, all balanced projects are characterized by different commercial, industrial or former military utilization, subsurface contamination, management of the preparation of the construction site and planned utilization. The following remediation techniques come into operation for subsurface contamination:

- Biological remediation techniques
- Soil washing (stationary/mobile)
- Air-sparging/bio-sparging

- Thermal treatment of contaminated soil
- Excavation of contaminated soil, recycling or disposal of the soil
- Hydraulic protection/remediation

In many of the projects examined, several of the above measures were used in parallel on the site.

The building measures caused by the demolition of buildings are also balanced with the program "environmental balancing".

5 Results

5.1 Introduction

Every examined redevelopment project is characterized by special conditions. Individual brownfield redevelopment projects - and in particular the subsurface remediation - are only partially comparable. Nevertheless tendencies and results that permit basic predictions appear in all examined cases.

The decision for a methodology, e.g. the selection of a remediation technique or a disposal method, is often limited by the timeframe of the investor, who is under economic pressure due to capital charges to begin with the development of his property. As time-consuming remediation can delay the new construction of buildings, such methods are rarely used for redevelopment projects.

Contaminated soil areas are often excavated and the contaminated material is used/disposed in waste dumps (as construction material). Due to the very low costs for disposal on waste dumps, this method has become considerably important in Germany, whereby the main environmental impacts are caused by construction work (demolition of buildings, excavation) and transportation.

However, during surface recycling projects implemented at the beginning of the 90's, alternative remediation technologies were frequently used. At the time, the recycling of contaminated soil material had priority over the disposal. Therefore, the objective of the projects was to clean and reuse the material, especially for the construction of roads and dykes.

5.2 General results

5.2.1 Transportation

In the brownfield redevelopment projects examined, the transportation processes (transportation of soil material and building debris to off site soil treatment plants, to recycling/disposal or building debris assortment plants) assumed a high proportion of the total energy demand, to which corresponding secondary environmental impacts are associated. Since, due to the economic situation, transportation is very cheap, long distances to soil treatment plants or building debris assortment plants are deemed acceptable (e.g. longer than 400 km).

However, in the case of excavation and recycling/disposal of material, the distances to nearby dumps are usually short, so that the impact on the environment through transportation remains small here.

These transportation processes result in a substantial consumption of fossil resources and contribute to a large proportion of the total emissions of the project, which is depicted by the impact categories acidification and summer smog.

It is important to consider not only the transportation to the soil treatment plant, but also the last place of utilization for the treated (cleaned) soil material as well as the disposal of waste products of the utilization. In one concrete example, the transportation distance after soil cleaning achieved similar orders of magnitude as did the distances to the treatment plant.

Transport by road dominates in the examined brownfield redevelopment projects.

5.2.2 Units and pumps

The long-term use of pumps and units, e.g. for soil vapor extraction or hydraulic protection of a contaminated area, results in a high cumulative energy demand and corresponding emissions. The secondary environmental impacts can achieve orders of magnitude equal to those for a complete excavation and following disposal in a dump. The impact on the environment is a result of the energy consumption for lateral channel sealers and pumps with long running times. The use of construction material (HDPE-lines, concrete) is negligible.

One project showed that long-term in-situ methods (air-sparging/bio-sparging) have nearly the same impact on the environment as the excavation of the contaminated soil in combination with an on-site remediation (biological soil treatment) of soil contaminated with aromatic hydrocarbons/mineral oils.

In-situ methods are suitable especially for the remediation of sites where the former buildings should be preserved.

5.2.3 Refilling with clean soil

An environmental impact can result from the use of clean gravel and sand as filling materials/replacements for contaminated soil material. This results in high land consumption. If filling material from gravel pits is specially mined for the replacement of soil, then this "consumption" can result in both sharply rising energy consumption (mining, processing, and transportation) and land consumption.

5.3 Critical analysis of the results

When applying the software [1], it must be noted that the results delivered on the secondary environmental impacts contain inaccuracies caused by the application. These inaccuracies are caused by the use of averaged data which allow a wide range of applications for the tool. Yet the quality of the generic data is unsatisfactory in those cases where processes are not yet exactly examined and calculated and where numerous assumptions must be made. Therefore, the documentation specifies that a significant difference between two remediation methods is given if the difference in the same impact category is larger than 100 %.

Within the life cycle assessment context, investigations show that it is difficult to balance the secondary environmental impacts resulting from the operation of construction machines [7].

6 Recommendations

6.1 Avoid or optimize transport

When conducting brownfield redevelopment projects, material flows should be coordinated and transports, especially for bulk materials (building rubble, excavated soil), should be held short or avoided.

The use of on-site washing equipment or the reuse of building rubble as secondary raw materials on the construction site constitutes an optimum concerning the avoidance of transports. This makes particular sense when very large masses are to be treated and sufficient space for the equipment is available (e.g. biological procedures) and also practiced.

Transport by ship is preferable to transport by rail. Road transportation show the worst performance with respect to the resulting secondary environmental impacts.

6.2 Recycling of building materials

The recycling of building materials is very desirable in demolition measures and has positive effects on the secondary impact on the environment: recycled building materials can substitute natural materials such as sand and gravel. If appropriately free of contamination they can find use as filling materials. In this context it should be pointed out that controlled demolition measures can lead to a mass reduction if "pure" materials can be brought back into the material/economic cycle.

6.3 Prefer on-site remediation techniques

An efficient way to preserve resources is to use on-site remediation methods (mobile soil washing plants, biological procedures). The necessary footprints, which must be available for the operation of a system, pose a limitation. Likewise, reintegration and/or on-site re-use of cleaned soil material should be possible in order to avoid the transportation of materials to other sites. Among on-site techniques, biological remediation methods have proven favorable for soil contaminated with MKWs.

6.4 Uses of effective in-situ remediation techniques

In one case, in-situ remediation techniques (air-sparging/bio-sparging) were balanced, with the result that under environmental criteria these measures do better than excavation of contaminated soil. Short remediation times are important in this context. If longer operation periods, e.g. for the use of a lateral channel sealer, are necessary, this advantage is quickly lost. Here, innovative measures like thermally enhanced soil vapor extraction or in-situ surfactant enhanced extraction could be an advantage.

6.5 Adjustment of the land development

A first step towards minimizing secondary environmental impacts of remediation can be made if the subsurface remediation technique is adapted according to the planned future use (living or trade). Another possibility is the adjustment of the subsequent land development (type and location) to the subsurface contaminations, e.g. parking-lots as surface sealing/protection. Similarly it is efficient under environmental criteria to use the excavationed material for the establishment of the new building. This co-ordination should take place in the planning phase of a project.

7 Conclusions

7.1 Specific evaluation approaches for brownfield redevelopment

On the one hand, brownfield redevelopment can lead to the reuse of land and to the removal of negative environmental impacts (subsurface contaminations, groundwater-loads). On the other hand, resources are consumed. A necessary requirement for brownfield redevelopment projects is that natural resources must be preserved to a large extent and that measures with as small an effect as possible on the environment should be used.

A useful evaluation approach for brownfield redevelopment projects is to determine the consumption of land caused by the measures. It is quite obvious that the direct use of gravel or sand to refill excavation pits resulting from a soil exchange causes an expansion of the gravel or sand pits. This leads to a direct consumption of land. Using the size of the gravel pits and pertinent areas as well as the quantity of mined gravel, a value for the landscape consumption per mass unit of gravel mined can be determined. Using data from Olschowy [8], Frischknecht et al. [9] calculated that a land consumption of 1,8 cm² occurs per kg of sand and gravel. These values are based on the annual average production of gravel and sand of 200 million tons, which corresponds to an area of 3000 hectares. 20 % area loss is added by the necessary border distances, development and technical installations. Olschowy [8] grants that due to the statistic boundary conditions, the real usage must be twice as large.

In one examined case, the brownfield area regained through recycling and preparation for new construction was overcompensated by expansion of gravel pits to provide gravel and sand for the soil exchange. An alternative in this case would be the refilling of excavated pits with cleaned soil, excavated soil material or recycled building materials.

The second direct consumption of land results from the dumping of contaminated soil, which is usually substantially utilized on dumps. This however usually constitutes filling up the remaining space of a dump. The mining and the processing of fossil fuels lead to a further consumption of land. These forms of consumption are considered in the generic data records used (see Frischknecht et al. [9]).

7.2 Result: A new definition of brownfield redevelopment

The results of the investigations lead to the following statements: As a basic principle, brownfield redevelopment should lead to a reduction of the consumption of agricultural land for housing and infrastructure, as it brings derelict land back into the economic cycle. As a consequence, natural areas can be preserved. In this context, it is useful to conduct an evaluation of brownfield redevelopment under ecological criteria (keyword eco balancing) as well as other criteria (e. g. energy consumption) to enable a comparison of gained vs. consumed area. The conclusions of this thesis should be reflected as follows in the ITVA [10] definition of brownfield redevelopment commonly used in Germany: "The recycling of derelict land is the useful reintegration of such properties into the economic and natural cycle which have lost their previous function and use – e.g. abandoned industrial plants and commercial enterprises, military properties and traffic areas – by means of planning, environmental, technical and political-economic measures" which are sustainable under ecological criteria."

8 Perspectives

The previous investigations have shown in principle that the determination of the secondary impact on the environment of brownfield redevelopment projects can form a basis for the optimization of the measures and for first recommendations for action. The software used must be further developed in order to be able to illustrate individual building processes and process steps more exactly.
In the coming years, further project evaluations will take place and, in particular, an evaluation approach will be developed that considers the utilization of land according to the measures implemented.

Acknowledgements
This project is financed by the Federal State of Baden-Württemberg

References

[1] LfU-Landesanstalt für Umweltschutz Baden-Württemberg, *Umweltbilanzierung von Altlastensanierungsverfahren.* CD-ROM Version 1.0 Rev.16., Karlsruhe, 1999.
[2] Doka, G., *Ökoinventar der Entsorgungsprozesse von Baumaterialien. Grundlagen zur Integration der Entsorgung in Ökobilanzen von Gebäuden.* Untersuchung im Rahmen des IEA BCS Annex 31: Energy Related

Environment Impact of Buildings. Forschungsprogramm „Rationelle Energienutzung in Gebäuden". Bundesamt für Energie, Zürich, Februar 2000.

[3] Doetsch, P & Rüpke, A., *Revitalisierung von Altstandorten versus Inanspruchnahme von Naturflächen: Gegenüberstellung der Flächenalternativen zur gewerblichen Nutzung durch qualitative, quantitative und monetäre Bewertung der gesellschaftlichen Potentiale und Effekte*. UBA-Texte 15/98, Selbstverlag, Berlin, 1998.

[4] ScanRail Consult, HOH Water Technology A/S, NIRAS Consulting Engineers and Planners A/S and Revisoramvirket/PKF (eds). *Environmental/Economic Evaluation and Optimising of Contaminated Sites Remediation. Method to Involve Environmental Assessment.* EU LIFE Project no. 96ENV/DK/0016, Executive Summary, February 2000.

[5] Diamond, M. L., Page, Cynthia, A. P., Campbell, M., McKenna, S. & Lall, R., *Life-Cycle Framework for Assessment of Site Remediation Options: Method and Generic Survey.* Environmental Toxicology and Chemistry, **Vol. 8, No. 4**, pp. 788-800, 1999.

[6] Volkwein, S., Hurtig, H.-W. & Klöpffer, W., *Life Cycle Assessment of Contaminated Sites Remediation.* Int. J. LCA **4 (5)**, pp. 263–274, 1999.

[7] Bundesamt für Umwelt, Wald und Landschaft BUWAL (eds). *Schadstoffemissionen und Treibstoffverbrauch von Baumaschinen – Synthesebereicht*, Umwelt-Materialien Nr. **23** Luft: Bern, 1993.

[8] Olschowy, G., *Bergbau und Landschaft. Rekultivierung durch Landschaftspflege und Landschaftsplanung*, Parey-Verlag: Hamburg, Berlin, pp. 40 – 42, 1993.

[9] Frischknecht, R., Hofstetter, P., Knoepfel, I., Dones, R., Zollinger, E., *Ökoinventare für Energiesysteme. Grundlagen für den ökologischen Vergleich von Energiesystemen und den Einbezug von Energiesystemen in Ökobilanzen für die Schweiz.* Bundesamt für Energiewirtschaft: Zürich, 3. Auflage, Dezember 1995.

[10] ITVA (eds). *Flächenrecycling. Arbeitshilfe – C 5-1.* Selbstverlag, Berlin. 1998.

Brownfields contribution in managing urban sprawling and soil consumption in Brescia

N. Marchettini[1], E. Tiezzi[1] & R. Ridolfi[1]

[1]Dept. of Chemical and Biosystems Sciences, University of Siena, Via della Diana 2/A, 53100 SIENA, Italy.

Abstract

The industrial district of Brescia is one of the most active in northern Italy. The scheduled reconversion of a large industrial site with a very central position (the so-called "Comparto Milano") has given the opportunity, during the last year, to assess the general potential contribution of brownfields reuse in the urban dynamics management of the region. The results were very reassuring about the chosen direction of an intensive reuse of idled areas, but very alarming about enduring soil consumption and urban sprawling trends.
In many areas landscape identities are irreversibly lost, and the forecast doesn't suggest significant changes: such a scenario addresses the opportunity of considerate urban reuse as the main answer to still growing urban market demand.

1 Introduction

With his 200 hectares of surface, the historical centre of Brescia is one of the largest in northern Italy. So large that, from the walls building at the end of the XII sec. until the end of the XIX, the city never spread out the ancient perimeter.[1]
Since the first years of the last century, however, the steel and pipe industry sited the largest plants just outside the ring of the demolished walls. In the south-west, nearby the way to Milan, rose the largest industrial quarter.
The crisis of Italian steel and iron industry in the early eighties gave way to a general process of relinquishment of industrial sites, but for the city of Brescia the phenomenon was particularly dramatic since it involved the core business of its strong industrial district.

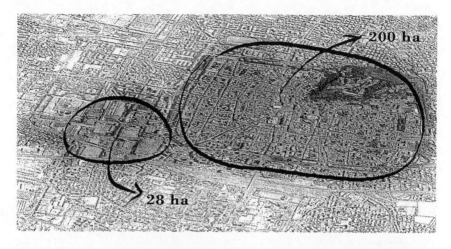

Figure 1: The idled industrial site and the ancient centre of Brescia.

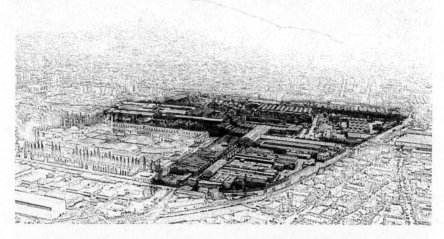

Figure 2: The so-called Comparto Milano in the urban landscape (from West).

As a result a unique site (for location, relative dimension, history) was made available for a project of urban redevelopment capable to affect deeply the face of the city. Beyond the specific issue of the so-called Comparto Milano, what is of interest here is that such a large event, increasing the level of public attention on this subject, was a good opportunity to reconsider the whole theme of brownfields in the territory of the province.

This paper shows the results of an inquiry that tried to assess the potential role of former industrial areas on the general scenario of current soil consumption dynamics.

2 Methodology and instruments

Figure 3: A projection of orthophotographs on a digital terrain model of the
province of Brescia.

To evaluate the general soil consumption, historical cartographic series has been
digitalized within a GIS system. The first modern topographic survey for the
province was made in the 1885, and, until the 1990, all the maps were on paper.
Digitalizing urban areas from traditional cartography (in a 1:25.000 scale) it has
been possible to obtain 7 layers of soil consumption: 1885, 1913, 1931, 1955,
1971, 1981 and 1991.[2]

After that date no further maps were made available, and to update our
information we have used two series of orthophotographs, respectively visible
RGB files and panchromatic scans: both have a resolution of 1 meter/pixel, and
are as accurate as a 1:10.000 scale in traditional topography. This more flexible
and accurate instrument made us able to distinguish industrial areas within
urbanized surfaces, even if just in a "typological" sense. The results of this
interpretation have also been projected over the historical layers, under a work
hypothesis of typological persistence, to obtain an estimate of the industry-
generated soil consumption.

To achieve a general assessment about the consistency of the abandoned or idled
industrial areas, all local authorities have been asked to compile a questionnaire,
considering entities larger than 10.000 square meters, or smaller areas with a
particular strategic relevance.[3] The 61% of the municipalities have completed the
set of questions, but - from a proxy with commercial data: i.e. presence of large-

mid economic units in mature sectors - only the 19% of the defaulting ones were potential owners of brownfields. At this general level we should ascribe to our results a tolerance of 10-15%, which is acceptable for the general purpose of this survey.

As a second step, the province has been divided into nine homogeneous subsystems: in this paper we'll briefly report the results of the central subsystem (numbered as #1), including the city area.

At this second level we have considered more detailed data, both from the city administration statistics and regional GIS data from ERSAL, the regional office for land and soil studies.

3 The Province scenario: soil consumption and abandonment

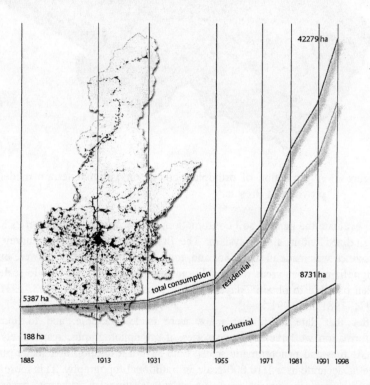

Figure 4: Historical trends of soil consumption, distinguishing residential and industrial settlements typology. On the background: present time urbanized surfaces.

The total amount of dismissed industrial areas over the whole province is turned out to be of 215 hectares, counting the abandoned military area of Montichiari, for a total of 499.400 m^2 built surfaces.

The 26% of the reported areas are within parcels smaller than 10.000 m^2, the 18% are in the dimensional interval from 1 and 2 *ha*, and the 24% is between 2 and 10 hectares. Two sites, with the mentioned powder magazine, are larger than 100.000 m^2.

In this first survey it seems that, even if relevant in absolute terms, the idled industrial areas couldn't have a serious weight in the general soil consumption balance of the province: 215 *ha* are less of 2.5% of industrial use destined surfaces, and even just a 0.5% of the total amount of urbanized areas. At the current rates of the still growing demand of the real estate business, a complete re-use of these areas could, in theory, satisfy the market for a few months. Looking beyond these raw quantitative considerations, our subject could become more interesting if we consider the potential *strategic* role of these areas, sometimes due to their centrality, or in other cases for their attitude to industrial reuse.

To achieve a closer sight we have divided the provincial system in several, quite homogeneous areas, or subsystems, which we have chosen considering orographic parameters, economical structure and the different historical attitudes.

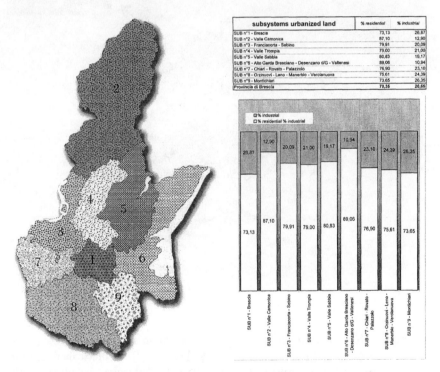

subsystems urbanized land	% residential	% industrial
SUB n°1 - Brescia	73,13	26,87
SUB n°2 - Valle Camonica	87,10	12,90
SUB n°3 - Franciacorta - Sebino	79,91	20,09
SUB n°4 - Valle Trompia	79,00	21,00
SUB n°5 - Valle Sabbia	80,83	19,17
SUB n°6 - Alto Garda Bresciano - Desenzano d/G - Valtenesi	89,06	10,94
SUB n°7 - Chiari - Rovato - Palazzolo	76,90	23,10
SUB n°8 - Orzinuovi - Leno - Manerbio - Verolanuova	75,61	24,39
SUB n°9 - Montichiari	73,65	26,35
Provincia di Brescia	79,35	20,65

Figure 5: The nine subsystems and the respective amounts of industrial areas.

From the point of view of the potential demand, abandoned industrial areas should be considered in different ways in the valley floors (subsystems 2, 4 e 5), where the older steel and iron plants can likely be converted only into alternative industrial destinations, and in the central district or near the lakes, where the market could sustain the additional costs of the re-conversion even for offices, residence or tourist trade. Here we have to adopt the conditional mood because, without a specific action of local administrations, and with the systematic definition of new building-suitable areas made by local plans, any private investor is induced in undertaking the risks and the additional costs of a complete reclamation of a site.

4 The city of Brescia and the central district

The only exception to the observations above could be found in the proximities of the central business district of Brescia. Here the usual lack of suitable lands for new buildings, that grows together with centrality, is become a fruitful background for innovative projects.

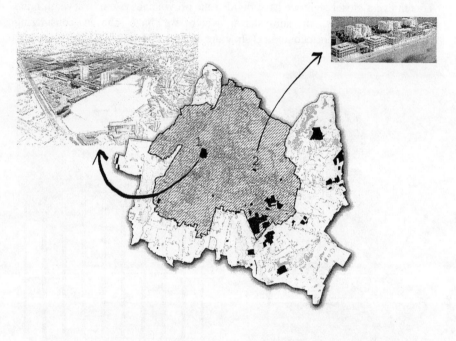

Figure 6: The two main urban-reuse projects within the city of Brescia: the Comparto Milano (1), and Borgo Wührer (2). In black there are the subsystem's "brownfields" in a very extended sense, including marble quarries and gravel pits.

Within the city boundary there are 69 parcels completely or partially abandoned or idled, for a total amount of 52.2 *ha* of area and 25.8 *ha* of built surfaces.
The 80% of the units is smaller than a hectare, and in the largest 7 units is concentrated the 78% of the total area. We have to stress the fact that only the 30% of the idled city areas can be strictly classified as "brownfield", due to the

fact that the others were been generic warehouses or were dedicated to relatively low-impact manufacturing industries.

The situation is quite different in the outer part of the subsystem: in the belt of the municipalities around the city we have found several different situations, also with specific and sometimes heavy problems of reclamation, but the total weight of these brownfields is relatively insignificant if compared with the pursuing urbanization processes, reaching here the highest values in the province. The statement could be different if we considered the open marble quarries and the gravel pits, but this theme will put us far beyond the initial purpose of this work.

5 Conclusions

Beyond any consideration about specific situations and some isolated positive examples, what really emerges from this study is the hugeness of soil consumption in the province under study: a dynamic that is not directly related with an increase in population, nor with the number of families or with any increase in the actual number of workers.

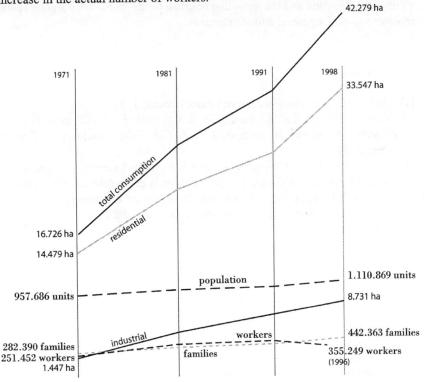

Figure 7: The increase in soil occupation in the last three decades vs. demographic trends.

We can say that a general increase in personal pretensions (considering that the threshold of a room for each person is already surpassed) and speculative pressures are the real driving forces of the system. Also interesting is the fact that, from a series of contacts and interview with privileged actors, public administrators, local planners and real estate consultant there is no real perception (nor interest) of the problem. A continuous, growing process of soil consumption is generally considered just physiological.

The role of abandoned or idled areas could be significant only in the context of a strong policy of reduction of the loss of ecologically productive soils. In that case the large units, often in central position, left by the industrial relinquishment, could be strategically fundamental, if appropriately adopted as catalysts, or attractors to govern the urban processes.

Beyond this, the surveyed increasing soil-wasting is so dramatic that even the option of re-use even residential areas should be considered, giving the owners the opportunity, with permissions and incentives, of re-build low-density quarters with different typologies, trying to concentrate the settlements in "urban" and more closed forms.

Unfortunately the actual political and cultural trend in northern Italy is going in an opposite direction and the sprawling of urban forms seems to be a good spatial representation of a general attitude of minds[4].

References

[1] Nardini, F., *Brescia e I bresciani*, Grafo: Brescia, 1984.

[2] Dominico, D., Olivari, A., Rosini, M. & Vavassori F., *A GIS analysis of soil consumption trends in Brescia*, Ufficio GIS della Provincia di Brescia: Brescia 2002.

[3] Studio Consolati., Indagine sulle aree industriali dismesse, *Documenti del Piano Territoriale di Coordinamento*, Provincia di Brescia: Brescia 1994.

[4] Paolillo, P.L., (ed). *Terre Lombarde. Studi per un eco-programma in aree bergamasche e bresciane*, Giuffrè editore: Milano 2000.

Cost-effectiveness in polluted site sampling campaign

C.I. Giasi & P. Masi
Department of Civil and Environmental Engineering.
Polytechnic of Bari, Italy

Abstract

Soil pollution is a difficult subject to investigate, cause to problems connected to uncertainty in polluted areas optimal sampling designing and in right conceptual model determining. After a preliminary sensitive investigation, an appropriate step-by-step approach could allow to reach representative results. After a description of the main codified sampling designs, the present paper is supply a sample case study, which could give some rough guidelines for an effective oriented approach to the subject. In literature there are a lot of indications, but few studies on the necessary sample number to reach a fixed confidence level in operative approaches. In such complex problems, in fact, the uncertainty level knowledge during the investigation is fundamental to attribute the correct significance to results. Coming from a probabilistic modelling of different over-threshold areas shape, the herringbone sampling pattern seems to give good indications about sampling design and number, since from a primary approach to the area, thanks to the possibility of percentage error a priori determining.

In the sample site, in Southern Italy, only the "analyte" contamination has been considered and, based on a herringbone sampling pattern, has been possible to demonstrate, on a kriged prediction map, that few optimal locations could have been chosen for a preliminary sampling campaign.

1 Introduction

The perception of environmental damage impact is more and more growing in the common sense, so that, almost in the last years, new environmental studies and specific professions are developing. Another aspect is that a great deal of studies in environmental matter is oriented to define a correct use of environmental resources, and to reach effective standards in order to perform polluted sites characterizations and reclamations. The last two aspects confirm the socio-economic consequence of the mentioned impact.

In fact, despite of the employ of economic resources, in a great deal of site characterization and reclamation projects a true reliable result isn't reached. The goal of the present study is to demonstrate how to employ geostatistical and grid efficiency based preliminary study optimization in a polluted site characterization plan in order to reach more effective economic employ without lacking the significance of results.

2 Sampling designs

The first step to design an efficiency – oriented sampling campaign consist in a survey of good and lacking aspects of the commonly applied sampling designs in order to decide which is the most effective to reach the objective of a reliable result in a site characterization plan. Overdimensioned employed procedures, in fact, come principally from the lack of clearness on the goal of the different sampling design and on the applicability limits of each one. So, as a first step it has been considered to check the applicability of each sampling design in order to decide which one can supply the best results on the examined site.

In the international literature there are seven sampling design which seems to be the more frequently applied:

- Judgemental sampling;
- Simple random sampling;
- Stratified sampling;
- Systematic grid sampling;
- Ranked set sampling;
- Adaptive sampling;
- Composite sampling.

Being the base of more complex sampling campaign the first three sampling procedures will be further described in order to highlight their good and lacking points.

Judgemental sampling can be used either as a stand-alone sampling, or as a first step in a more complex sampling design. The choice of number and sampling location is based on the previous knowledge of the examined site. The higher the knowledge on the site the better the results supplied by such a design criterion application. Such a design, whose effectiveness is linked to the site information quality, determines a difficult application in case of lack of historical or subsoil knowledge on the examined site. Another subject that can limit such procedure applicability regards the data interpretation, which can be only based on the designer judgement. In a great deal of such a criterion practical applications a GIS based analysis of the sampled concentration data determine a wrong result based on good quality data. In fact such an analysis contains the limit of being generally based on information with a different quality degree regard the contamination sources and targets. An example could be a judgemental sampling campaign designed in order to find over threshold polluted locations on a contaminated site; in this case despite to respecting a regular location spread on the examined area, a very common practice is to locate some boreholes near pipe trenches, where, thanks to the greater soil permeability, a preferential subsoil contamination migration path is allowed. This way of proceeding, which allows investigating high probably over threshold locations, while supplying a good economic effectiveness, determine a different accuracy degree in describing sources, paths and targets of a migrating contamination. Only the designer judgement can take into account of these different characterization accuracy degrees in order to estimate the real site conceptual model, while the application of a migration model could supply an incomplete analysis being based on different detailed target, path, and contamination sources data.

Due to the complex site geology, the application of this sampling design to the case-study-examined area would have determined wrong data interpretation coming from a partial knowledge of the subsoil situation.

In the *simple random sampling*, samples number and locations are based on a random criterion. Once a list of probable hot spots compiled, it consists of:

- casually extracting some numbers corresponding to the previous listed locations;
- on the basis of them positioning the boreholes in the corresponding places.

This way of proceeding has to be preceded by the determination of the sample number in order to be sure about its significance. If neglected the last aspect leads to not representative sampling campaign and to a bad estimation of the origin population parameters. A second source of mistakes comes from the necessity of a sample dimension accorded to characterization purpose (i.e. the sample number in order to find the mean contamination value on the site has to be smaller than the sample number to hit a determined number of hot spots with a certain confidence level). In the last case a sampling campaign dimensioned to respect the decree 471/99 could even be underdimensioned.

A further consideration can be based on the frequent confusion between judgemental and simple random sampling, which contribute to a frequent overestimation of the mean site contamination values. Due to the judgementally directed sampling location selection, in fact, the data are shifted to high concentration values. The application of a simple random sampling to the below reported case study, according to the goal of searching for the number of hot spots and their locations, would have determined the undesirable effect of a high borehole number, which couldn't agree with the searched economic efficiency.

The first step in a *stratified sampling* campaign consists of dividing the examined site in sub-areas on the basis of expected contamination type and site geology. By overlapping the different obtained "strata", it's possible to subdivide the site into homogeneous areas characterized by the same kind of contamination and geology. In the following fig.1 (EPA Quality staff (2000)) by means of the mentioned sampling design application, 9 homogeneous strata have been determined in an area surrounding a smokestack, on the basis of prevalent wind direction and of site geology. A good point of the mentioned design is the well-structured way of proceeding. Cause to the greater homogeneity of the sampled sub-areas, such a structured sampling design contributes to a reduction, for example, of the contemporary analyzed variable. In order to avoid a great deal of strata, cause to the different contamination sources and kinds, on the below examined site it has been better not to apply such an approach.

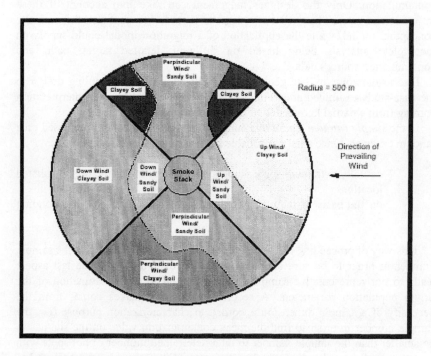

Figure 1: Stratified sampling

3 The examined site

The examined site, which covers an area of about 100 hectares, is located in south of Italy; in the present paper it will be referred to as "Plant". Cause to an accident a great deal of high-pressured contaminant (which will be referred to as "analyte") based products has been leaked, and spread on the surrounding soil.

The characterization campaign was based on a regular grid sampling; some hundreds of sampling locations have been defined, relevant to a sampling grid dimensioned from 40×40 m to 25×25 m in the installation area, and to a grid of 100×100 meters in the surrounding areas; it has been reached a sampling grid dimension of 2×2 m to describe with full particulars high punctual concentrations with respect to single pollutants.

For each borehole 3 representative samples was analyzed and the presence of over threshold concentration of metals and organic substances have been checked.

The site characterization produced the following results:
the soil surface interested by concentration of pollutants over the threshold has been determined in few percentage points the total plant area;
8% of contamination is due to the "analyte".

4 The proposed approach

When the present study has been conducted all the pollution data in each environmental matrix had yet been measured and known.

Based on the consideration that a well dimensioned soil characterization is a structured procedure that allows managing step by step few highly representative data, a more cost-effective approach to the subject has been carried on using more advanced design tools. In order to illustrate the proposed method only the analyte contamination at the depth of 1m (see the below reported location map) has been considered (fig.2).

By means of known contamination levels at the sampled points the contamination level at a 10000 locations has been predicted based on geostatistical tools which main features are reported in the further.

5 Geostatistical tools

Geostatistics provide a set of applicable effective spatial techniques for statistical interpolation of spread observed random functions based on the explored data structure (Rosenbaum and Nathanail (1996). The main geostatistical tool is the semivariogram, which can be computed as half the expected square of the difference in value between each pair of observations a given distance apart. The squared differences plot versus distance (or lag h) gives a sharp idea of the random functions correlation distance and structure. In case of *p(h)* paired observation $z(x_\alpha)$, $z(x_\alpha+h)$, $\alpha=1,2,...,p(h)$, the experimental semivariogram can be estimated as:

$$\hat{\gamma}(h) = \frac{1}{2p(h)} \sum_{\alpha=1}^{p(h)} \{z(x_\alpha) - z(x_\alpha + h)\}^2 .$$

Webster and Oliver (1993) showed that a reliable estimation of semivariogram values requires at least 150 data, and larger samples are needed to describe anisotropy, however this doesn't means that geostatistics can't be applied to smaller data sets. Previous knowledge and ancillary information coming from better sampled areas can be used in order to supply a lack of information about the examined one.

A good knowledge of the study area combined with such a semivariogram spatial description can improve our understanding of the physical mechanism controlling the spatial patterns, but the description of a spatial pattern is rarely a goal per se. By means of mathematical models, a semivariogram fitting is possible in order to predict the property of interest value at unsampled locations. Where the geological structure is suspected to have anisotropy, a supporting vector can be considered in order to take into account for minimum and maximum spatial dependence directions. In the last case the experimental semivariogram formula becomes:

$$\hat{\gamma}(h) = \frac{1}{2p(h)} \sum_{\alpha=1}^{p(h)} \{z_v(x_\alpha) - z_v(x_\alpha + h)\}^2 ;$$

being v the support.

On the basis of the variogram-investigated random functions spatial structure, a linear weighted moving average technique (kriging) can be performed. The most frequently used kriging form is ordinary kriging (OK), in which case each estimate is done as:

$$\hat{z}(V) = \sum_{\alpha=1}^{n} \lambda_\alpha z(x_\alpha) ;$$

where:

V is the support over which the estimate is made;

λ_α are the weights assigned to the available observations with respect to the condition:

$$\sum_{\alpha=1}^{n} \lambda_\alpha = 1;$$

in order to ensure that the estimate be unbiased.

If any trend exists, a kriging procedure with a good approach degree it is need the data trend removal. In fact if a trend is exists in the data, it is the non-random (deterministic) component of the surface (Johnston et al., 2001). By removing the trend using a mathematical formula is possible to obtain random distributed data. Trend removal can help justify assumptions of stationarity.

6 Sampling pattern choice

Based on the results of the real data sampling campaign the location map and the relative kriged prediction will be considered in the present study as representing the "real" contamination spread on the examined site.

Before designing a new sampling grid the main question to be answered has been: which is the confidence level to be respected in each step of a sampling campaign? Regard the last subject a lot of studies have been produced which substantially agree that the confidence level must be related to the accuracy in the investigations, and can vary from 68 to 99% (McCoy and Associates, Inc., 1992). However, regard the majority of practical applications, all the consulted studies seem to agree on the confidence level of 95%. Beside to this substantial agreement there isn't any clear explanation on the way in order to obtain a 95% confidence level in practical sampling grid design.

To the purpose, according to Nathanail et al (1996) the coverage level of the examined site has been assumed as an index of the sampling grid accuracy. On the side a herringbone sampling scheme has been considered owing to the showed stationarity of the probability of hitting a hot-spot, regardless of its main direction (for elongate targets). Being confirmed through a Monte Carlo analysis of 14000 different coupled grid-target positions, such an indication highlights a first lacking point of regular grid located sampling patterns. These kinds of sampling designs, in fact, seem not to have a reliable behaviour in hot-spot hitting owing to the highly variable probability value found in the previous mentioned Monte Carlo simulation.

Other literature researches have strengthened this kind of idea. Coming from an a-priori study conducted in terms of error variance in geostatistical analyses, a variance minimization algorithm supply an important tool to our purpose. Despite to the implicit assumption that the variance of the errors is independent of the actual data values (a situation referred to as "homoscedasticity") the mentioned method seems to be a simple but effective a-priori approach to sampling grid designing.

As an example of such an approach, using spatial simulated annealing (SSA) a spatial sampling scheme can be optimized for minimal kriging variance. This kind of study, conducted by van Groenigen (2000) yields a sampling pattern similar to the previous mentioned herringbone scheme.

A more advanced design-oriented study might take into account anisotropy or the presence of possible high concentration areas based on expert judgement, but there is no doubt on the greater reliability coming from one of the mentioned sampling patterns.

7 A new sampling campaign and result comparison: a case study

In the present case study one hot spot in the considered Plant area has been hit through the real sampling campaign. In the studied area the sample number consisted of 176 samples and chemical analyses in order to detect the analyte. On the basis of the sampled real concentrations, the first step of the present study consisted in determine the "real" analyte spreading on the studied Plant area through kriging. The kriged prediction map has been reported in fig.2 in which the sampling locations have been highlighted.

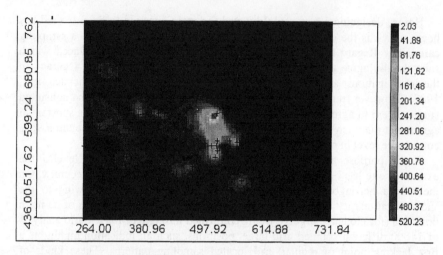

Figure 2: Analyte "real" prediction map.

On the basis of the previous "real" prediction map a new sampling campaign has been designed. In order to reach a better cost efficiency for the new campaign, the above mentioned effective sampling pattern has been used. New concentration values have been measured on the formerly generated kriged prediction map. Doing so, a new location map for the analyte has been derived (see fig. 3).

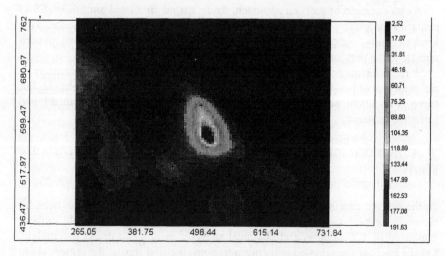

Figure 3: Analyte effective-sampling-based prediction map

The new prediction map uses something like one third of the total sample number and can be used for an effective very preliminary site survey in order to hit hot-spots with known dimension on an examined site. In fact, based on the last sampled concentration values, a new kriged prediction map has been determined, which demonstrated the same hot spot location and a similar coverage area. Despite to the lower concentration values it's possible to strengthen the quality of prediction map through a second sampling in the over threshold area.

8 Conclusions

On the basis of what above reported a main consideration should be done: a traditional based site characterization sampling design can't supply tools for the most effective economical employ. When the problem is to hit a certain hot-spot whose dimensions are known or predictable with a good approach degree, a variance minimization based sampling design can supply effective tools in order to map the pollutant concentration on a site area, and allows hitting a very preliminary area of interest to be further investigated. Such a way of proceeding, thanks to other advanced geostatistical tools, allows a gradually growing data set managing and a step by step decision making process. On the side the main direction in which the economical resources have to be employed could be stated in each step of the characterization campaign allowing a better effectiveness of the expenses.

References

[1] EPA Quality Staff (2000) "Guidance for choosing a Sampling Design for Environmental Data Collection". EPA QA/G-5S
[2] Johnston, K., Ver Hoef, M., Krivoruchko, K., Lucas, N. (2001) *"Geostatistical Analyst User Guide"* printed in United States of America.
[3] McCoy and Associates Inc. (1992) *"Soil sampling and analysis-practicies and pitfalls"* The hazardous Waste Contaminant, November/December.
[4] M.S. Rosenbaum and C.P. Nathanai (1996)l *"Petrophisical database for ground characterization"* Marine and Petroleum Geology Vol.13 n.4, pp 427 – 435.
[5] Van Groenigen, J.W., (2000) "The influence of variogram parameters on optimal sampling schemes for mapping by kriging" Geoderma 97, 223 – 236.
[6] Webster, R., Oliver, M.A., (1993) *"How large a sample is needed to estimate the regional variogram adequately?"* In Soare, A. (Ed.) Geostatistics Troia '92, Kluwer Academic Poblishers, Dordrecht, pp. 155-166.

Section 4:
Development issues

Section 4:
Development issues.

Brownfields development issues

A. Bogen
Down To Earth, LLC, United States

Abstract

Brownfields, underutilized or abandoned industrial and commercial sites are a worldwide legacy. Regulations, resources and conditions vary from site to site but turn on universal concerns. The interconnectedness of health, economic and ecological issues become ever more apparent on, over and in the Earth. Successful Brownfields reuse means addressing these fundamental and connected problems.

An area with many Brownfields is the northeast United States. Connecticut, is a small state in that region with highly proactive policies. Four elements unique to Brownfields development are discussed below in the context of that state. The elements are Time, Regulatory Knowledge, Money and Will.

1 Time

Timelines are critical to the management of any real estate development project. Permits, financing rates and windows, seasonal concerns and tenant access impact all projects but are unresolved until the environmental matters of a Brownfields site are brought into focus. There are two blocks of time discussed below as *lead-time* and *data collection* time.

1.1 Lead-time

Some of the increased lead-time is due to negotiating access agreements necessary to perform the environmental assessments. Among other time consuming areas is encouraging a seller's willingness to have testing results revealed since there is the perception that developing the data could trigger enforcement action. This is a real concern. The developer will have to determine the probable types of risk associated with the site, under the regulations. He or she will then have do determine who will pay for the resolution of the liability questions. Although the owner is responsible, the resources able to resolve the issues may belong to the developer and the public.

A generic or initial response may be sought from the regulator as to what is required on the site. This is good opportunity to use a consultant as an emissary since no client identification is necessary. Legal assistance is needed to arrive at what indemnification is possible under the law. Liability definitions have been steadily modified as Brownfields projects accumulate. It should be known who is responsible for reporting test results to regulators and what immediate actions those reports could trigger. The inherent nature of a Brownfields site is one of some complexity and difficulty. The developer needs to determine with the regulator what their concerns are to find a way to work from the perspective of resolution rather than enforcement. The developer also needs information on operations and gap insurance, assuming that previous experience is limited. This field too has many emerging and important products.

Keep in mind the continuing relationships with community, regulators and funding sources including public funds. The process of project advocacy will stir dormant community memories and issues, which need to be understood and receive responses. Sometimes Brownfields projects serve as lighting rods for political, cultural and economic issues. Banks and other interim managers such as municipalities need satisfaction that their involvement is without penalty. It may be possible to set up the site so that the seller perceives this as the best choice to free the asset or liability from their portfolio.

1.2 Data collection time

Once sampling is underway the public and press will want to receive notices of results. There needs to be definition about who ought to receive what and what the responses will be in order not to unjustly mischaracterize the site conditions. Place reports in draft form with conditional conclusions where ever possible. Initial findings may only bring increased alarm, which may not be properly understood or managed until additional testing is done. There are health concerns and civic duty involved so a considerable degree of openness is required.

The perception or reality of health issues may require public meetings over a period of time to achieve consensus and invaluable political support. This is usually where the press focuses. There are underlying mutation fears and stories. The credibility of assurances is critical to forestall delays. Engaging the highest authority and building partnerships early with health regulators can address these concerns. House or interested party staff will not have the same credibility even if there is concurrence on results and actions. Along the way, schedule open report reviews to give air to concerns and invite comment. This will enable issues to be addressed in a continuing fashion rather than them cropping up as periodic occurring obstacles to the project.

Site development plans are more complicated to draft and implement when they need to incorporate contamination data. These plans contain authorized yet conditional resolutions. The elements vary according to ground water quality controls, aquifer protection criteria and risk abatement. Prepare these as a visual aid to help people overcome their concerns about contaminates managed on site. This process will also define the likelihood that the project will receive abiding public support, which is critical to defining remediation costs and development timelines.

2 Regulatory knowledge

Health and regulatory officials need to agree on jurisdictions and standards, which can be in conflict among local, state and federal agencies. It will be important to have negotiated a common protocol supported as much as possible by all interested parties. The outcomes need this support to achieve community and lender acceptance. This determination can be in the format of written questions answered by the regulators or technical professionals wherever possible in the development of the Scope of Work. Subsequently, the contractors bidding on the assessment work will provide data within that framework. This helps to support the efficient use of funds. It may also uphold a well-founded claim against the contractors insurance should there be a serious error or omission. In effect the Brownfields manager is asking for a written scope of work from the technical and enforcement people. This helps to force a conclusion, at the start, of what needs to be known and what needs to be done as a result of that knowledge. Otherwise random and reactive efforts can create cost, time delay and not render certainty.

The Scope of Work needs to be a utilitarian document. Only proceed with tasks as far as you have to at the time. The initial sampling and testing should give a sense of what contaminates are present. Field labs can reduce costs by allowing for some immediate re-sampling or expanded sampling if probable areas of concern are immediately identified. Continue the work under the scope as the information develops but get as broad a collection of information at the start as is possible. This approach provides protection against going in too deep on a dangerous project – which is one that may be significantly inverted in value. That is the cost of cleanup far exceeds the market value of the property. An enormous public investment and return would be required for the reuse of that site.

Utilizing *in-situ* remediation techniques can be very cost effective if the conditions allow. Human health exposure and mobility matters are usually the first criteria to be satisfied. Some contamination may not migrate off site and can be incorporated under structures, parking lots or berms. Some contaminates may be treated on the site over an extended period of time but allow for concurrent development and use. Suggesting ways to manage contamination without removal and disposal can be very cost effective. Good coordination among designer and environmental technicians as well as marketing people is required to present acceptable alternatives.

At times a site is made up of multiple parcels. The environmental matters may be more readily addressed if the parcels are treated as one for final disposition purposes. This would allow the movement of materials within the parcels to concentrate problems in a section as opposed to have scattered complex matters all over the site. There are regulations that allow for treatment of a large site with certain exclusions that acknowledge the existence of "widespread coal ash contamination", as an example.

Environmental land use restrictions allow the non-removal of contaminants under a structure if that structure is to be reused or more specifically if the slab is not removed. There can be site specific resolutions such as the temporary removal of some of the floor when a new floor replaces it. The human health exposure criteria must be in compliance.

An early and essential part of the development of the scope of work is communication with sophisticated insurance providers. Gap insurance can make a risk acceptable to the long-term financier. Carriers rely on their experts who may suggest protocols that are acceptable and definable to their satisfaction reducing the possibility of a costly conflicted claim. Federal clean up loans require remediation standards that will bring liability releases only when followed completely. All parties need to agree on what the outcome will be at the start.

Sustainable development design elements are still unfamiliar to many regulators and need to be fit into the discussion. A problem such as controlling off site water runoff provides an example. Gray water reuse in cooling and heating, in fountains and in landscaped retention ponds can reduce the need for sewer expansion. Permeable pavers will allow water to percolate; roof gardens will provide environmental amenity and water control function. These approaches may require more initial capital, which will be offset over time with savings from operating efficiencies.

Phytoremedial techniques are still experimental but proving themselves in many situations. The regulatory community in the US has shown itself willing to consider these options in a range of applications if the human health exposure and pollutant mobility criteria can be met. Long term cleanup of shallow metal contaminates, for example, can make a project feasible if there is no cost for expensive excavation and removal. Some areas don't lend themselves to soil removal resolution due to the shallowness of groundwater or proximity to a water body. The regulations usually address the standard to which a site must be cleaned – not necessarily the pace. Poplar trees, sunflower and mustard plants as well as zucchini may replace excavators at sites with patience for and knowledge of Nature's ways.

3 Money

As with all real estate development decisions, reuse of Brownfields are dependent on the projected economic returns to the investor. Brownfields must rely on estimates of the additional line item of projected environment expense to determine feasibility. The narrowing of the probable range of those environmental expenses is critical to determining the risk and bid for a site. Before expensive detailed documents are created, the bidder must arrive at a go/no go decision point. Essentially this is a determination of the acquisition price. If that price is unreasonable, after concerted negotiation, the project is not market ready. Determining the realistic acquisition cost requires some information developed in a process, which follows.

3.1 Acquisition cost

The site acquisition cost must reflect an appropriate discount to absorb the additional assessment, cleanup and process expenses. To determine the range of or the limit of the site acquisition cost, consider the following:

- What are the comparative market prices for "clean" sites
- What is the probable time and professional cost to determine a baseline of environmental information? There are numerous tasks such as file searches to obtain background information that is often located in multiple departments and which may take hours. Also valuable are site walks and notes on the preliminary inventory of areas of concern. There may be structural integrity questions, which may require determinations of safety before the environmental review of the interiors can proceed. Visits with both local and ultimate decision making regulators to quantify their concerns can render surprising results better known before more significant expenses are incurred. There are also the initial drilling and sampling costs.
- What legal costs will be incurred to negotiate and complete access agreements to perform the sampling?
- What is the value of the additional process time to gain approval of the proposed environmental remediation plans
- What is the estimated cleanup cost
- What is the opportunity cost or desired rate of return on these seed funds, which may not be recovered if the project is deemed unfeasible.
- There are tax benefit programs available, which allow immediate write off of cleanup costs. These may only be considered an entrepreneurial reward for the more complicated project and not be included in this estimate.
- Public funding may be available to reduce the hard costs of assessment or cleanup. These often require building, capital improvement or employment commitments that the developer may not want as an obligation or may not be able to meet.
- Traditional lending sources may not provide funding until the site receives the appropriate regulatory releases. Therefore there can be the increased cost of short- term borrowing needed until the liability concerns are satisfied and long-term financing is secured. This may require significant capital capacity.

Astute risk managers may be disparaged as "bottom feeders", but the costs are real and must be factored into any offer. There are other softer costs that can occur. One example is the cost of any number of community meetings to determine the local acceptability of the proposal. An industrial re-use may not be well received. These sites often have a stigmatized reputation, which can require heavy lease advertising and additional remedial costs to overcome.

Diagnostic funds must be used with great efficiency. Traditional engineering approaches often start with limited testing. These accumulate into more and more data until there is reasonable certainty of a detailed site assessment. However this approach is expensive and lengthy. The prospective developer needs a reliable quick cut at the expense of dealing with the contamination. The environmental mess is after all only a line item in the development of the proforma. This is where Brownfields can seem overwhelming. It is not easy to resolve the complexity and diversity of the suspected contaminates. Breaking the facade down into comprehensible and practical units is an early task. This requires some dialogue and visioning.

Beginning at the proposed end use is valuable. Envision the reuse. Determine where roads and structures are likely to be placed. Where can landscaping berms and plantings be deployed? Where are there water concerns and abutting property considerations? Sketch these in on the site plan.

Then take an overlay of the preliminary areas of environmental concern and look at how the two coincide. Can the contamination in an area safely be incorporated into a berm or under a structure or parking lot so that removal costs are reduced? Often some areas of contamination can remain in place. Remediation may be restricted to soil removed for required invasive procedures such as utility trenching or foundations. These exercises can lead to a conceptual remedial action plan. The regulators can respond to the vision of the reuse rather than just tables and problems.

3.2 Public funding

Connecticut has a variety of funding programs, which are successfully meeting the different aspects of Brownfields. The following shows most of the types of assistance currently available:

- Special Contaminated Property Remediation Insurance Fund – A loan program for assessments on sites where there is a proposed use. Can incorporate demolition costs.
- Economic Development and Assistance Act – Loans for construction, demolition, remediation, renovations, equipment acquisitions and training.
- Connecticut Development Authority – Provides a variety of loans and loan guarantees including a new Brownfields program that will loan money to clean up and prepare a site for reuse with repayment through incremental tax payments.
- Enterprise Zones – A package of tax abatements for particularly distressed areas.
- Utilities Assistance – Special incentives for conservation planning.
- Job Training – Grants for training employees.

- Environmental Protection Agency Pilot and Brownfields Programs – A wide coverage of grants for assessment, loans for clean up, innovative technology applications and partnerships with other federal agencies assisting redevelopment.

Adequate seed capital to perform the assessment tasks detailed previously can be obtained from public sources such as those above. This public funding develops information that remains in the public realm and to its benefit. The studies may be utilized in a later project should the first attempt stall. Federal and state assistance can also support remediation and infrastructure modification when those are deemed comprehensive and required to reuse the site. The overall benefit to the community must be made clear while the project is being formulated to receive the political support. Tax increment financing means some surrender of public income. The value of that exchange must be made clear. There is often the contrary argument that the community would be giving up a tax stream to the partial benefit of the developer. It must be explained that this public contribution may provide the only economic basis for the deal. The site can't be returned to productive use unless there is this abatement. The tax stream is not being given up for someone else's gain – rather it is being established because of the public contribution. The future benefit is in concept continuing. However, this is sensitive political matter requiring discussion within the context of all the benefits and sacrifices all parties make to repair a Brownfields site.

4 Will

Brownfields reuse demands perseverance. The developer must have a realistic understanding with his prospective tenants. Their financing needs to be committed for a window that allows the Brownfields therapy to take place.

Some professionals believe that many of the "best" sites have been redeveloped in areas where Brownfields activities have occurred for nearly a decade. What remain are the truly difficult sites. Some of these sites were bypassed before liability was better resolved and before the larger sums became available for assessment. Another layer of sites should become reusable. Then too, contaminated sites are always emerging as soil and ground water testing reveals issues around property transfers. Rather then let these sites fall into disrepair, ways could be found to utilize some public assistance to make deals possible. The responsible party's obligations and the avoidance of windfall returns need to be puzzled out. The main concern is to incorporate reuse of these sites into long-term planning.

Brownfields are also not just market driven projects. The host communities and public funding sources are more prepared than ever to lift sites. The legacy of contaminated industrial sites is no longer acceptable. Open space programs supported by a public willing to be taxed for quality of life parallel the desire to reuse Brownfields. The swap in bonding to provide cleanup costs rather than infrastructure costs makes a simple sense. Many Brownfields sites have utilities, transportation and permitted uses in place. This reuse supports in fill and Smart

Growth principles. Sometimes the new on site access roads can encapsulate contamination, which allows existing government development money to accomplish both aims simultaneously.

And yet, delays and increased costs due to unforeseen problems are part of the process. There can be more or new and different types of contamination than estimated. The community may demand the application of a higher standard of remediation than the regulations do and get that higher standard applied through political pressure. Understanding such possibilities early lessens the likelihood of discouragement on the part of political allies and funding sources. That anticipation builds a reserve of will and commitment for project managers.

The patient vision and considered responses of the developer must lead and support the community, regulators, financiers and staff. Brownfields are not projects for novices. The experienced eye must look at the risks carefully before committing. A track record will be needed to demonstrate how these site problems can be broken down. The use of public funds requires an open process. Neighborhood health concerns and historic resentments against the blighted site will be allayed only if brought forth and met.

5 Results

Experience helps manage the additional risks inherent in Brownfields. The world community is learning from projects all the time. Improved returns and more success stories will be forthcoming. The gap in economic rationality that is often a Brownfields site must be made up with public funding. That funding returns cash and many ancillary benefits as in the following forms:

- Tax streams from personal property, capital invested in equipment, real estate, income and wages, sales from products, value-added levies, employee purchases, etc.
- Health benefits – The health risks to human and other species may be incalculable. The insidious nature of long term exposure to contaminants is showing up in expensive medical care as well as well as in habitat degradation. Acknowledging this can reverse a community objection to the reuse. An obstacle can become an asset. The site will not continue to threaten that neighborhood.
- Sustainable development – The reuse of the site means that much less pressure on open space. Infrastructure will be reused. That neighborhood will be relieved of that particular blight making it more desirable and something less important to leave behind for families. The site is much less likely to become a future Brownfields if the economic use falters later.

Examples of Connecticut Brownfields projects that provide illustrations of managing costs and the development of protocols follow:

- Derby, Downtown Revitalization District – A combination of grants from state, federal and community foundation sources was utilized. Phase I and II studies, a Conceptual Remediation Plan, preliminary engineering and planning studies were prepared for the 22-acre site at a cost of about $170,000 US. Multiple parcels were envisioned as one for redevelopment goals to allow for moving contaminated materials within the site. The redevelopment plans indicate a probable mixed-use project valued at approximately $50,000,000 US where there was little tax stream and abandoned and decaying structures.

- Waterbury, Bunker Hill Park – This urban site was owned by a church, which had grave concerns over liability for suspected contamination. Neither the City nor state, which planned to make improvements after the site was transferred to the City, could accept title with unknown environmental conditions. The area residents were very concerned about health issues because four generations had played there. The regional EPA funded Pilot group provided a $15,000 grant for a combined Phase I and II study that defined the matter enough for the renovation process to continue.
- Willimantic, American Thread Mills – This site had a number of abandoned, old mill buildings dominating the center of the community. A phased approach resulted in the demolition of some structures and the cleanup of others. The wonderful architecture of granite walls with high ceilings with huge windows and thick wooden posts and beams were cleaned. The site now houses a number of businesses. Each section is renovated when a tenant commits. The project is funded from a number of sources and is ongoing for about a decade.
- Several Sites – Millions of square feet of new retail construction have been opened on significant Brownfields sites across the state. These sites had abandoned industrial structures determined to be uneconomic for redevelopment. They were razed, contamination removed or encapsulated and in one instance a small river was relocated to a site margin so that the site could be most efficiently reused. These projects have produced tax streams where there were only health concerns and blight.

6 Conclusions

The projected outcome must be envisioned and shared from the start to build trust, commitment and a framework for operating decisions. The project flows out from the orderly portrayal onto a spreadsheet. The team members can see their tasks and see issues when they develop. This reduces time and cost. There is the reward of working competently and in coordination.

The rewards can be spiritual. The blight of a Brownfields site depresses the people who see it every day. That same image causes people to shun the area and seek living quarters elsewhere. Restoring the site can help restore area pride. The site will no longer feature graffiti, broken windows and risk to children. Incorporating neighborhood concerns into the design can contribute to a sense of acceptance and partnership. Working through diverse issues creates a respect among regulators, developers, elected officials and residents.

Ecological benefits are sometimes obvious. The tanks and pits, the asbestos covered pipes and lead paint surfaces are no longer a visible problem. The groundwater is protected. Surface water runoff does not carry contaminant plumes off site. Problematic dust no longer invades households or lungs. But then too creatures and birds no longer ingest or track contamination to their homes and families.

The economic returns for Brownfields sites vary. Not every site produces expense covering revenue. In many instances, the projects are enormous undertakings like the construction of stadiums or malls. The public leverage used may bring slight and difficult to measure returns. However it can be seen that any positive return is worthwhile when compared to the costs of inaction and avoidance.

Smaller sites pose different problems and are not receiving the same attention as the larger ones. The potential for returns is restricted since the cost of remediation may end up being higher per square foot. It is necessary to empower regional development capacity to assist the entrepreneurs who eye these sites for reuse. That site alone may meet the eternal best location requirement of successful development.

The obstacles can be overcome. Indeed they must be overcome. Brownfields can no longer be disregarded. With will, time, money and knowledge they are manageable.

Integrated concept for groundwater remediation INCORE

T.Ertel & B. Schug
UW Umweltwirtschaft GmbH, Stuttgart, Germany

Abstract

Pressure on drinking water resources due to urban industrial areas is a common problem in European cities. State-of-the-art remediation of contaminated groundwater is based on single-site approaches neglecting cost-effective regional groundwater management and urban planning aspects. Ongoing research by a consortium of technology developers, problem holders, economists and administrators is focussed on a comprehensive approach to tackle these problems. Technology development include the integrated investigation of groundwater contamination at regional scale, fingerprinting and on-site analysis at single site scale and the implementation of innovative remediation technologies. Problem oriented implementation strategies include administrative procedures, urban planning aspects, and real estate market needs. Key goals to be met are the conservation of groundwater resources and economic revitalisation of urban industrial areas

1 Introduction

Most European cities are located in river basins and use groundwater from local shallow aquifer systems. Rapid industrial development in this century has been and is still generating groundwater pollution within city boundaries, exceeding legal requirements. Within the last decades changes in land use have resulted in complex contamination patterns, such as heterogeneous distribution of contaminants, different contaminants and large Landfill areas (see fig. 1).
State of the art approaches for site investigation and remediation regard only site specific solutions (see fig.2). But in general all site owners are declaring that contamination has been caused either by their neighbours or a former user of this site. Traditional investigation methods in most cases need very long time for the identification of the real polluter or even fail on this matter.

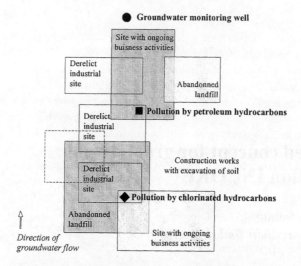

Figure 1: Actual situation on the investigation site

Today, legal treatment of soil and groundwater impurities is targeted on the specific endangering situation caused by a specific polluter. All measures aim at a fast reduction of environmental damages so that dangers from a single property for public security do not exist any more. This procedure fails in extensive polluted areas with different property owners and complex pollutant situation:

1. Evaluation of groundwater pollution is based on the sampling of single wells downstream the polluted site. But underground heterogeneity and spatially varying input of pollutants causes an inhomogeneous pollutant spreading in the groundwater.
2. Identification of polluters is possible only with high expenditure if different inputs of similar hazardous substances overlap.
3. Remediation of single hot spots contributes only rarely to a sustainable improvement of groundwater quality.

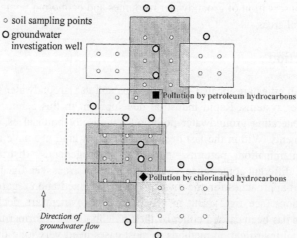

Figure 2: Classical investigation strategy

Today, large amounts of private and public money are spent to identify and assess point sources of contamination without being able to reliably quantify their impact on the groundwater quality; numerous remediation schemes are operated without an economical evaluation of their long-term performance.

Four European cities, Stuttgart Linz, Strasbourg and Milan, which share the same groundwater problems within their industrialised urban areas, have committed themselves to jointly develop such solutions. Within the different project areas specific local conditions vary with respect to groundwater conditions, existence of public and private monitoring wells, type of pollutants, size of areas, rising groundwater problems etc., and therefore provide a representative range to be expected at EU scale. In order to achieve the project goals in a cost-efficient way, different parts of the anticipated tool set are applied and evaluated at different levels of detail in the four selected areas.

2 Innovative approach

The proposed strategy for investigation, remediation and revitalisation of industrial areas is based on an integral quantification of total contaminant emissions, by considering entire industrial areas instead of single sites, in order to obtain a high level of investigation certainty.

An innovative cyclic approach is being proposed which starts with the screening of groundwater plumes at the scale of entire industrial areas, and ends with the remediation of individual source areas or the containment of plumes. The major advantage of this approach is that the number of local scale sites or the size of the area to be considered, respectively, is reduced step by step from one cycle to the next. Consequently, a large potentially contaminated area is screened but only a small area may ultimately be remediated. Figure 4 presents the scheme of this new approach.

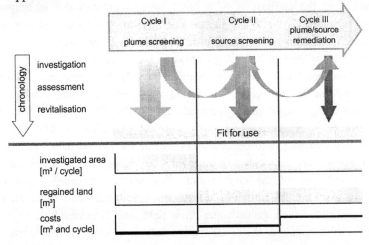

Figure 3: INCORE cyclic approach

It exhibits that the investigation / assessment / revitalisation cycle is repeated three times at a different scale:

(a) In cycle I of the new approach, the groundwater quality is screened downstream of the potential source areas. This is done by employing an innovative flux-based groundwater investigation approach.

(b) In cycle II only those sites are being re-considered where the groundwater quality is not acceptable. There, analytical methods are used to backtrack and identify the contamination source and determine its extent. Based on a site-specific risk assessment, the decision whether or not to proceed to the third cycle is made.

(c) In cycle III source zones are being considered for emission oriented remediation or implementation of monitored natural attenuation. For this purpose a comparative cost-benefit analysis is performed based on the available technologies. Again, the future land-use and the corresponding tolerable emissions need to be taken into consideration in order to make a final decision on the most appropriate technology.

The proposed "fit for use" –approach finally provides a cost-efficient set of tools, meeting both technological and administrative demands, for optimised investigation, evaluation and management of contaminated ground-water and land in industrialised urban communities.

3 INCORE results

3.1 Cycle I - plume screening

At the beginning of the project investigations are emission oriented and shall focus on groundwater contamination. Quantification of contaminant emission is obtained from the application of a novel integral groundwater investigation method, which yields the total pollutant mass flux and the mean and maximum pollutant concentrations originating from contaminant source zones.

The basic idea of the new integral groundwater investigation technique is that the total contaminant mass flux downstream of potentially contaminated sites is covered by the capture zones of one or more pumping wells, which are positioned along control planes perpendicular to the mean groundwater flow direction (see fig. 4) Analyses of multiple groundwater samples obtained at the wells during pumping yield concentration time series. Within these time series, concentrations vary as a consequence of the increasing capture zones and the spatial distribution of the contaminant mass within the aquifer. The total contaminant mass flux (emission) and the mean and maximum concentrations within the undisturbed groundwater flow field are determined through the mathematical inversion of the concentration time series using a particle tracking based inversion algorithm and a numerical flow and transport model of the investigation area. These tools are also used for determination and the delimitation of boundaries of potential polluted source zones. With the means of numeric-stochastical flow and transport model-ling approaches based on geostatistical simulation techniques parameter uncertainty can be described. This yields to estimations of probabilities of contaminant concentrations exceeding regulation limits within large areas. The size of these areas under consideration depends on the transport and degradation behavior of different contaminants.

Figure 4: Principle of integral groundwater investigation approach

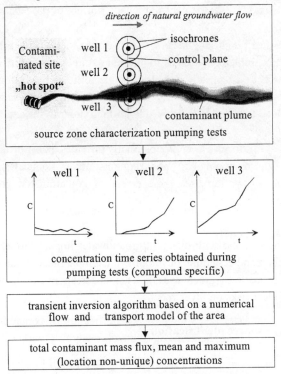

Final result is a map (see fig. 5) of the investigation area distinguishing between areas with different level of groundwater impact. This allows a ranking of these areas with a distinct level of confidence. This map enables the administration to set priorities for further activities. Focussing the efforts on the red areas helps to concentrate man-power resources on these sites which cause major impacts on groundwater pollution. This leads to maximum effectiveness in administrative activities.

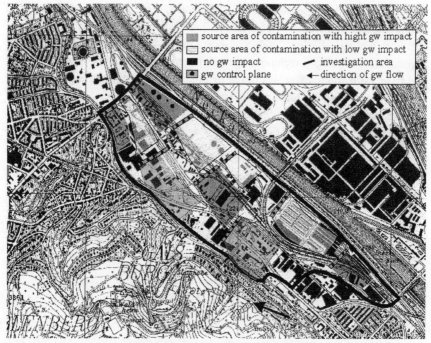

Figure 5: Cycle I - plume screening

The mapping of areas with different groundwater impact further identifies areas in which urban and economic development can take place without any hindrance by pollution matters (green areas). This ensures development and structural change with low risks on investment.

3.2 Cycle II - source identification

The results of the integrated investigation give a rough localisation of suspected source areas. However, more precise identification of the contamination source area is needed in order to apply the polluter pays principle. With the means of cost-effective laboratory and on-site analytical systems as well as isotopic finger-printing techniques the backtracking from the control plane along the path line of the plume yields to a precise localisation of the source of contamination (see fig.6).

The results of cycle II verify the sources of groundwater pollution and identify the polluter with a very high degree of probability. This secures the application of the "polluter pays principle".

These results also help to avoid law-suits which leads to an acceleration of investments and to faster administrative procedures.

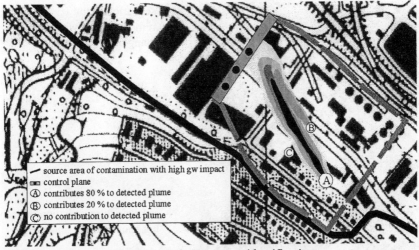

source area of contamination with high gw impact
control plane
(A) contributes 80 % to detected plume
(B) contributes 20 % to detected plume
(C) no contribution to detected plume

Figure 6: Cycle II - source identification

3.3 Cycle III - remediation strategy

Monitored natural attenuation as well as emission-focussed in-situ remediation combined with passive remediation techniques are major options which have to be considered in the management of contaminated groundwater and land in urban industrial areas.

The work deals with dual solutions of source and plume remediation, taking into account natural attenuation as a part of a comprehensive remediation approach, depending on the remaining/tolerable contamination levels and extent. The basic idea is to find the most efficient hot spot treatment technology for a given hydro-geological and contamination situation, and to treat the remaining plumes with passive remediation technologies.

These alternative or partly combined remediation scenarios are actually under Remediation schemes lead to joint remediation strategies for several sources of contamination. The application of "private-public partnership" and innovative technologies will result in cost-efficient and faster treatment of polluted sites. By this means the costs for industrial settlement as an important locational factor for industry will be lowered, which strengthens international competitiveness.

consideration in feasibility studies (see fig. 7).

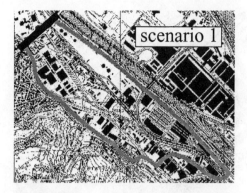

scenario 1: joint
remediation of whole area
using one reactive barrier

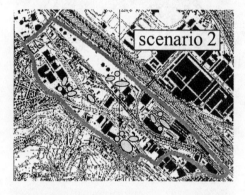

scenario 2: remediation of each
hot spot / source separately

scenario 3: combined
remediation of clusters of
related hot spots

Figure 7: Cycle III - remediation strategy

4 Outlook

In the final working phase of INCORE the activities are focussed on three main questions:
   Does the methodology of Cycle I and II ensure a strict application of the "polluter pays principle"?
   How cost-effective is the "fit for use" approach of INCORE?
   What kind of administrative measures will be able to support the implementation of the INCORE-methodology in future practical application?

5 Acknowledgement

The project receives funding from the European Commission in the specific research and technological development program "Energy, Environment and Sustainable Development". The author is solely responsible for the content of this publication which does not represent the opinion of the community. The community is also not responsible for any use that might be made of data appearing in this publication.

Smart growth and prosperity

C. Milburn
Phalen Corridor Initiative, East Side Area Business Association, USA

Abstract

This paper will outline the concrete steps and challenges of one of the most comprehensive brownfield redevelopment projects in the USA. A road to be constructed in 2002, gave rise to a vision capturing $237 million of investment to date. The paper includes historic and current photos, maps, charts, and newspaper stories. The Phalen Corridor Initiative is a complex public-private collaborative where corporations learn to work with neighborhood groups and every level of government. This project addresses environmental justice as African Americans, Hmong, Somali, and Hispanics have moved to the East Side of Saint Paul. However, the once thriving plants and decent housing have been replaced by absentee owned rental houses and polluted land. Successes of the project include...

1) Williams Hill Business Center, a converted brownfield and winner of the Minnesota Economic Development Award for 2000. The site includes 650 jobs, and the companies hire through a job bank the collaborative established. 2) Medium density, mixed income, sustainable housing designed by a partnership with fourteen corporations (3M and Andersen Window, etc.), community groups, Habitat for Humanity and the City. 3) A retail center, built on a marsh, was torn down in the nation's first example of retail re-conversion to a wetland. Now, a bank, a 62 million-dollar state agency, over one hundred units of senior housing and a national grocery are being built surrounding the new wetland. 4) A multi-modal transit corridor that includes a new road, bicycle/pedestrian trails, storm water ponding, a large bridge, rail, a $60 million interstate interchange, and bus transit. This paper will include a complex funding strategy with corporate, federal, local and state dollars.

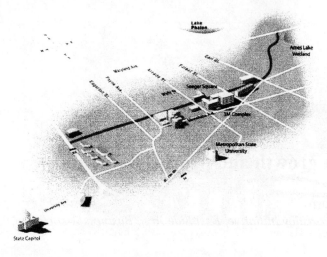

The Phalen Corridor

1 Introduction

As Americans continue to flee the poverty and blight of the inner city, suburban sprawl consumes millions of acres of dwindling farmland and green spaces and the abandoned urban core becomes more and more distressed. In his groundbreaking article, "Competitive Advantages of the Inner City" [1] Michael Porter of the Harvard Business School acknowledges that inner city areas in America have powerful impediments against redevelopment. Porter, however, calls on a partnership involving each sector in the American economy to work in consort to foster wealth creation for even the most blighted urban areas. Porter states that, "Overcoming the business disadvantages of the inner city as well as building on its inherent advantages will require commitment and involvement of business, government, and the nonprofit sector" (Porter 1995, 10).

The Phalen Corridor Initiative converts the detriments found in the urban core into advantages. The high unemployment rates in the city, translates into an available workforce. Polluted and abandoned brownfields are redeveloped into productive use. The Initiative is rebuilding the area's transportation infrastructure and reverses spatial disconnect by building decent homes and reclaiming green space closer to the population centers.

Since its inception in 1994, the collaborative has brought reinvestment and a renewed sense of hope and pride to some of Saint Paul's most distressed areas.

Theodore Hamms, 1850, photo courtesy of Minnesota Historical Society

2 History

Beginning in the 1850s, the rail lines that crisscrossed the East Side of Saint Paul helped bring raw materials to industries including breweries, foundries and manufacturing plants. By 1910, favorable manufacturing conditions helped to entice the burgeoning Minnesota Mining and Manufacturing Company (3M) to relocate to the area. The industries hired eager immigrants new to America. Waves of Swedes, Norwegians, Italians and Mexicans worked at the plants, occupied hastily constructed housing, and created a bustling social and commercial community. This era of industrial expansion was at its pinnacle in 1955, when 3M opened the world's largest abrasives manufacturing plant as part of its sprawling East Side campus. [2].

Swede Hollow, photo courtesy Minnesota Historical Society

3 Decline

By the 1980s, the economic bottom began to fall out. Manufacturing techniques in America had changed. The old multistoried buildings with low ceilings and multiple columns could not accommodate the need for larger advanced equipment or new distribution techniques. Companies began to shutter plants in search of cheaper labor, which could be found either in the southern states or off shore. In the early 1980s, the once bustling Whirlpool Seegar Square washing machine plant shut down, laying-off over 1,000 employees [3]. Over the next twenty-two years, the crippling spiral of layoffs, plant closings, and abandonment of polluted sites continued unabated. Griffith Wheelworks, the nearby American Hoist Plant, Cannon Conveyor, and Globe Roofing all closed their doors. On Thanksgiving Day 1997, the Stroh Brewery (formally Hamm's) closed, laying-off the last 365 employees. In a final blow to the community, in March of 2002, 3M announced the layoff of their last 500 manufacturing jobs making a total of over 3000 lay offs in just over 20 years.

3M Plant, photo courtesy of Minnesota Historical Society

In the 1990s, portions of the area suffered from 17 percent unemployment, almost 25 percent of families were receiving welfare payments (compared with 8 percent city-wide), and between 33 and 50 percent of men were working halftime or less [4].

4 A collaborative forms

According to Michael Welch in, <u>Environmental Business Research</u> [5], the right mix of partners can make or break a redevelopment project. In "Jobs and the Urban Poor," Beth Siegel and Peter Kwass assert that successful redevelopment projects aimed at finding people employment must, "develop a collaborative process to bring a wide range of key actors in the sector together including industry, labor and the public sector..." [6].

The right partners for Saint Paul's East Side began to coalesce in 1994. A local banker, an executive at 3M, a city planner, the director of a community development corporation, and an executive at Stroh Brewery met and began the Phalen Corridor Initiative. A steering committee was formed involving elected officials from every level of government, representatives from many of the area's community and nonprofit groups, and executives from area corporations. The collaborative resisted the urge to incorporate as a nonprofit, favoring a strategy of sharing power. (One partner acts as a "sponsoring agency," performing the fiscal responsibilities.)

Since most of the former job generators had been large-scale industry located on rail tracks, it was reasoned that a road should be built to connect the string of abandoned sites to two interstate highways, thus making them attractive to modern industry. Local, state and federal sources were tapped to begin to raise the $69 million needed to build a road to gain access to theses areas. Committees included an Environmental Impact Statement (EIS) Taskforce to design the road, a Workforce Development Taskforce to foster job development programs accessible to neighborhood residents and to be run at the new industrial sites, and a Communications Committee charged with disseminating information on the advantages of the project to the neighborhood and the public at large.

In 1995, Saint Paul became one of the first 16 US Environmental Protection Agency "Showplace Communities." While other areas of the city squabbled over redevelopment options, the four neighborhoods of the East Side brought in comprehensive proposals to local, state and federal agencies. The initial projects included endorsing the development of a new industrial park on the west end of the corridor (Williams Hill), a job bank, and a proposal to tear down much of a blighted shopping mall and convert it back into the wetland that had once been on the site (Ames Lake). The third project was the first time in the US that a wetland was reclaimed to create an amenity that might eventually attract reuses to the surrounding area.

Williams Hill, pre construction

Williams Hill Business Center today

During the first few years, the Initiative followed a reactionary model, responding to the needs and dictates of the partners. Whenever an opportunity was presented, the Steering Committee would exert what influence it could to urge elected officials to support the specific project. In the case of Williams Hill, (a 27-acre site of 30 meters tall polluted construction fill and abandoned dump site) members of the partnership and staff were recruited by the development agency of the City (the Saint Paul Port Authority) to attend legislative and city hearings and to lobby behind the scenes. The Phalen Corridor Initiative joined an even larger state-wide collaborative involving churches and community groups to secure $21 million in each biennium for brownfield clean up dollars for a state-wide fund. As the project became more complex, a new way of coordinating agencies and organizations was demanded.

5 Change leads to prosperity

The East Side community has not only changed economically, but the demographic shifts are profound. Between the 1990 and 2000 Census, the percentage of White Americans has been reduced by 23 percent, replaced by a 225 percent increase in Asian/Pacific Islanders (most often the Hmong ethnic group from Laos/Cambodia) and an almost 200 percent increase in both Hispanic and African American people. [7]. While many US "rustbelt" cities have shrunk, Saint Paul actually grew in population. The East Side alone accounted for 61percent of this growth over the last decade.

As the community began to change, more people of color joined and supported the Phalen Corridor Initiative. When the first projects along the Corridor became realized, these neighborhood leaders and funding organizations looked to the Initiative to tackle more and more issues. A concern often expressed was that if a community member got a job in the new industrial park and yet the area was still plagued by substandard housing, high crime, and degraded green space, what was there to keep this person (and their wealth) from leaving the neighborhood? Also, as companies moved to the area, what would there be to entice their employees to move closer to their new plants? Soon new schools, recreational facilities, enhanced retail opportunities and improved transit, were being attracted to the area in hopes to reinvent the whole community.

In 2000, the staff of the Phalen Corridor Initiative began to introduce housing concepts to an area called Railroad Island, the first neighborhood along the western edge of the Corridor. In this small community, 70 percent of the housing was absentee-owned rental, drug sales were rampant and the community was clearly in disrepair. The neighborhood group had not received financial or political support for a housing program, which focused on building new, single-family homes on large lots with detached garages [8]. The Phalen Corridor partnership began to introduce plans for over 140 units higher density housing meant to attract various income and age groups [9]. An architecture firm, the City of St. Paul, a contractor and a local community development nonprofit comprised this new partnership advocating energy efficient housing built to last well over a century. Entrenched attitudes were challenged and innovative concepts were scrutinized through a series of community meetings. The community adopted many of the concepts and the first phase of the construction began in May 2002.

Signage for the Railroad Island housing development

Because the housing will require substantial public subsidy, corporations, including Andersen Windows (the largest window manufacturer in the World) and 3M have been solicited to provide products at reduced cost. Over fourteen companies have also agreed to help build nine Habitat for Humanity Homes in which the future low income owners are supported in every step of the building and purchase of their houses.

In another community along the Phalen Corridor, early plans called for the removal of over 70 homes in a blighted area to be replaced by a new industrial park and new housing. The community balked at the scope of the redevelopment. Rather than engage in a protracted battle, the Initiative partners met repeatedly with the neighbors until the residents themselves helped develop plans for a new industrial park (to be built in 2004) and a greatly reduced housing development.

This new kind of partnership has fostered the formation of the Phalen Corridor Project Team, including city staff involved in roads, housing, commercial development, parks and even industrial development. Three municipal agencies have signed the first ever "development agreement" where funding is shared between agencies and across Phalen Corridor projects, a strategy unheard of in the "turf sensitive" municipal government of the past. This team approach model has brought a value added advantage to the Initiative as the expertise of each of the partners is shared and the project is enhanced exponentially.

Phalen Shopping Center, before wetland redevelopment

Ames Lake wetland redevelopment with new construction of bank and BCA in background

6 Conclusion

Through creative and nimble partnerships,$ 237 million of funding has been secured for industrial redevelopment, roads, two new schools, a new YMCA, a new transit facility, workforce development programs, green space, housing and commercial development. Over 1,000 jobs have been brought to the community to date with triple that number expected in the future. In August of 2002, the first third of Phalen Boulevard will begin construction to be completed in 2003. Even prior to the building of this road, a vision has captured the imagination of neighborhood residents and decision makers alike. Redeveloping brownfield sites has led to a community-wide reinvention of blighted areas and brought pride back to a distressed community.

References

1] Porter, Michael E., "The Competitive Advantages of the Inner-City," Harvard Business Review
    May-June 1995, pp55-71

2] Huck, Virginia , Brand of the Tartan - The 3M Story, 1955, Minnesota Mining and Manufacturing
    Company Library of Congress Card Number, 55-8359, p57

3] Bartsch, Charlie, Senior Policy Analyst, Northeast Midwest Institute, "St Paul, Minnesota:
    Integrating New Urbanism," Brownfield Blueprints (A Study of the Showcase Communities
    Initiative), Superfund Brownfield Research Institute, 1998, pp315-318

4] Phalen Corridor Redevelopment Proposal, March 1994, City of Saint Paul (1990 US Census), pp9

5] Welch, Michael, "Notes from the Brownfields Front," Environmental Business Research, Number 1,
    1997, pp1-6

6] Siegel, Beth and Kwass, Peter, Jobs and the Urban Poor - Publicly Initiated Sectoral Strategies, Mt
    Auburn Associates, November 1995, Executive Summary pvii

7] 2000, US Census, Saint Paul Planning District 2, 4, 5

8] Railroad Island Small Area Plan and 40 Acre Study, 1994, City of Saint Paul

9] Deegan Jessie, and Milburn, Curt, "New Housing Ahead", East Side Review, February 28 2000,
    pp8-10

The city of Muskegon Heights, Michigan

A. Wentz[1] & K. Lynnes[2]
[1]*U.S. Environmental Protection Agency, Region 5, Chicago, Illinois*
[2]*Williams & Beck, Inc., Consultants to the city of Muskegon Heights*

Abstract

The City of Muskegon Heights, Michigan is a classic example of the inextricable link between environmental justice and brownfields. Muskegon Heights is a small, predominantly African-American city near the shores of Lake Michigan. The City had a thriving industrial job base from the early 1900's until the early 1970's, when many plants closed their doors and moved to the southern United States or other countries. The economic devastation caused by the City's high unemployment rate, the loss of high paying manufacturing jobs, and environmental degradation cannot be overestimated.

Despite these significant obstacles, the City is deeply committed to urban revitalization. Muskegon Heights' brownfields program is a key part of this community-based revitalization process. A decades-old wastewater treatment plant caught a national spotlight in the United States through the efforts of the City. The City operated the plant from the 1920's until the early 1970's when it connected to a regional treatment system. The plant was then leased to a succession of hazardous waste treatment companies that used it to treat electroplating and oil wastes until 1990. The current corporate successor of the lessee, the Safety-Kleen Corporation, is in bankruptcy. The site has been unused for the last decade due to concerns about heavy metal and solvent contamination and cleanup liability. Instead of being discouraged by the regulatory and financial hurdles, the City has approached the project as an opportunity to create both needed housing and a prototype for other low-income, minority, urban communities that want to attract middle-income families back to their community. In 2001 this project was selected by the U.S. Environmental Protection Agency (U.S. EPA) as one of five national Resource Conservation and Recovery Act (RCRA) Brownfields pilots.

Regulatory framework

The national RCRA program requires corrective action be taken at permitted hazardous waste treatment, storage and disposal facilities.   The State of Michigan, which is authorized to execute this national program within the state, has never elevated this site to a high priority status, partly because sampling data is lacking and partly because enforcement would be futile against a bankrupt company and a low-income, already-disadvantaged municipality. The safe redevelopment of this property is, however, a high priority for this community. The complex hazardous waste regulatory scheme and the current blighted condition of the site create a number of barriers to redevelopment.

The barriers

The first barrier is the City's exposure to liability for RCRA corrective action. Although the City never treated hazardous wastes at the plant, they have RCRA liability because they own the property. The second barrier is the issue of liability protection for a potential buyer. Michigan's statutory liability protection provision for innocent buyers of brownfield properties does not extend to hazardous waste sites. The third barrier is the prohibitively high cost of demolition for the plant buildings and treatment structures. Current, conservative estimates of the cost are over $1.8 million. The fourth barrier is obtaining funding for remediation when the City's budget is tight and the former hazardous waste treatment operator is in bankruptcy.

A vision

Despite these problems, City Manager Melvin Burns and Mayor Rillistine Wilkins have a vision for the 51-acre site, which includes middle-income housing, playgrounds, and a hiking trail. The principals of Mona Terrace L.L.C., a local development company, are community leaders who share the City's vision. They all know that the higher wage jobs that are being created as other brownfield sites in the City are redeveloped will attract middle-income families back to the City and that these families will help revitalize the community.   Mona Terrace, L.L.C., approached the City last year about using the plant site for a "new urban" residential development. The preliminary design includes a combination of single family homes, duplexes and townhouses. The housing density is higher than a typical suburban development because the site plan incorporates an existing natural area as community open space. The site is near an elementary school and is served by public transportation.   Before Mona Terrace's vision becomes a reality, the site must be investigated, existing structures demolished and cleanup strategies developed and implemented.

The starting point

In the early months of 2001, the City of Muskegon Heights applied to the U.S. EPA for pilot status under the Agency's initiative to streamline and improve the RCRA corrective action program. A December 22, 2000 memorandum from U.S. EPA's

Office of Solid Waste stated that the RCRA Brownfields Pilot program "would test a variety of innovative approaches that expedite the cleanup of brownfield properties subject to RCRA and to use the information gathered to create improvements in the administration of the RCRA program at contaminated sites." The Muskegon Heights application was one of five selected nationwide in 2001. By applying to the RCRA Brownfields Pilot program the City hoped to bring all stakeholders together to focus on creating an innovative, results-oriented, and protective approach to the redevelopment of the site.

Attendees at the organizing meeting held in Michigan included representatives from the Michigan Department of Environmental Quality (MDEQ), Safety-Kleen Corporation, U.S. EPA's offices in Washington D.C. as well as the regional office in Chicago, Mona Terrace LLC., and the City of Muskegon Heights. Sixteen people in all attended. The first task for the group was to establish a process for communicating with each other through regular conference calls and e-mail. The group then focused on creating an initial list of short term goals, which included: a comprehensive history of the site through all it's changes in management, finding funding for additional analysis (in order to fill in the data gaps on what contamination may be at the site), a community outreach program for local residents so that they might understand both the level of contamination at the site and how it will be cleaned up to make it safe for residential development, and, of course, a liability guarantee for the prospective purchasers so that they would be able to secure bank financing for the project.

Progress thus far

A project of this complexity requires performing multiple tasks simultaneously while maintaining a constant focus on the "big picture". For example, the prospective purchaser agreement negotiated among the developer, EPA and MDEQ must not only protect Mona Terrace from liability for the pre-existing contamination, it must be easily assignable to individual home buyers so that they would not be burdened by RCRA requirements. Another example is the need to coordinate the environmental investigation and cleanup activities with site plan development. Because it will not be possible to remove all contamination to allow for unrestricted residential use, the City's consultant and the state and federal regulatory agencies must work closely with the developer to ensure that any engineering and institutional controls are integrated into the design.

Two of the initial short-term goals have been accomplished. The City and the developer have organized a community focus group to facilitate community outreach. The EPA has provided grant assistance to help prepare outreach materials. A comprehensive site history has been completed based upon a review of the voluminous MDEQ files, information provided by Safety-Kleen and interviews with local residents. The information gleaned from the development of this chronology is being used to identify data gaps remaining from prior investigations, prepare demolition specifications, and research historic insurance policies.

Progress is also being made toward overcoming the four development barriers described above. The City's consultant has prepared a site investigation workplan based upon the site history and the proposed future use that is designed to fill data gaps and identify any environmental "deal killers" (contamination problems that would be too expensive or time-consuming to address). Funding for this investigation is being provided by the City with assistance from the EPA regional office in Chicago. The investigation is scheduled for late spring of 2002.

The momentum resulting from this project has lead to increased public attention to the historical sediment contamination in Little Black Creek, which runs through this site. The state has known since the mid-1970's that the water and sediment in this creek is contaminated by heavy metals and other hazardous substances. The creek runs through two City parks and adjacent to an elementary school before it reaches the former wastewater treatment plant site. Although the shores of the creek are posted with warning signs, children still play in the water. The Community Foundation of Muskegon County was recently awarded a grant from the C.S. Mott Foundation for a preliminary evaluation of water quality concerns in the Mona Lake watershed, which includes this site. The scope of work under the grant will include sediment and water sampling in Little Black Creek. A local university has applied for additional funding from EPA for further, more comprehensive sediment and water sampling. The MDEQ, is responsive to the community's concerns about the health hazards posed by the contamination in the creek and the City plans to use the data from the C.S. Mott grant to encourage the MDEQ to use state funds to remove contaminated sediment.

The Future

The next critical step in this project is finding funding for demolition. The Michigan Economic Development Corporation (MEDC), the state's economic development agency, rejected the City's initial application for demolition funds. MEDC was concerned that the site might be too contaminated to develop and that the project had too many unknowns to commit funds from the state's limited budget for brownfield redevelopment support. The site history and the data from the site investigation described above will be used to provide MEDC with the documentation needed to reconsider the City's request for demolition funds.

The prospective purchaser agreement that will be negotiated for this project will be one of the first in the United States for a RCRA brownfield site. Developers nationwide will be able to benefit from the hard work in Muskegon Heights. This project will also be one of the first in the state to use a new lender liability protection program developed by EPA and MDEQ to encourage lenders to finance brownfield redevelopment projects at former hazardous waste sites.

None of this progress would be possible without the City's determination to revitalize Muskegon Heights, the willingness of Safety-Kleen to take a risk, and EPA's and MDEQ's commitment to innovation. Two years from now when this blighted site is transformed into a thriving neighborhood, Muskegon Heights will be proof that working together can make property re-use a reality.

Section 5:
Financial and insurance aspects

Value creation in the redevelopment of environmentally impaired properties

S.M. Hollingshead
Renova Partners LLC, USA

Abstract

The essence of value creation is the application and coordination of a broad range of capabilities to the complex process of contaminated asset redevelopment. This process includes careful risk analysis and containment, optimal land use and infrastructure planning, financial engineering, targeted real estate marketing and access to a myriad of governmental entities.

The goal of value creation through an integrated site-planning model is to:
- Align the interests of all stakeholders
- Contain risk through a variety of risk control measures to achieve certainty
- Minimize remediation costs
- Identify the true market for an asset
- Shorten cycle time

Coordination of these objectives is demonstrated in a review of two redevelopment projects conducted by the author.

1 Introduction

While the economic prospects for redeveloping contaminated properties vary widely, a significant portion of these assets have remedial requirements the cost of which overwhelms the future value of the property. The result of this unfortunate equation is that these properties continue to sit idle and are only addressed to the extent that public or private funds can be used to offset the negative value of the property. Practitioners within the brownfield redevelopment field have looked for other tools that might bring greater value to these properties and expedite their redevelopment and reuse.

The concept of "value creation" is a new paradigm for solving the ubiquitous and divisive environmental challenge of cleaning up and reusing contaminated sites. Against the backdrop in the United States of recent legislative initiative promoting brownfields redevelopment, voluntary clean up programs and liability protection, the value creation paradigm redefines traditional clean up strategies by expanding the environmental goals to encompass real estate development, land revitalization, public-private partnerships, liability protection/containment and alliances with all levels of government. The essence of value creation is the application of a broad range of capabilities to a highly complex process. This process includes careful risk analysis and containment, optimal land use and infrastructure planning, financial engineering, targeted real estate marketing and access to governmental assistance programs.

2 Variables that affect value

What are the variables that affect value? While these change from site to site, key among them are:
- Nature and extent of the contamination
- Time required to conduct clean up
- Clean up methodology and technologies available to expedite the process and reduce costs
- Future use considerations including:
  - Present zoning
  - Community and neighborhood concerns
  - Dynamics of local economy
  - Opportunities to reposition property through enhancement of infrastructure
  - Unmet requirements in the local or regional market place
- Available funding mechanisms for clean up and vertical development

3 Integrated site planning

What is the process then by which we manipulate these variables to improve value? The term that Renova Partners uses to describe this process is "integrated site planning". Integrated site planning includes:
- Identifying the true market for an asset
- Aligning the interests of all stakeholders
- Shortening cycle time
- Containing present and future risk through a variety of risk control measures to achieve certainty

3.1 Prior practice

Historically most corporations and public entities have approached brownfield redevelopment through a series of largely unconnected steps as depicted in the following diagram:

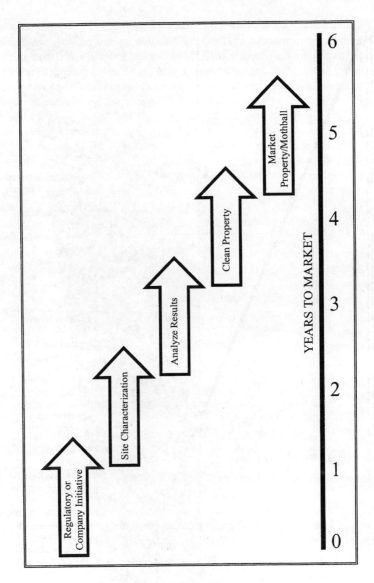

Figure 1: Historical redevelopment process.

3.2 New practice

Integrated site planning changes the model to conduct many of these steps simultaneously and to examine the appropriate exit strategies for responsible parties, third party interim ownership and other stakeholders:

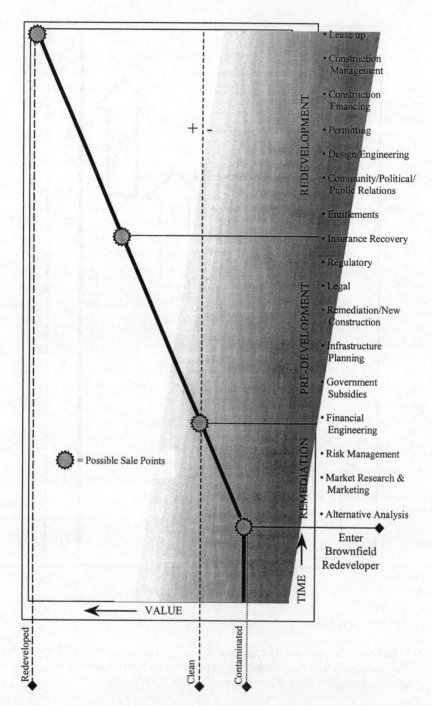

Figure 2: Integrated redevelopment process.

3.3 Identifying true market

The standard real estate practice of identifying that a property is for sale for a given price and letting the market drive the transaction and future reuse of the property is unsatisfactory for many brownfield properties. In order to maximize value, the point at which increased remediation intersects with increased value must be understood. While the highest and best use of a property may seem to be industrial, commercial, retail, residential (the types of residential i.e., high density versus single family are more important considerations than merely the residential designation), or recreational, the interaction of the cost to reach an acceptable level of clean up and elimination of exposure pathways must be calculated and optimized if best value is to be obtained. Further, many brownfield properties are located in areas of transition or may be sufficiently isolated as to make any future reuse unclear. Therefore, rather than waiting for the market to drive redevelopment, it is necessary to determine what needs are least served in that locality or region and evaluate the potential to meet that need in the redevelopment of the property. The following case studies illustrate how hidden some of the reuse options may be. They also demonstrate that without stakeholder participation reuse to achieve increased value may be impossible to pursue.

3.4 Aligning the interests of stakeholders

As discussed in 3.3, stakeholder participation can be key to creating a reuse for the property that maximizes value. Aligning the interests of stakeholders interacts with value in many other ways:
- Avoids lengthy disputes over reuse options that increase costs and/or nullify the project
- Recognizes the potential to serve multiple interests and potentially draw upon more diverse sources of funding to support project
- Identifies opportunities for and facilitates public funding

3.5 Minimizing the remediation costs

There are three key considerations to minimizing remediation costs in a way that maximizes value:
- Achieving a clean up level that supports best value when analyzed along side potential next use
- Achieving clean up within a time frame that supports best value when analyzed along side capital costs and development delays
- Planning clean up so as to use demolition and redevelopment as the remedy wherever possible or to conduct these activities as one effort (for example, site grading and excavation of contaminated soils) rather than multiple tasks.

3.6 Shortening cycle time

In most cases cycle time affects the impact of ongoing exposures, the ability to meet a window within a given market, and the cost of capital. Therefore any method to reduce the timing of a project must be considered in order to achieve maximum value.

3.7 Containing present and future risk through a variety of risk control measures to achieve certainty

By coordinating each of the following elements, risk can be managed more thoroughly and usually at a lower cost:

- Remediation
- Long-term operations and management
- Engineering controls
- Institutional or land use controls
- Environmental insurance

In too many cases, if remediation precedes land use planning, engineering and/or land use controls may be implemented that preclude appropriate redevelopment, increase cost of environmental insurance or provoke future issues amongst stakeholders.

4 California case study

4.1 Overview

The owner of this 220 acre operating refinery at the North East end of the San Francisco Bay had decided to close their operations at the site.   Despite being of relatively new construction,   the refinery   needed extensive retooling to meet current additive requirements. The site was not suitable for commercial or retail development because of its steep slope (45% grade to the Bay). There   were   two potential reuses for the

Figure 3: California site, before.

site. It could be sold as an asphalt manufacturing facility, which could operate using the existing infrastructure and would not require retooling. This alternative provided only a minimal land value. The community did not favor the continuing presence of an operating refinery of any type given prior air emissions issues and the proximity of many residences to the site's Northern boundary. The site could also be sold for residential development. Residential development created a much higher land value but that transaction required waiting until an extensive remediation could be performed. The owner was concerned that the cost of remediating to a residential standard might overwhelm the increase in land value. Residential development also posed the threat of future liability arising from bodily injury or property damage claims from owners of residences built on the site. Neither alternative addressed the owner's goals, which were:

- A significant increase in land value
- Avoidance of the direct costs of remediation
- Elimination or limitation of future liability from third parties

4.2 Application of integrated site planning

The owner contacted a brownfield redevelopment firm with the notion that the firm might be willing to take title to the property, conduct the remediation and shield both the owner and developer from future liability arising from the clean up of the site. The redevelopment company was interested but unwilling to assume the entitlement/land development risk. Further, both parties conditioned their interest on the ability of the redevelopment company to acquire significant limits of environmental stop loss and real estate pollution legal liability coverage on a full occurrence basis. The owner was extremely risk adverse and wanted long-term protection from future

Figure 4: California site, after.

claims arising from the residential use of the property. From the beginning of their discussions, both parties focused on the owner's goals and their ability to frame a transaction to meet those goals. The owner was willing to compromise on many points not related to the achievement of its three goals. Ultimately a transaction was structured to include:

- The site would be purchased on an "As Is, Where Is" basis. The buyer would contractually assume all liability for clean up and any liability to third parties arising from the contamination conditions in perpetuity.
- The buyer formed a single purpose entity to jointly acquire the property with a national developer. The redevelopment company retained all of the environmental liability and the developer all of the entitlement risk.
- The remediation and site development plans were integrated so that considerable savings could be achieved by conducting contaminated soil excavation/burial simultaneous with site grading for development.
- The clean up costs and potential future claims were secured by $50,000,000 (50 million) in environmental insurance policy limits. Six policies were written to address various concerns.
- The seller discounted the property value with consideration for clean up costs, demolition, certain uninsurable risks and stigma.

The transaction closed in 1998. Surficial clean up, and entitlement were completed in 2000 and vertical development begun in 2001.

5 New York case study

5.1 Overview

This site, a contaminated former oil terminal, was located in an economically depressed area of the Hudson River Valley. The site offered no potential for commercial or residential development. Stakeholders concerned with the future of the river were opposed to industrial redevelopment both because of the visual impact from the river and because of the potential for additional contamination from new industrial operations. Given little or no market in the area for this type of property, estimated cleanup costs were greater than perceived land values. The site had been

Figure 5: New York site, before.

idle for a number of years and no clean up was taking place, an additional problem for stakeholders concerned with impacts to the river.

5.2 Application of integrated site planning

This case is an excellent example of the value created by detailed market research and the integration of end use and clean up. Market research revealed that due to the extensive environmental clean up taking place along this part of the Hudson, there was certain demand for a new soil incineration facility. The state was anxious to site and permit such a facility and the potential fees for soil incineration supported a value for the property that could include the cost to clean up. Further, construction of the foundation of the facility provided and opportunity to address part of the clean up, capping of contaminated soils, through the redevelopment thus saving considerable clean up expense. The warehouse type construction, as compared to tanks or other industrial development, satisfied stakeholders. The construction was bland in appearance and the covered soil incineration posed no pathways for contamination to the river. A buyer was identified who was willing to acquire the property, conduct and pay for the clean up, and construct and operate the soil incinerator.

Figure 6: New York site, after.

6 Conclusion

Value creation through integrated site planning is an effective method for conducting brownfield redevelopment throughout the United States. While only two cases were examined here, there are now several hundred similar projects.

Economic aspects of polluted soil bioremediation

J. Troquet[1] & M. Troquet[2]
[1]*Biobasic Environnement, Pollution control through biotechnology, France*
[2]*Institut des Sciences de l'Ingénieur de l'Université Blaise Pascal, France*

Abstract

Industrialisation in developed country has left many type of pollution during the last 150 years. The end of pollutants course is generally soil and groundwater. In the main part of west countries, authorities must list the contaminated sites and have to consider their clean-up.

In the same time, the use of biotechnology for the removal of organic pollution from soil and groundwater is in progress and offer new possibilities of cost savings. The main characteristics of bioremediation processes are their capacity to be implemented *in-situ,* that is to say to avoid great civil engineering works.

Nevertheless before to choose this innovative technology, it is necessary to define the best treatment strategy, which is in close relation with local conditions. In this paper we will investigate some specific cases for which it is possible to implement an effective bioremediation process. In addition, the main process variables will be identified in order to keep the competitive price of this biotechnological approach. In this way, it will be emphasised particularly about the whole knowledge of the quality and quantity of the pollutants to be treated and both physical and biological factors active on the bioprocess in the first stage. Secondly, pilot lab experiment will be conducted to identify the different parameters of oxidizer and nutriments supplementation and micro-organisms growth kinetics.

This strategy appears surprising, because the real evaluation of bioremediation cost cannot be established without the two steps described above, and this first operation could represent 5 to 15 % of the global cost of rehabilitation.

1   Introduction

With the economic and industrial development, all countries in the world faces an enormous task in cleanup hazardous waste. If we promote today clean technologies for industries, it is also necessary to rehabilitate old industrial sites, particularly in urban zone, and remediate rapidly after ecological disasters. Soil, groundwater and marine coast are often subject to contamination by various organic and mineral pollutants from a wide variety of sources. Organic pollutants, like crude oil, are one of the most important pollutant in marine environment with the mediated oil spills like the Exxon Valdez and Sea Empress in the United-States and other Amoco Cadiz or Erika in France. It has been estimated that worldwide somewhere between 1.7 to 8.8×10^6 tons of hydrocarbons impact marine waters and coasts each year [1]. Around the coast of the UK alone, between the years of 1986 to 1996, 6,845 oil spills were reported [2]. In France 281 pollution reports in 2000 and 347 in 2001 were notified with respectively only 83 and 62 pollution sources identified [3]. This data show the difficulties to really quantify the marine pollution.

For soil and groundwater, the problem is the same but for other reason, especially because the pollution is quite very diffuse and no visible. In different countries, the regulation organisation, national administration or industries try to draw up inventories of polluted site through historical studies. Among those, the US Environmental Protection Agency (EPA) and US Air Force in United-States, British Gas in United-Kingdom, the French Agency for Environment and Energy Management (ADEME) and the Ministry of Industry in France are the most active and they have promoted various demonstration programmes for testing land remediation. Economic assessment is not only a key factor for decisions in the field of site rehabilitation, but also one of the integrating tools for damage assessment. Indeed, economic cost induced by accidental, chronic or historical spills, belong to two distinct categories:

- costs associated with decrease in services rendered by the natural active assets;
- costs of rehabilitation or replacement of this asset.

Treatment methods are divided into those for soil remediation and for surface and groundwater remediation. Further categorization results in the consideration of biological, chemical and physical treatment technologies. Eighteen different methods have been reviewed that pertain to remediation of soils [4]. Bioremediation appears the simplest and cost effective method for large volumes of contaminated soil by toxic chemicals [5, 6], probably because this process naturally occurs in soil and groundwater and can be easily enhanced by different techniques.

2   The market evaluation

The pollution amount in soil and groundwater is very extensive, accumulated during 150 years of industrial activities. In United-States, thousands of sites have been identified throughout the country, with an estimated cleanup cost over $ 1.7 trillion using existing technologies [7]. On the other hand, it has been estimated that the bioremediation component of environmental biotechnology will be some $500 million by the year 2000 [8]. This two values show the difficulty to make an inventory of contaminated sites and to estimate the remediation cost.

The US Air Force has estimated that it has 2,000 underground tank sites to clean up [9] by this method. In 1986, conservative estimates indicated that they were approximately 3.5 million Underground Storage Tanks (USTs) in the United-States, of which roughly 200,000 to 400,000 are leaking or showing signs of deterioration. By 1994, the estimate of leaking tanks had grown to over 600,000. The average cost of a non-hazardous UST cleanup is estimated at $70,000 [10]. Other evaluation talk about 6 million USTs, with 15-20% of them are leaking. The associated cleanup costs are estimated to average about $180,000 per site [11]. Always in United-States, the national market for bioremediation has been estimated to be up to $1 billion. Currently there are over 200 firms in the US that offer bioremediation services [12]. Nevertheless, many other evaluations exist in different countries. In United-Kingdom, it has been estimated that there are some 100,000 sites which will cost between £10,000 and £20,000 million to clean up [8]. For example, British Gas has a major programme of contaminated land remediation involving approximately the 1,000 former gaswork sites, and up to £200,000 can be saved on each site with bioremediation [13].

In France, 896 contaminated sites have been identified in 1996 by different type of pollution [14] : 41 % are contaminated by petroleum hydrocarbons, 17 % by lead, 16 % by polycyclic aromatic hydrocarbons (PAHs), 13 % by halogenated organic compounds, 6 % by polychlorobiphenyls (PCBs), 6 % by other volatile organic compounds (VOCs) and lastly 5 % by mercury. The regulated market is annually evaluated at €45 million [15]. In this last country, the global market is estimated between €7,500 and €10,500 million for the next ten years for 200,000 to 300,000 sites potentially contaminated [15].

In Germany, 110,000 polluted sites are really identified among 200,000 suspected sites for a total remediation cost of €200,000 million, but actually, a greater financial effort is consented with regard of other European countries [15].

3   Methods of soil treatment

Various technologies are currently available to treat contaminated soils by hazardous chemicals, including excavation and containment in secured landfills, vapor extraction, stabilization and solidification, soil flushing, soil washing, solvent extraction, thermal desorption, vitrification and incineration [4,16]. However, many of these technologies are either costly or do not result in complete decontamination of sites and in the actual communication especially devoted to bioremediation, only comparison for treatment costs will be reported.

4   Bioremediation capability

Bioremediation technologies can be broadly classified as *ex-situ* or *in-situ*. In both cases, micro-organisms are used because they have an extensive capacity to degrade synthetic compounds; therefore, bioremediation can be applied to various sites contaminated with a wide variety of chemical pollutants. The first bioremediation technologies that have been developed were *ex-situ* technologies, i.e. biotreatment of excavated soil in contrast to *in-situ* technologies which aim is the treatment without excavation and often is taking care of both groundwater and soil pollution [17].

The current practice and perspective of bioremediation have been recently reviewed in terms of processes, efficient micro-organisms and molecular degradation activity, through the molecular microbiological point of view [18]. Before evaluating bioremediation potential, it is necessary to explore the different processing ways and the vocabulary associated to this biotechnology. The most commonly techniques include [19] :

- *in-situ* **bioremediation,**
- **landfarming,**
- **slurry-phase bioreactors.**

4.1   *In-situ* bioremediation,

relies on biological cleanup without excavation and include many processes:

- **Intrinsic bioremediation, or bio-attenuation,** which is natural attenuation of organic pollutants by micro-organisms present in soil an groundwater. No actions are taken to enhance the biodegradation of contaminants beyond the existing capacity of the system. Intrinsic bioremediation is implemented by demonstrating that the native microbial population has the potential to reduce contaminant levels. It is less costly than conventional engineered technologies, but site remediation to regulatory standards generally cannot be accomplished in short time frame. Before intrinsic bioremediation can be implemented at a site, an evaluation is required to determine if it is a viable remediation method. Intrinsic bioremediation is expected to cost around $200,000, whereas the cost of pump and treat for the same pollution configuration can easily reach several million dollars [20].

- **Biostimulation,** consists in the stimulation of microbial growth *in-situ* by providing extra nutrients: Nitrogen and Phosphorus sources are added in relation with the hydrocarbons concentration according to the ratio carbon 100, nitrogen 10 and phosphorus 1. An electron acceptors such as oxygen is also supplied. Other substrates can be added as phenol, toluene or methane. In Japan, the effectiveness of *in-situ* biostimulation by methane injection into trichloroethylene contaminated groundwater was demonstrated by small-scale field experiments funded by the Environmental Agency [21].
  It is also possible to supplement the endogenous microbial population after enrichment of this specific microbial community in a Biological Activated Carbon (BAC) reactor for example [22].

- **Bioaugmentation,** is the addition of extra micro-organisms to enhance a specific biological activity when indigenous bacteria are not sufficiently active. As for biostimulation, the cost of bioaugmentation will vary with the nature of compound, of soil, and with the inoculum type and quantity [23].

Determination of the potential success of bioaugmentation requires an understanding of the bioavaibility of the pollutants, the survival and activity of the added micro-organisms and the general environmental conditions that control soil bioremediation rates. The use of indigenous strains is not subject to notification requirement since they already exist in nature. However, researchers are showing an increased interest in the use of genetically-engineered-micro-organisms (GEMs) for remediation of recalcitrant contaminants [24].

- **Biosparging,** is a process increasing the biological activity by improving the supply of oxygen by air or pure oxygen sparging into the soil layer. Generally, air injection was tried at first, but was often replaced by pure oxygen in order to increase the biodegradation rates [8]. Although biosparging has been used for soil and groundwater remediation since the mid-1980s, and the review of published literature reveals only a limited number of field investigations. The main encountered problem is that the injected air probably moved through the subsurface in channels and there was not really all the volume of soil that was oxygenated [25].

- **Bioventing,** is defined as the use of induced low volume air movement through the unsaturated soil associated with vapour extraction. A vacuum is applied at some depth in the contaminated soil which drawn air down into the soil from wells drilled around the site and sweeps out any volatile organic compounds [26]. The use of bioventing depends on specific conditions but could result in substantial cost savings as compared to current traditional methods of contaminant source treatment. Costs can vary dramatically depending on specific conditions. However costs from Air Force have been estimated for bioventing at $3 to $80 per cubic meter: $80 for 380 cubic meter and less, $3 for pollution volume greater than 15,000 m^3.

- **Biofiltration,** is particularly well adapted to groundwater bioremediation. This application involves sand mixed with non-indigenous micro-organisms into a trench in the subsurface ahead of the contaminant plume [27].

- **Phytoremediation,** consists in the use of plants for contaminants and metals removal from the soil. Many studies have been recently reported on the effects of vegetation on the transformation or stabilisation of organic compounds [28] and also extraction of metals from the environment [29]. These studies show that there are many reasons to investigate the interactions that occur among hazardous compounds, micro-organisms, plants, soil air and water. The plant root zone (rhizosphere) has significantly larger numbers of micro-organisms than soil which do not have plants growing on them; this appears to enhance bioremediation of organic compounds. Phytoremediation has made tremendous gain in market acceptance in recent years. Cost comparisons to other remediation technologies have been reported and the consensus cost of phytoremediation has been estimated at $45-$180 per cubic meter of soil treatment, and $0.15-$1.50 per cubic meter for treatment of aqueous waste stream [30]. The main technical advantages of phytoremediation in comparison with other methods is that it is far less disruptive to the environment and it has potential versatility to treat a diverse range of hazardous materials. Finally, the process is relatively inexpensive, because it uses the same equipment and supplies that are generally used in agriculture.

- **Associated technologies:** in order to enhance the remediation of contaminated soil, surfactants and biosurfactants can be used successfully to increase contaminants solubility, and consequently their bioavailability. The main factors that should be considered when selecting surfactants include effectiveness, cost, public and regulatory perception, biodegradability and degradation by-products, toxicity to humans, animals and plants. Those specifications should be established by previous experience or by laboratory studies prior to the field-scale demonstrations [31].

4.2   Land farming,

is an established technique for the remediation of hydrocarbons contaminated soil. Given half-lives of the order of one year, it would take about 7 years of treatment to remove 6,400 ppm of hydrocarbons down the cleanup goal of 50 ppm [19]. The time required for land farming remediation is an important factor in costing of the project. When the remediation can be promoted in ideal conditions, i.e. no liner and the remediation site adjacent to the contaminated one, the cost reported can be very low. For example, in Australia, 2,700 cubic meter of hydrocarbons contaminated soil have been treated in a 12 months operation; the pollution decrease from 4,600 ppm to near 100 ppm, costed $A10,30 per ton, i.e. $A13,40 per cubic meter (respectively $US5,5 and $US7,10). This cost can reach $A101 per ton ($US54) for an other contaminated site (5,300 cubic meter) [32]. Relative costs associated with land farming in USA are reported to vary from $US10 to $US100 per ton [17].

As *in-situ* bioaugmentation, extra micro-organisms can be added. For example, a mixed culture obtained from land farming and introduced in a sandy sediment polluted by light Arabian has increased by 31% the bioremediation efficiency of the heavy fraction [33].

This technology generally used *ex-situ* can be formally divided in 3 classes:

- **Landfill,** is the oldest method of *ex-situ* disposal waste; initially most landfill were unsealed so that any leachate derived from the degradation of the contents could be dispersed in the surrounding soil and groundwater. The main problem of landfill disposal is finding suitable sites in term of geology. Today those sites will need to be lined with an impermeable barrier to avoid contamination of groundwater. This method is particularly used for storage and anaerobic biodegradation of domestic wastes. The problem with landfill is that, although the biodegradable materials is broken down, the degradation process is very slow [8]. Environmental regulations can considerably increase the cost of landfill technique by the hardening of specifications, but also by long-term costs for disposal at licensed site and new levy, as in the UK since 1996, both evaluated at £16 per cubic meter. On a direct cost basis the cost of landfill becomes more expensive [13].

- **Composting,** is the low-tech land farming technology and can be somewhat upgraded by mixing the soil with fresh organic residues (compost). Elevated temperature and increased microbial diversity and activity improve the reaction rates. Moreover, specific co-substrates could favour co-metabolism. The US Environmental Protection Agency (EPA) has conducted many demonstration programmes in bioremediation activity for cost and performance comparison. In the Dubose site (Florida), composting was believed to be approximately equal in cost to *in-situ* biological treatment: $266 per ton of soil treated (19,705 tons). This cost is relatively high because of the relatively large quantity of soil excavated (58,559 tons) [34].

- **Biopile**, is an advanced technology for composting method by the fact that all parameters are monitored in a controlled area. Biopiles refer to the pilling of the material to be biotreated by adding nutrients and air into piles usually to a height of 2-4 m. This method is particularly well adapted to mathematical modelling for parameters analysis [35]. Five biopiles performed in field scale (40 cubic meter) during 5 months, with addition of organic matter, show a decrease of oil pollution from approximately 2,400 to 700 ppm [36].

4.3   Slurry-phase bioreactor,

In this case, excavated polluted soil, or sediments lagoon are treated under controlled optimal conditions, ensuring effective contact between contaminants and micro-organisms. The latter are, in most cases, specific cultures of adapted micro-organisms.

A slurry phase bioremediation application performed at the French Limited Superfund Site, in Crosby, Texas, by the US Environmental Protection Agency (EPA), was a larger full scale application. The aim of the project was the remediation of a lagoon polluted by poly-nuclear aromatic hydrocarbons (PAHs), chlorinated organics, and metals. In 11 months, an innovative oxygenation system has permitted to remediate *in-situ* approximately 300,000 tons of tar-like sludge and subsoil from the lagoon. The global cost for this slurry-phase bioremediation system including technology development, project management, EPA oversight and backfill of the lagoon were approximately $49,000,000, i.e. $163 per ton, which correspond to $90 per ton for the treatment only, $55 for the before treatment and $18 for post-treatment activities [37]. Those last values emphasise that the indirect costs are very important: 45 % of the total cost.

On other *ex -situ* application with four closed-top reactors of 680 cubic meter, for remediation of the South-eastern Wood Preserving Superfund site, in Canton, Mississippi, EPA demonstrate the great potentiality for this technique. Actually the majority of the biodegradation of 8,000 ppm of PAHs occurred during the first 5 or 10 days of treatment. The total cost for activities directly attributed to treatment corresponds to $170 per ton ($300 per cubic meter) for 14,140 tons of soil and sludges treated. In addition the cost of post-treatment activities is $35 per tons ($46 per cubic meter) [38].

5   Discussion

Table 1 summarize the results encountered in literature for the cost of contaminated soil remediation. We can observe a large distribution depending for a part, with the amount of soil treated and for other part, with the type of contamination. It appears clearly that the cost is also function of environmental regulations, definitions of cleanup levels [10] and future progress in research. The first operations reported have supported the cost of development and those indirect costs can represent 45 % of expenses [37]. Unfortunately, many values, in euros equivalent, date from the 90s. Actual costs will be impacted by time and competition but the relative costs given in Table 1 and the research perspectives will give an idea of where bioremediation falls. In our company, we develop also new generation of oxygenation systems, which permit to increase *in-situ* and for different conditions the biodegradation rates. This new process will allow us to treat soils and groundwater *in-situ* at a cost of €7.5 to 15 per cubic meter.

Remediation method	Range of cost in € per cubic meter	References	Increasing expenses factors
Incineration	*305*	*4*	*energy*
Chemical immobilisation	*28-70*	*4*	*transportation, monitoring*
Soil Washing	*42*	*13*	*transportation, solvent disposal*
Basically Bioremediation	35 34-215	13 10	monitoring, time course of treatment
Bioventing	3.5-90	9	monitoring
Landfarming	20.5-205	32	monitoring
Landfill	41 130-215	13 10	excavation, transportation
Composting	542	34	excavation, transportation
Biopile	56-102	4	excavation, transportation
Phytoremediation	51-203	28	monitoring
Slurry-phase	185-392	37-38	excavation, transportation

All cost are converted in € with quotation at the April 4th 2002, assuming 1 cubic meter =1.8 ton, when it is not indicated in the text.

Table 1: Cost comparison for bioremediation and classical methods.

6   Conclusion

Bioremediation methods will play a significant role in the near future for remediation of contaminated soil and groundwater. Bioremediation and associated technologies are environmentally friendly technologies, are promoted *in-situ* and are often more cost effective than conventional treatment methods that are very energy consumer. Nevertheless, the contaminated soil cleanup market ask today numerous questions and the actual limited growth of this market is without common measure with the requirement in term of public safety. Therefore, as key environmental technology for the next century, the technical and regulatory communities need to continue to ensure that bioremediation will reach its full potential as a treatment method.

References

[1]   National Academy of Sciences, *Oil in the Sea: Inputs, Fates and Effects.* Washington DC: National Academy Press, 1985.
[2]   Department of the Environment, *Transport and the Regions: Digest of Environmental Statistics*, n° 20, London: HMSO, 1998.

[3] CEDRE, wwwcedre.ifremer.fr// .

[4] Hamby, D. M., Site remediation techniques supporting environmental restoration activities, a review, *The Science of the Total Environment*, **191**, pp. 203-224, 1996.

[5] Levin, M.A. and Gealt, M.A. (eds), *Biotreatment of Industrial and Hazardous Waste*, McGraw-Hill, New York, 1993.

[6] Mills, C.H., Bioremediation comes of age, *Civil Engineering*, **65**, pp. 80-81, 1995.

[7] Timian, S. J. and Connolly, D. M., The Regulation and Development of Bioremediation

[8] Scragg, A., *Environmental Biotechnology*, Pearson Education Limited, Edinburgh, England, 1999.

[9] ITRC, *General Protocol for Demonstration of in-situ Bioremediation Technologies*, Final Report, September, 1998.

[10] Lin, G-H., Sauer, N. E. and Cutright, T. J., Environmental Regulations: A Brief Overview of their Applications to Bioremediation, *International Biodeterioration & Biodegradation*, pp. 1-8, 1996.

[11] Kovalick, W. W., Jr., Removing impediments to the use of bioremediation and other innovative technologies , *In Environmental Biotechnology for Waste Treatment*, Sayler. G. C. Foix, R. and Blackburn, J. W., (Eds), Plenum Press, New York, pp. 53-60, 1991.

[12] Chengalur-Smith, I. N. and Tayi, G. K., Trade-off Analysis in the Assessment of Pollutant Concentrations: a Simulation-based Approach for Bioremediation, *Socio-Econ. Plann. Sci.*, **32 (3)**, pp. 211-231, 1998.

[13] Day, S.J., Morse, G.K. and Lester J.N., The cost effectiveness of contaminated land remediation strategies, *The Science of the Total Environment*, **201**, pp. 125-136, 1997.

[14] Ministère de l'Aménagement du Territoire et de l'Environnement, *Gestion des sites potentiellement pollués*, Editions du BRGM, Paris, 1997.

[15] French Academy of Sciences, Colin, F., (Ed.), *Pollution localisée des sols et sous-sols par les hydrocarbures et les solvants chlorés*, Editions Tec & Doc, Paris, 2000.

[16] US Environmental Protection Agency, *Technology Screening Guide for Treatment of CERCLA soils and sludges*. EPA/540/288/004, Washington DC, 1988.

[17] Skladany, G. J. and Metting, F. B. Jr., Bioremediation of contaminated soil, in Metting, F. B. Jr. (ed), *Soil Microbial Ecology: Application in Agricultural and Environmental Management*. Marcel Decker, New York, pp. 483-513, 1993.

[18] Iwamoto, T. and Nasu, M., Current Bioremediation Practice and Perspective, *Journal of Bioscience and Bioengineering*, **92 (1)**, 1-8, 2001.

[19] Vandevivere, P., Verstraete, W., Environmental Applications, *in Basic Biotechnology*, Ratledge and Kristiansen (eds), *Soil remediation*, Cambridge University Press, UK, pp. 549-557, 2001.

[20] NFESC Environmental Department, http://enviro.nfesc.mil/ps/projects/biorem.htm

[21] Yagi,O. and Nishimura, M., Environmental biotechnology, the Japan Perspective, pp. 201-207. In Sayler, G. S. (ed.), *Biotechnology in the sustainable environment*. Plenum Press, New York, 1997.

[22] Li, G., Huang, W., Lerner, D.N. and Zhang X., Enrichment of degrading microbes and bioremediation of petrochemical contaminants in polluted soil, *Water Research*. **34 (15)**, pp. 3845-3853, 2000.

[23] Vogel, T. M., Bioaugmentation as a soil bioremediation approach, *Current Opinion in Biotechnology*, **7**, pp. 311-316, 1996.

[24] Timmis, K.N., Rojo, F. and Ramos, J. L., Prospects for laboratory engineering bacteria to degrade pollutants, *In Environmental Biotechnology: Reducing Risks from Environmental Chemicals through Biotechnology*, Plenum Press, New York, 1988.

[25] Hall, B. L., Lachmar, T. E. and Dupont, R. R., Field monitoring and performance evaluation of an *in-situ* air sparging system at a gazoline-contaminated site, *Journal of Hazardous Materials*, **B74**, pp. 165-186, 2000.

[26] EPA Manual, Bioventing Principles and Practice, *Volume II: Bioventing Design*, EPA/625/XXX/001, DRAFT September, 1995.

[27] Warith, M., Fernandes, L. and Gaudet, N., Design of in-situ microbial filter for the remediation of naphtalene, *Waste Management*, **19**, pp. 9-25, 1999.

[28] Macek, T., Macková, J. and Kas, J., Exploitation of plants for the removal of organics in environmental remediation, *Biotechnologies Advances*, **18**, pp. 23-34, 2000.

[29] Garbisu, C. and Alkorta, I., Phytoextraction: a cost-effective plant-based technology for the removal of metals from the environment, *Bioresource Technology*, **77**, pp. 229-236, 2001.

[30] Flathman, P. E., Lanza, G. R. and Glass, D. J., Phytoremediation issue. Soil and Groundwater Clean-up, **2**, pp. 4-11, 1999.

[31] Mulligan, C. N., Yong, R. N. and Gibbs, B. F., Surfactant-enhanced remediation of contaminated soil: a review, *Engineering Geology*, **60**, pp. 371-380, 2001.

[32] Line, M. A., Garland C. D. and Crowley, M., Evaluation of landfarm remediation of hydrocarbon-contaminated soil at the Inveresk Railyard, Lauceston, Australia, *Waste Management*, **16 (7)**, pp. 567-570, 1996.

[33] Del'Arco, J. P. and de França, F. P., Biodegradation of crude oil in sandy sediment, *International Biodeterioration & Biodegradation*, **44**, pp. 87-92, 1999.

[34] US Environmental Protection Agency, Cost and Performance Report, *Composting Application at the Dubose Oil Products Co. Superfund Site*, Cantonment, Florida, 1995. www.clu-in.org/

[35] Jorgensen, K.S., Puustinen, J. and Suortti, A.-M., Bioremediation of petroleum hydrocarbon-contaminated soil by composting in biopiles, *Environmental Pollution*, **107**, pp. 245-254, 2000.

[36] Mesania F. A. and Jennings, A. A., Modelling soil pile bioremediation, *Environmental Modelling & Software*, **15**, pp. 411-424, 2000.

[37] US Environmental Protection Agency, Cost and Performance Report, *Slurry-Phase Bioremediation at the French Limited Superfund Site*, Crosby, Texas, 1995. www.clu-in.org/

[38] US Environmental Protection Agency, Cost and Performance Report, *Slurry-Phase Bioremediation Application at the South-eastern Wood Preserving Superfund Site*, Canton, Mississippi, 1995. www.clu-in.org/

Section 6:
Risk management

Sustainable management of groundwater resources with regard to contaminated land. RTD needs

J. Grima[1], J. López[1] & B. Ballesteros[1]
Valencia Regional Office, Geological Survey, Spain

Abstract

Environmental policies for water resources have evolved over time. Some years ago, only emission of hazardous substances into surface waters was controlled. Now, with the enforcement of the Water Framework Directive, water resources are considered as a whole resource that must be protected at catchment scale. The good ecological status of groundwater is a requirement of the Directive, and to address this goal, Member States should implement a programme of measures over the coming years.

The close link between soil and watercycle demands an overall strategy for the management of natural resources. In most industrialised countries brownfields are an origin of groundwater pollution and, in the opposite way, water pollution can be an important source of soil contamination. In order to reach the goal of successfully protecting water resources from point source contamination, some issues should be addressed, like obtaining public awareness of groundwater contamination, permitting activities potentially contaminating, identifying point sources and improving groundwater remediation techniques. Research and technological programmes are vital to achieving groundwater protection. Finally, it is important to involve the stakeholders by means of appropriate, evident and transparent criteria.

1 Introduction

As pointed out, water is the pillar of life. Its quality and quantity determines the quality of our lives and the places where we live, access to clean drinking water is a human right. According to EUROGEOSURVEYS (Eurogeosurveys [1]), of the total fresh water amount on Earth (35×10^6 km^3), only 0.3% comes from lakes, rivers and swamps. Groundwater is 30.1% of the rest, 69.6% of the remaining water being in the form of ice and snow. Having in mind these figures and taking into account that European countries are facing a significant pollution of groundwater resources caused by former industrial activities, it is easy to understand the increasing awareness of European citizens in relation to water for drinking and bathing, water in rivers, lakes and coastal waters and, in general, water environment.

With the aim of obtaining a sustainable management of groundwater resources effective measures should be taken to prevent and/or reduce the effect of contaminated land on water resources. To date, many industrialised countries have adopted specific legislation to protect groundwater. In the European Union, the Water Framework Directive has come into force on 22 December 2000, and has a strong quantitative component, i.e. a long-term balance between abstraction and natural recharge. On the other hand water pricing will be a major element to conserve adequate supplies (Blöch [2]).

It demands the development of cost effective in-situ treatment technologies. The Member States must prepare a programme of measures to attain good surface water and groundwater status by the end of 2010. To reach this ambitious goal, not only the application of innovative technologies for the remediation of contaminated groundwater must be promoted, but also field demonstrations, bench studies and technology evaluations are needed (Grima and Lopez [3]). Active treatment technologies and passive containment technologies must be investigated for use in cleaning up contaminated groundwater

The Water Framework Directive and future regulations derived from it establishes the necessity of remediation of groundwater masses but, in practice, achieving groundwater cleanup objectives is not possible due to the extent and persistence of contamination. In these situations of technical impracticability even if a cleanup approach is technically feasible, the scale of the operation (EU wide scale) may make it impossible so a more **risk-orientated approach** may be used. In this sense the concept of Risk Based Land Management (CLARINET [4]) was developed as one of the main products of CLARINET Concerted Action (Contaminated Land Rehabilitation Network for Environmental Technologies. The project was funded by DG Research of the European Commission under the 4th Framework Programme for Research and Technological Development, with representation of sixteen European countries.

2 Interdependence water-soil in contamination processes

The environmental problem of polluted sites has been recognised on a European level many years ago. While on the side of water resources, Water Framework Directive has provided a set of principles for the management of water pollution at catchment scale, the EU has not developed a specific soil protection policy (European Commission [5]). Although Water Framework Directive will provide a legislative driver for the remediation of contaminated land (Darmendrail and

Harris [6]), there is a real need of an integrated assessment for the management of contaminated sites.

In such an assessment land represents a geographical area, and also includes the soil, surface water and groundwater beneath the surface of the land, adding a third dimension to the traditional spatial planning interpretation of land (Vegter [7]), (Kasamas and Vegter [8]). On these broader basis as traditionally used in the context of soil contamination is that arises the concept of Risk Based Land Management as an output form CLARINET. This is in line with the requirements of the Water Framework Directive, as the management of contaminated land can contribute on a long-term basis to the protection of groundwater and superficial water in a sustainable way.

As water is in soil and soil is the receptor of many pollutants, there is a close link between soil and watercycle. It is, therefore, necessary make an integration of groundwater in decision support systems for contaminated land.

From the scientific side to assess the risk of groundwater pollution particular attention should be paid to pollutant transfer from soil to groundwater. Due to the complexity of the processes involved many idealisations and simplifications are usually required (Fergusson, et al [9]). It means that there are many scientific and technical uncertainties in contaminated land decision making. Managing these uncertainties means not only the introduction of concepts like probabilistic approaches, but also involving stakeholders in the management of risks.

Finally, a number of issues have close relation with contaminated land business and may affect management and decision support systems. Relevant topics include the following:

- Intensive agricultural land use practices can be the origin of diffuse contamination (mainly due to nitrates) and lead to high levels of consumption. It may originate the drinking water not to meet the standards and eutrophication of surface waters (Kasamas, et al [10])
- Erosion can lead to desertification and so affect the quality of soil
- Overexploitation of groundwater can produce seawater intrusion in coastal aquifers, especially in Mediterranean countries
- Rising groundwater levels in urban areas may originate an interaction of groundwater with overlaying urban contaminated lands. It sometimes produces a flushing effect as well as physical-chemical changes that mobilises contaminants
- Interaction of seawater with contaminated soils. Some immobile or scarce-mobile contaminants, such as heavy metals, may mobilise as a consequence of salinity changes in the physical medium

3 Legal framework

The approval of the Council Directive 96/61/EC of 24 September 1996 concerning integrated pollution prevention and control (IPPC) [11] and the Directive 2000/60/EC of the European Parliament and of the Council establishing a framework for the Community action in the field of water policy or, in a short way, the EU Water Framework Directive [12] published in the Official Journal (OJ L 327) on 22 December 2000, expresses the significant efforts that the European Union is facing up to prevent and control the industrial contamination and to ensure the good quality and quantity of water.

The purpose of the IPPC Directive is to achieve integrated prevention and control of pollution arising from industrial activities listed in Annex I of same. The way to tackle it is summarised as follows:

1. Integrated proceedings of permits, granting authorisation to operate all or part of an installation, subject to certain conditions, which guarantee that the installation complies with the requirements of the Directive
2. Establishment of emission limit values expressed in terms of certain specific parameters, concentration and/or level of an emission, which may not be exceeded during one or more periods of time
3. Transparency of the proceedings through the access to information and public participation in the permit procedure
4. Exchange of information between Member States and the industries concerned on best available techniques, associated monitoring, and developments in them.

With regard to the new authorisations or integrated permits, above point 1, the emission limit values of contaminant substances specially those listed in Annex III, should be specified. These new authorisations should also include the prescriptions to guarantee soil and groundwater protection.

The main goal of the EU Water Framework Directive (WFD) is the establishment of a protection framework of continental groundwater and superficial water, as well as coastal and estuary water in the European Union, through several objectives like: drinking water and other economical needs supply; environmental protection; diminishing the effects of drought and floods.

Some of the principles in which the WFD is based on to achieve this objective, also in agreement with the IPPC Directive, are the following:

1. Expanding the scope of water protection to all waters, surface waters and groundwater. The protection of human health, water resources and natural ecosystems has a priority status
2. Achieving "good status" for all waters by a set deadline
3. Water management based on river basins
4. "Combined approach" of emission limit values and quality standards
5. Getting the prices right
6. Getting the citizen involved more closely
7. Streamlining legislation

From the above paragraphs it may be inferred that the aim of the environmental legal framework in the EU is clearly focused on protecting the quality and quantity of water resources in order to ensure human and economical supply, as well as obtaining a high quality status for all the masses of water in the EU. Other key subject is also the effort to generate a transparent legal framework through easy access and exchange of information, and the involvement in awareness of the citizens in the environmental problems.

However, no new emission limits of contaminants have been established in these two Directives, and it seems to be one of the main goals where the future legislative efforts should be focused on. Until that date, the applied limits will those specified in the Directives listed in the Annex II of the IPPC Directive.

Within this framework, the sustainable management of water resources and already contaminated lands appears to be one of the most powerful tools in the future environmental policy of the European Union.

In addition new legislation in relation to management of soil and water resources is being developed in the E.U. Examples of it are:

- EC Draft Directive [13] on prevention and restoration of significant environmental damage (environmental liability). Fault-based liability is also proposed for any other activities, which cause damage that affects the favourable conservation status of biodiversity
- A daughter Directive on quality of groundwater and the standards to be set is being drafted in 2002
- A Soil Directive is being designed. As a first step, a Commission Communication on soil issues is expected to be developed until mid 2002

4 Technological development in groundwater protection and remediation

A coherent RTD strategy is needed in order to obtain cost-effective methods and improve sustainability of groundwater remediation. The remediation time frame for a groundwater remedy should be kept within reasonable limits. The right application of the existing techniques, as well as the development of new ones to be applied on problematic aquifers (low permeability, fractured formations, high depths), is shown to be a main goal for future investigation programmes. Comparative studies to determine the effectiveness of multi-technique sequences where biological, chemical and physical methods are combined are also lacking.

4.1 Development of new technologies

Before a specific remedial technology has been selected, some investigation must be carried out, to determine the extent of the contamination and pollutant fate and transport. In this context the relationship between surface water and groundwater is an important issue to be studied. The following issues may be addressed:

- Development of simple (non-intrusive) methods of site investigation.

- Methods to asses the natural potential of soil and the unsaturated zone to attenuate contaminants, and techniques to monitor the processes.
- Key processes controlling the quality of groundwater/surface water and their interactions.
- Interactive metabolism of contaminants in aquifers.
- Free phase fate and transport.
- Modelling of aquifers paying special attention to fractured and non-homogeneous ones.

4.2 Improving groundwater remediation techniques

We are far from having a set of techniques able to decontaminate every kind of aquifer in a sustainable way (Arctander [14]). In this regard, research needs for improving the effectiveness of groundwater remediation techniques have been short-listed.

- Remediation in low permeability formations and those aquifers where low hydraulic conductivity hinder the use of classic techniques. Low hydraulic conductivity does allow neither air nor solutions injection as well as hinder contaminated groundwater removal
- Influence of rising groundwater tables in urban areas where there is a land contamination.
- Methods to assess interaction of seawater with contaminated soil in coastal aquifers.
- Remediation techniques for inorganic substances and compounds, since most of modern day techniques are specific for organic contamination.
- Genetic information needed by specialised microbes to produce the required enzymes in order to degrade specific contaminant substances, as well as effectiveness of genetically manipulated organisms.
- Vulnerability of microbes to certain substances that produces inhibition of bioremediation techniques.
- Toxicity of by-products generated by the application of remediation techniques.
- Development of new non aggressive methods in order to increase the solubility of contaminants to enhance their movement and removal, avoiding the destruction of the basic aquifer structure as well as the undesirable presence of residual reagents.
- Improvement of methods for dissolving heavy metals in their metallic state, present in the pores of aquifers.
- Optimisation of remediation multitechnique sequences.
- Analysis of geochemical stability systems in order to determinate the dissolution / precipitation potential of metals according to the Eh-pH changes produced in aquifers during the application of remediation techniques.
- Degradation processes of contaminated vapours in the vadose zone, as sub-products of remediation.

- Investigation on new plants with potential phytoremediation application, as well as genetic engineering to improve their natural capabilities.
- Determination of processes of accumulation and degradation through plant metabolism, in order to determinate the enzymes that breakdown complex organic molecule into simpler CO_2 and H_2O ones. The goal of this investigation should be the synthesis of those enzymes.
- Recovery of metals from enriched plant material in phytoremediation techniques, intended for their removal from the environment and / or the food chain.

4.3 Monitoring of remedial performance

It is essential to verify success of aquifer cleanup operations as well as to detect changes in environmental conditions, control the presence of toxic transformation product and verify possible undesired spreading of the plume. A facility should monitor until the groundwater cleanup levels are met at the point of compliance for both protection (new pollution) and remediation of water resources (past pollution). Furthermore, to evaluate data and support decision-making, statistical methods should be also improved.

Basically, systems of groundwater quality control are focused in the definition of a monitoring network, and precise detection techniques of pollutants. For a proper definition of a control network hydraulic and hydrogeological characteristics of the aquifer should be determined prior to remedial activities, so concentration, distribution and movement parameters of contamination in the subsurface can be modelled. To achieve this goal, new investigation programmes on hydrogeology and aquifer modelling should be carried out.

Once the monitoring network has been designed and performed, sufficiently accurate analytical detection techniques should be employed in order to detect small concentration changes. Development of new techniques and improvement of previous ones should be achieved.

5 Implementation of risk based land management

During last decades pressures in land use have strongly increased. In the European Environment Agency report, the Dobris Assessment, it is mentioned in its chapter on soil degradation, that data on contaminated sites are not suitable for aggregation in a consistent manner. In addition, European countries have different legislative and procedural approaches to the problems of groundwater protection and remediation of groundwater contamination (Darmendrail [6]). On the other hand, there exist a number of similarities in the management of contaminated land in most European countries. For example, management is handled at a regional level. Moreover, Water Directive provides a common legal framework for European countries and can be a driver for the remediation of contaminated land. Implementation requires an integration of differences and consideration of environmental and spatial planning perspectives. The former is focussed on the impact on contamination on human health and environmental quality, and the latter deals with the management of impact on contaminated land on the way land is used, for example regenerating industrial areas, or increasing agriculture use, or for creating a nature area.

For the protection and remediation of groundwater quality in relation to contaminated land sites risk assessment approaches are used in most European countries. The main principles that underlay risk assessment are the definition of the sustainability of the resources, prevention of new pollution and remediation of past pollution if necessary to protect human health or the environment (Darmendrail and Harris [6]) (Ferguson [15]).

Related to the implementation process itself, some conditions must be fulfilled for the legislation to have the desired effect. First, the legislation must be appropriate for the protection and remediation of water resources. Secondly, responsible authorities must be capable of driving the process, and finally, financial instruments must be provided. As stated by the Ad Hoc International Working Group on Contaminated Land (Ad Hoc International Working Group on Contaminated Land [16]):

> "Most industrialised countries have developed laws on the protection of groundwater. Following the principle of precaution, most of these laws require the maintenance of the multifunctionality of all groundwaters. These requirements are mostly very stringent, but their implementation pose enormous financial problems, as complete decontamination and the aftercare measures to ensure continued effectiveness of the long-term in cases of only partial decontamination or containment of the contaminants are very expensive. It is therefore necessary to examine these laws from the point of view of the contaminated sites management and to consider another philosophy of groundwater protection"

This philosophy must consider longer time perspective of sustainable environmental management, and can be characterised by three elements (Vegter [7]):

- Suitability for use. It focuses on the quality of the land for uses and functions. In relation with water resources Water Framework Directive considered groundwater as a resource to be protected from any change. In fact, one of the goals is achieving a good ecological status of groundwater masses.
- Protection of the environment. Environmental protection of soil and water resources with the aim of preserve, protect and improve their ecological quality and to establish a sustainable way on utilisation of these natural resources
- Long-term care. Reduction of aftercare is needed in order to avoid solutions that need control and maintenance during long periods of time.

Taking account of these factors brings up a requirement to improve the ideas on to date contaminated sites clean up and to improve risk management strategies. Natural Attenuation is not a new strategy (Müller [17]) and, moreover, it is not a "do nothing" approach. Hopefully, evolution of remediation techniques will bring new procedures to define acceptable levels of residual pollution and remediation objectives on a site-specific basis.

6 Conclusions

Interaction between water and soil is an important environmental problem, so successful implementation of Water Framework Directive has to involve land management. Although Water Framework Directive will provide a legislative driver for the remediation of contaminated land, there is a real need of an integrated assessment for the management of contaminated sites. In such an assessment land represents a geographical area, and also includes the soil, surface water and groundwater beneath the surface of the land.

The Water Framework Directive is highly precautionary in its approach to preventing new pollution and issues the need of remediation of damaged groundwater masses to attain good quality status by the end of 2010. Reaching those groundwater cleanup objectives in time is fairly unattainable due to technical and scale impossibilities. However an orientated approach may be achieved, developing new remediation technologies, optimising the application of existing ones and using proper monitoring techniques. Natural Attenuation although not widely accepted must be investigated as the only viable remediation option in many cases.

Investigation should be focused, among other, on problematic aquifers where present day applied techniques have failed due to their heterogeneous hydraulic behaviour or high operating depths, as well as the application of new technologies such as genetic engineering in order to improve bio and phytoremediation techniques. Typifying proper remediation multitechnique sequences shows to be a main objective in orders to optimise the existing techniques.

It is likewise important to improve the Knowledge State of aquifers to be cleaned up, by mean of hydrogeological investigation and modelling, prior to the application of remedial activities and the establishment of monitoring networks. The latter is essential to verify the success of cleanup operations, and therefore extra effort should be carried out to improve the design, the data processing and the analytical pollutant detection techniques.

References

[1] www.eurogeosurveys.org
[2] Blöch, H. European's Commission approach for water and soil protection in the EU. Sustainable Management of Contaminated Land, pp 2-4, 2001

[3] Grima, J., López, J. RTd for improving groundwater remediation technologies. Sustainable Management of Contaminated Land. Clarinet Final Conference, pp 38-41, 2001

[4] CLARINET. Contaminated Land Rehabilitation Network in Europe: http://www.clarinet.at

[5] European Commission. EC Communication "Towards a Thematic Strategy for Soil Protection". Brussels, 16.4.2002, COM pp 179 final, 2002

[6] Darmendrail, D., Harris, B. Land Contamination and Reclamation.EPP Publications, Vol. 9, n°1, pp 1-6,2001

[7] Vegter, J. Land Contamination and Reclamation. EPP Publication,Vol. 9, n°1, pp 95-100, 2001

[8] Kasamas, H. And Vegter, J. Scientific and Research Needs for Contaminated Land Management. Land Contamination and Reclamation. EPP Publications, volume 9, 79-85, 2001

[9] Fergusson, C. et al. Fundamental concepts of Risk assessment. Risk Assessment for Contaminated Sites in Europe, vol. 1, Scientific Basis, pp 7-21, 1998

[10] Kasamas, H. et al. Sustainable management of Contaminated Land, an overview. Final report of CLARINET. In preparation

[11] Council Directive 96/617EC of 24 September 1996 concerning integrated pollution prevention and control, Official Journal L 257, 10/10/1996

[12] Directive 2000/60/EC of the European Parliament and of the Council of 23 October 2000 establishing a framework for Community action in the field of water policy. L 327; Official Journal of the European Communities; 22/12/2000.

[13] EC Draft Directive on prevention and restoration of significant environmental damage (White Paper on Environmental Liability). Issued by DG Environment on 30/07/2001.

[14] Arctander, E. State of the Art in Remediation Technology in Europe. CLARINET Network, 2000

[15] Ferguson, C. Assessing Risks from from Contaminated Sites: Policy and Practise in 16 European Countries. Land Contamination & Reclamation, 7(2), pp 33, EPP Publications, 1999.

[16] Ad Hoc International Working Group on Contaminated Land: http://www.adhocgroup.ch/

[17] Müller, D. Common ground of Risk Based Land Management and Water Policies. Groundwater Protection in Selected Countries:Point Sources. Environmental Report, 1999.

Identification of subsurface environmental barriers and brownfield development

K. Schejbalová [1] & M. Vacek [1]
[1] KAP, spol. s r.o., Czech Republic

Abstract

Restrictions and risks resulting from subsurface environmental aspects has become a significant part of the documentation during the preparatory phase of brownfield development. Brownfield foundation conditions, soil, soil gas and ground water contamination, contamination of building constructions and wastes disposed on site and other site-specific subsurface environmental barriers determine the future commercial success of the brownfield development. Developers should acknowledge these aspects during the early stages of the brownfield decision-making process, since these may impact on technical design, costs and time schedule of the investor's intention. A step-by-step approach has become a normal procedure for the site assessment to identify subsurface environmental barriers. Application of the step-by-step approach is considered to be the most efficient for sites characterised by a long - term industrial operational history. It is proposed to apply the step-by-step approach divided into: Phase I-Review of present level of knowledge, identification of uncertainties involved; Phase II- Site investigations; and Phase III- Risk assessment, feasibility and cost-benefit study. The identification requires significant effort in terms of human resources, time, technical sources and costs. Results and outputs of the process can profoundly change the idea of future brownfield economical capacities, which may even require their conversion. Therefore, it's the author's opinion that it would be highly practical to prepare a "Road map" to enhance understanding of subsurface environmental barriers identification and management, similar to that prepared by US EPA.

1 Introduction

Identification of restrictions resulting from subsurface environmental barriers has become a significant part of documentation during the preparatory phase of brownfield development. Definition of the subsurface barriers represents a key aspect to be considered for future technical design and environmental protection measures, which may reflect in the commercial success of preparation, construction and operation of the future development zone.

At the initial stages of brownfield decision-making process it is critical to understand   the prevalent subsurface environmental barriers, including:

- general environmental, geological and hydrogeological conditions;

- foundation conditions;

- soil, soil gas and ground water contamination, contamination of building constructions, wastes disposed on site; and

- other, site-specific subsurface environmental barriers (undermined areas, methane emissions, radon emissions, cultural heritage, buried constructions and ammunition etc.).

2 Methodological approach

To identify subsurface environmental barriers a step-by-step approach is recommended during execution of the site assessment. However depending on site character, a single Step approach is not excluded. A single step approach seems to be appropriate for greenfields and/or for selected brownfields, which demonstrate a clearly documented short-term operational history and are located on non-complicated geological structure.

Otherwise, application of the step-by step approach seems to be more appropriate for sites characterised by:

- long - term industrial operational history (typical brownfield);

- complicated geological structure and hydrogeological conditions;

- extensive historical non-documented demolitions; and

- significant public interest.

It is proposed to apply a step-by-step approach divided into the following Phases:

- Phase I: Identification of potential subsurface environmental barriers based on a review of available data and human knowledge;

- Phase II: Site investigations; and

- Phase III: Risk assessment, Feasibility and Cost-benefit study.

For each step, results and solutions of previous phase are assigned to consequent phase in terms of its need, objectives, level of detail, methodological approach and type and extent of work performed.

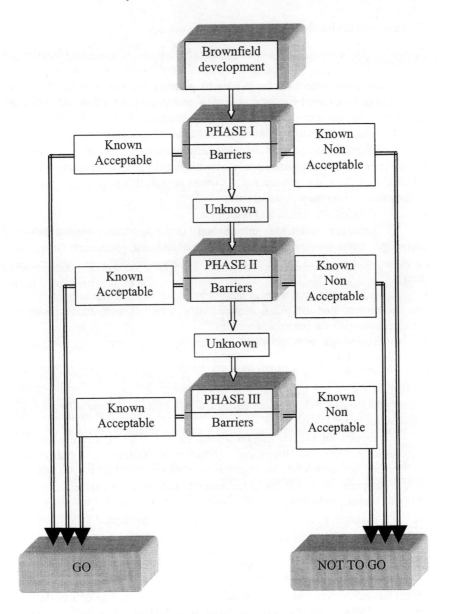

Figure 1: Decision-making flowchart

The decision-making flowchart is shown in Figure 1.

2.1 Phase I

Phase I conceptual objectives are to identify potential for existence of subsurface environmental barriers by means of review and description of present state of knowledge. The outputs of Phase I should also include:

- identification of unknowns and uncertainties;
- identification of need for consequent phase; and
- proposal and justification of objectives and extent of Phase II, once the potential for subsurface environmental barriers are identified.

Information sources particularly are:

- detailed site review and inspection;
- generally available information (e.g. location, geomorphology, hydrology, hydrogeology, climate, water management, protection areas and zones, seismology, land slides, radon and methane emission areas, undermined areas);
- information available at state administration and municipalities (e.g. past and present land use plans, administrative decisions, potentials for cultural heritage, potentials for buried ammunition);
- interviews with previous site management, witnesses, neighbours, permanent inhabitants;
- review of site history (e.g. in terms of land use, use of individual production shops and halls, use and management of hazardous materials and wastes, changes in technologies, constructional history, decommissioning and demolitions, breakdowns);
- reports on site investigations and studies (e.g. engineering geology, hydrogeology, site contamination, groundwater monitoring, archaeology, methane and radon emissions measurements, subsidence studies, E.I.A); and
- reports on remediation, reclamation and clean-up performed on site and in its immediate vicinity.

2.2 Phase II

Phase II conceptual objectives are to prove/disprove suspicions and indication of the existence of subsurface environmental barriers and to minimise level and magnitude of unknowns and uncertainties identified during the course of previous phase. Individual types of works should be effectively used to collect data for purposes of all site investigations. Depending on the site characteristics given in the previous phase site investigations should be realised as follows:

- geological and hydrogeological investigation;
- engineering geological investigation of foundation conditions;
- site contamination investigation;
- construction of groundwater monitoring network;
- other site-specific investigations (e.g. subsidence measurements, methane emissions, radon investigation, salvage archaeological investigation).

The following types of work are usually performed:

- geophysical measurements;
- drilling;
- field tests;
- sampling works;
- laboratory analyses;
- surveying; and
- other site-specific measurements, if necessary.

The outputs of Phase II should bring sufficient site-specific information on location, quality and extent of subsurface environmental barriers. Nevertheless, identification of remaining uncertainties and evaluation of data and its sufficiency should be an inevitable part of Phase II outputs as well.

Provided that collected data and information are sufficient and complete and Phase II can finish the process of identification of subsurface barriers, provided one is able to propose the following parameters:

- definition and assessment of risks arising from subsurface barriers;
- definition of acceptable level of risks;
- conceptual technical proposal of measures to achieve acceptable level of risk; and
- cost estimations for proposed measures.

Special approaches should be adopted for definition and assessment of risks and definition of acceptable level of risk of identified site contamination. In spite of risks arising from other subsurface barriers, these cannot be standardised and can vary depending on many factors and scenarios. The recommended approach is described in the following chapter.

Evaluation of affects of remaining uncertainties and unknowns, identification of need for Phase III and justification and proposal of essential objectives of Phase III and its technical proposal should be involved in outputs of Phase II as well.

2.3 Phase III

Phase III conceptual objectives are fully dependent on conclusions of Phase II and the relevance of remaining uncertainties. Output parameters listed in the previous chapter cannot be sometimes defined due to site-specific circumstances. Certain critical uncertainties may occur with some heavily contaminated sites even after Phase II is completed. This often happens e.g. when groundwater contamination plume has not been sufficiently outlined because:

- groundwater contamination plume migrates beyond the site border; and/or
- the site was identified to be a receptor of contaminated groundwater migrating to the site from an upstream profile.

In such cases one can only undertake essential actions to complete data collection of the enlarged area of interest, including repeating Phase II, and sometimes even Phase I. Only when uncertainties are minimized it is possible to define and assess risks and acceptable level of risks arising from site contamination.

Contaminated site risk assessment based on source – migration pathway– exposure principles is usually divided into the following steps:

- specification of contamination hot-spots and sources;
- specification of migration pathways;
- selection and identification of potential/realistic scenarios of migration and potential/realistic exposed receptors;
- definition and evaluation of potential/realistic risks for receptors as shown; and
- definition/calculation of acceptable level of risk represented by acceptable contaminant concentrations (acceptable residual concentrations) on converse principle.

Conceptual technical proposal of measures to achieve acceptable level of risk and cost estimations related to these measures can be proposed once a risk assessment is completed.

Complex and high-budgeted technological measures are usually proposed for heavily contaminated sites. In these cases, it is recommended to identify the best available technical approach by undertaking a Feasibility and Cost-benefit study. Evaluations performed within these studies are usually as follows:

- identification of realistic technical options;
- non-economic aspects (risks of failure, efficiency and reliability);
- costs;
- selection of best available technology by means of integration of information stated above (costs versus non-economic aspects); and
- conceptual technical specifications of selected best- available option.

3 Conclusions

It is evident, that the process described above requires significant effort in terms of human resources, time, technical sources and costs. Particularly Phase III may face heterogeneous political, public and managerial pressures and conflicts. The entire process requires significant understanding, skilled communication, negotiations and decision-making with and between involved parties. Conclusions and outputs of the process of identifying subsurface barriers can profoundly change the idea of future brownfield economical capacities, which may cause conversion and changes of developer's intentions and even stop the development programme.

The issues related to identification of subsurface environmental barriers and their affects on future development seem to be very sophisticated. Consequently follows the process of risk management, i.e. remediation and cleanup, which have not been addressed in this paper. Risk management programme may also profoundly change the developers' intentions and, if failed, even stop the development programme.

There are many guidelines and approaches all over the world and in the Europe dealing with these issues and they do not differ from each other essentially. Based on the author's experience in Czech Republic, these guidelines cannot help to much to improve developers' understanding, since they are heavily focused on technical and special aspects, while managerial aspects fall into the shade.

Therefore, it's the author's opinion to initiate efforts to assist developers with a more effective understanding of these issues in a higher systematic level. In this respect, it would be highly practical to prepare a "Road map" to enhance understanding of subsurface environmental barriers identification and management, similar to that prepared by US EPA.

Proper risk management: The key to successful brownfield development

R. D. Espinoza[1] & L. Luccioni[2]
[1]GeoSyntec Consultants, Columbia, Maryland, USA
[2]GeoSyntec Consultants, Huntington Beach, California, USA

Abstract

Society benefits from the redevelopment of environmentally impaired properties, often referred to as brownfields. For most investors and developers, brownfield redevelopment projects are considered too risky and demand higher returns on the investment needed to cleanup and redevelop a contaminated property. This paper proposes a framework for hedging the risks associated with brownfield development and shows how the use of hedging mechanisms can positively affect the value of a brownfield investment opportunity, thus increasing the likelihood that the project will provide an attractive return on investment.

1   Introduction

Several incentives have been recently proposed and implemented to promote the development of brownfields (i.e., abandoned, idled, or underutilized environmentally impaired properties). These incentives consist of: (i) federal and state environmental regulations incentives (i.e., limitations on investors or developers liabilities); (ii) economic incentives (e.g., tax breaks, municipal and/or federal grants); and (iii) administrative incentives (e.g., faster review process of construction permit applications for brownfield projects).

Despite these incentives, investors and developers often find the development of greenfields (i.e., uncontaminated virgin land) more attractive than the development of brownfields. Investors and developers tend to identify the real or perceived risk related to environmental conditions as the main barrier to investing in brownfield development. Also, investors and developers require a considerably higher rate of return for brownfield redevelopment projects than for other real estate projects due to these environmental risks.

There are, indeed, several sources of risk associated with brownfield development. Environmental risks include the cost of remediation and third-party liability. Still, as with any other real estate development, the risks of property value change and time required to realize the investment (i.e., clean-up and construction duration) are also risks inherent to brownfield development. Consequently, identifying and quantifying these risks requires the integration of expertise from different fields such as economics, engineering, and finance.

This paper presents a framework for hedging risks and calculating the value of a brownfield investment opportunity so that investor's return can be maximized. This framework is presented in the following three sections. Section 2 focuses on the sources of risks, the quantifications of these risks, and hedging mechanisms. Section 3 focuses on the valuation tools used by investors/developers to make investment decisions. Particular attention is given to taking into account both risk and managerial flexibility in making investment decisions. To illustrate these considerations, both net present value and real option valuation are presented in this section. Section 4 discusses two examples to illustrate the effect of risk hedging on investment decision-making.

2   Sources of risks

2.1   General

From the investors' point of view, risk can be classified as either diversifiable (i.e., private) or non-diversifiable (i.e., market) risk. Risk is classified as diversifiable if it can be eliminated by holding several investments with uncorrelated or negatively correlated risk profiles. A classical example of diversifiable (i.e., private) risk is an insurance policy that covers car accidents; insurance companies diversify this risk by insuring a large number of drivers. On the other hand, risk that is correlated with the overall macro-economy may not be diversified. A classical example of a non-diversifiable (i.e., market) risk is the risk associated with the change in price of a market index, such as the Standard and Poor's index. The risks associated with brownfield development usually comprise both private and market risks. In order to make appropriate investment decisions when dealing with brownfields (e.g., buy or lease a contaminated property, amount and type of insurance required), both sources of risk (i.e., market and private) should be included in the valuation analysis. Descriptions of both private and market risks related to brownfield projects are presented below.

2.2   Private risks

Private risks associated with brownfield development consist of technical, legal, and regulatory risks. Typical examples of private risk are the extent of contamination, the outcome of new cleanup technology, the time to complete the project [1], the liability claim from a third party, and changes in regulation.

Private risks can be either endogenous, if the uncertainty only gets resolved as the developer invests in the project and additional information is obtained (e.g., cleanup cost, time to project completion), or exogenous, if the uncertainty is independent of the developer's decision of going ahead with the project (e.g., change in environmental laws). Private risk is difficult to quantify due to the absence of observable market prices. For example, the cost of cleanup of a contaminated property is difficult to predict with a high degree of confidence due to technical, legal, and regulatory uncertainty. The evaluation of expected cleanup cost is complicated because each contaminated property is unique, making it difficult to use standard statistical (i.e., actuarial) techniques commonly used to evaluate expected cost. Cost for third-party liability claims that may arise during the cleanup process resulting in costly legal battles are also difficult to estimate. Continuous changes of regulatory standards in response to scientific results and public pressure further complicate the valuation process. Nonetheless, a measure of the uncertainty concerning each of these issues can be obtained from technical experts on the subject (i.e., lawyers, engineers, toxicologists). The information thus obtained can be used in a systematic manner to provide a measure of the environmental uncertainty and the price associated with it.

Several insurance products have been recently created to help developers hedge against private risks. These insurance products fall into two main categories: (i) cleanup cost-cap products to hedge against risk associated with clean-up cost overrun, time to completion, and changes in regulation; and (ii) third-party liability products to hedge against legal risks. To quantify the cost associated with these insurance products, one can view buying insurance policies as buying (i.e., paying a premium) the right to sell a liability for a given price (i.e., the insurance contract amount) if the value of the liability increases. This view shows the parallel between buying an insurance policy and a put option (i.e., financial instruments that confer on the seller the right to sell a stock at a specified price) in the financial market. The usefulness of using financial-market concepts to price private risk will become more apparent as new financial products that allow trading of private risks (e.g., weather derivatives, emissions trading, etc.) are developed.

2.3 Market risks

The main market risks associated with brownfield development are the market value of the land and interest rates. Market risk is correlated with the general movement of the economy, and so there is a market where the prices of the underlying asset can be observed. Market risk is always exogenous (i.e., independent of the developer's decision of going ahead with the project). The value of the developed property changes with time depending upon economic conditions coupled with supply and demand of the real estate asset. A measure of the market risk regarding real estate prices is given by its price volatility (i.e., the standard deviation around a historical mean of the property). A proxy for market risk information for developed properties is tracked by regional Real Estate Investment Trusts (REITs). REITs are portfolios of real estate properties and are usually listed on the New York Stock Exchange. Put options on REITs located in the same region as the candidate brownfield property can be used to hedge the market risk regarding the value of land. Thus, if the market value of land in the target region decreases, the value of REITs decreases, and the value of the put option would increase. An example of this application is presented in Section 4.

3 Valuation tools

3.1 Overview

The primary objective of investors/developers is to maximize the return on their investments in real assets. To achieve this goal, investors and developers evaluate investment opportunities using valuation tools. The net present-value (NPV) method is the method most widely used by corporations for valuing investments. Recently, a superior valuation technique, referred to as real option valuation (ROV), has emerged [2]. ROV explicitly captures the value of management flexibility and also accounts for variation in risk through the life of a project. Properly accounting for changes in risk is essential, because investors and developers require a rate of return on investment that reflects the risk of the investment. This section presents both NPV and ROV valuation techniques.

3.2   Net present value method

In the NPV method, future cash flows are discounted and compared to the present value of investment cost. The difference between the discounted future cash flow and the present value of the investment cost is called the net present value. Projects with positive NPV are considered to be profitable whereas projects with negative NPV are generally considered to be unprofitable.

If the value of a remediated and redeveloped brownfield is V, the initial investment is I_o, and k is the discount rate (also known as the hurdle rate, usually calculated as the average cost of capital), then the NPV can be calculated as:

$$NPV = \frac{V}{1+k} - I_o \geq 0$$

The hurdle rate (k) accounts for the risk associated with the uncertain future cash flow of selling the redeveloped property in the future. Because the risk associated with redeveloping a property is greater than the risk associated with depositing I_o in a fixed-rate-of-return bank savings account, k is greater than the risk-free interest rate (r). To account for the environmental risk usually associated with redevelopment of brownfields, the discount rate is further increased ($k_b > k > r$). Although the application of NPV is simple and straight forward, there are several problems with the use of this approach:

(i)   Risk premiums are applied to k to account for environmental risks, even though environmental risks are private risks that can be diversified away.

(ii)   The NPV method uses a constant discount rate, even though environmental projects risk profiles changes with time. Using a constant discount rate does not take into account the fact that technical uncertainties such as the cost of cleanup and time to completion get resolved (and therefore risk associated with these variables reduces significantly over time).

(iii) The NPV method ignores the manager's ability to shape the outcome of the investment result (i.e., the NPV method assumes that the decision is made at time $t=0$ and that management does not have any ability to reformulate the project if the initial results are unfavorable or if market conditions change).

3.3   Real option valuation

The real option valuation (ROV) technique overcomes the limitations inherent in the NPV method as listed above. The ROV is based on the option pricing theory developed by Black and Scholes [3] to price stock options. Mathematically, the value of the option is represented by a partial differential equation [4]. Cox, et.

al.[5], introduced a simple representation of the evolution with time of the value of the underlying asset (Figure 1).

Like their financial counterparts, a real option is the right, but not the obligation, to take an action (e.g., changing cleanup technology, buying the neighboring land, building a smaller/bigger structures) at a predetermined cost called the exercise price, for the life of the option. In the case of a put option, the option is the amount of money, or premium, to buy environmental insurance. The value of real options depends on six basic variables [2].

1. *The value of the underlying risky asset (S).* In the case of brownfield development, S is the value of land (Figure 1). If the value of land goes up, then the value of the option also goes up.
2. *The exercise price (X).* In the case of brownfields, X represents the amount of money invested to exercise the call option (e.g., to sell the cleaned property) or the threshold above which a cost-cap insurance will be triggered (Figure 1).

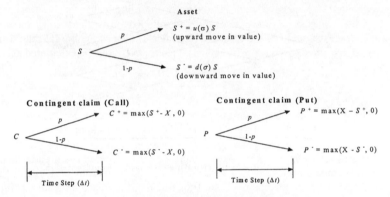

Figure 1 – Binomial Lattice Representation

3. *The time to expiration of the option (t).* This is the expected time to completion of the project or the duration of the insurance policy (Figure 2). As time to expiration increases, the value of the option increases.
4. *The standard deviation of the value of the underlying risky asset (σ).* The standard deviation represents the risk of the project and determines how high or low the asset value can be worth over the next period (Figure 1). The value of a call option increases with the risk of the underlying asset because the payoff depends on the value of the underlying asset exceeding the exercise price and the probability (*p*) of this occurring increases with volatility. For instance, the more volatile the cleanup cost, the higher the insurance premium.

5. *The risk free interest rate over the life of the option (r).* This is used to calculate the value of the option by discounting backwards to time 0. As the value of the risk-free rate goes up, the value of the option decreases. The proper selection of the risk-free interest rate (commonly based on the market for U.S. government debt instruments) is important. Usually, higher interest

rates are associated with longer-maturity debt instruments. The selected interest rate should match the time horizon of the cash flows for the project under consideration. For example, remediation expenses projected to occur in three years should be discounted using the yield of the three-year U.S. Treasury note. This discounting approach is consistent with U.S. Securities and Exchange Commission Staff Accounting Bulletin 92, and produces results that are suitable for financial reporting and disclosure.

6. *Cash flow from or to the asset during the life of the option.* The value of the option is affected by the amount of money that is continuously invested/received over the life of the option. For brownfields, this could represent the expenses associated with maintenance over the life of the option.

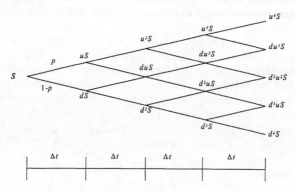

Figure 2 – Multi-step Binomial Lattice Representation

3.4 Decision management

Managers have long recognized that active management of projects that have a variety of types of flexibility adds value. In general, the greater the level of flexibility, the better chance a manager has to make a project profitable. This holds true for environmental projects, where recent changes in regulation allow for more flexibility while cleaning up and redeveloping a brownfield property.

In the past, cleanup of brownfields was driven by strict regulations with little regard for the potential use of the property or the general conditions of the surrounding environment. As a result, redevelopment of brownfields was limited by enforced strict cleanup. Recent changes in environmental regulations allow for more flexible approaches to achieving remediation goals that consider the future use of the property (i.e., commercial, agricultural, housing). Also, innovation in clean-up technology has allowed for faster and cheaper cleanups, providing greater flexibility to the remediation process. Risk based corrective action (RBCA) approaches to remediation have sparked redevelopment of contaminated properties. RBCA design can provide developers/owners with several alternatives for remediation depending upon the end use of the property and its corresponding exposure. These recent developments impact a greater amount of flexibility of brownfield project managers; therefore, brownfield projects are more likely to be profitable than in the past. To calculate an accurate valuation of a proposed redevelopment project, these new sources, and other sources of management flexibility should be incorporated in the economic evaluation method for the project.

Flexibility needs to be considered in investment valuations to accurately reflect the value added due to the different alternatives available to the seller/buyer. By understanding and quantifying the value of flexibility, the value of contaminated properties can be increased. Unwilling sellers may identify the circumstances under which divesting of brownfields is optimal, municipalities can identify the appropriate incentives that may spark redevelopment of brownfield that otherwise may languish; buyers can develop an optimal investment rule.

Once project-related uncertainty and flexibility has been properly accounted for by using the appropriate valuation tool, like ROV, managers need to track the project to ensure that the appropriate decisions (i.e., the ones that maximize profits/minimizes loses) are being made over time. For example, depending upon market conditions and the environmental liability, it may be more convenient to lease a contaminated land (and keep the environmental liability with the landowner) than to purchase the land (and assume the environmental liability). The terms of a business buyout, a merger, or a brownfield site redevelopment can be structured to take into account the results of a new remediation technology.

4   Examples

4.1   Introduction

This section introduces two simple examples to illustrate the importance of quantifying and hedging the main risks associated with brownfield development. The first example shows the importance of the "hurdle" rate in the decision making process, thus emphasizing the need to properly determine and account for the "hurdle" rate. The second example shows, first, how hedging the main risks alters the attractiveness of the project and, second, how to price risks using ROV. For both examples, the NPV is used as the project valuation technique to help managers in the "go/no-go" decision-making process. Finally, the decision process illustrated in these examples is simplified for the sake of clarity and to illustrate the focus of this paper, which is the role of risk management in the decision-making process to invest in brownfield properties.

4.2   Example I – Effect of risk/discount rate

Let us assume that the annual weighted average cost of capital (WACC) of Land4Sale, a real estate developer, is 20 percent; investment of $8.4 million is made at the beginning of the period; and that the average time to redevelop and sell a greenfield is two years. Land4Sale will develop this greenfield property only if the NPV is greater than zero, which means that the expected price of the redeveloped property needs to be $12 million or more:

$$NPV = \frac{\$12\,m}{(1.2)^2} - \$8.4\,m = \$0\,m$$

Now, assume that, for contaminated properties, Land4Sale uses a "hurdle" rate of 40 percent (i.e., a 20 percent risk premium over the WACC to account for the environmental risk). The NPV of the project is –$2.3 million, which results in a "no go" decision:

$$NPV = \frac{\$12\,m}{(1.4)^2} - \$8.4\,m = -\$2.3\,m$$

To be considered viable, the expected price of the redeveloped property needs to be at least $16.5 million compared to $12 million in the case of the uncontaminated property.

4.3   Example II – Effect of risk hedging

4.3.1   No risk hedging

Now, let us assume that Land4Sale is interested in buying a brownfield that can be redeveloped for commercial purposes. As in the previous examples, Land4Sale expects to sell the redeveloped property for $12 million in two years. The expected cost of cleanup is $5 million (for simplicity, assume that $2.5 million is paid at the end of the first year and $2.5 million is paid at the end of the second year) and the expected cost overrun cost is 20 percent (i.e., $1 million) of the estimated cleanup cost. The risk free interest rate is 10 percent. If the property is sold for $1.5 million, should Land4Sale invest in the brownfield?

The net present value of the project is:

$$NPV = \frac{\$12\,m}{(1.4)^2} - \frac{\$2.5\,m}{1.1} - \frac{\$2.5\,m}{(1.1)^2} - \frac{\$1.0\,m}{(1.1)^2} - \$1.5\,m = -\$0.5\,m$$

The expected value of the property is discounted at 40 percent because of the environmental risks present in addition to the usual real-estate risks. The expected cost of cleanup is discounted using the risk-free interest rate because the uncertainty in the cleanup cost is accounted for by the cost overrun variable. The expected cost overrun is also discounted using the risk-free rate because the uncertainty associated with cost overrun is already accounted for through the 40 percent discount rate applied to the expected value of the property. Based on NPV, Land4Sale should reject this investment opportunity.

4.3.2   Risk Hedging

Land4Sale management is now evaluating the investment opportunity presented in Section 4.3.1 under the scenario where the main risks (i.e., cost overrun and price of land) are hedged. To hedge the risk associated with cost overrun, Land4Sale decides to buy a cost-cap environmental insurance policy. The insurance premium may be estimated as follows: the estimated cost overrun (i.e., $1.0 million) is assumed to follow a stochastic process represented by the binomial distribution shown in Figure 3. For simplicity, two time steps (one year each) are selected. After two years, the possible outcomes of the cost overrun are shown by Figure 3a. The insurance premium P (i.e., $0.28 million) is then computed using ROV (with a risk neutral probability $p = 0.52$) by working backwards (i.e., right-to-left) through the binomial tree shown in Figure 3b and using the risk free interest rate (i.e., $r = 10$ percent) to discount through time.

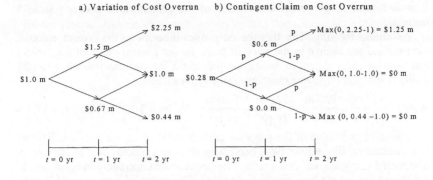

Figure 3 – Lattice Representation of Cost Overrun

To hedge against changes in the price of land, Land4Sale decides to buy a put option to protect itself against a drop in land prices below $12 million. The price of put option is determined as follows: the price of land is assumed to vary stochastically with time (Figure 4a shows the different values that land price can take over a period of two years). The value of the put option (i.e., $0.6 million) is estimated using ROV from Figure 4b, $p = 0.82$ and $r = 10$ percent.

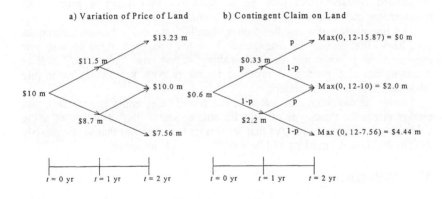

Figure 4 – Lattice Representation of Land Price

As in the previous examples, Land4Sale expects to sell the redeveloped property for $12 million in two years. The expected cost of cleanup is $5 million (for simplicity, assume that $2.5 million is paid at the end of the first year and $2.5 million is paid at the end of the second year) and the expected cost overrun cost is 20 percent (i.e., $1 million) of the estimated cleanup cost. The risk free interest rate is 10 percent and the value of the contaminated property is $1.5 million.

Assuming for simplicity that the only risks inherent to the project are cost overrun and the change in the price of land, the net present value of the project can now be recalculated, including the premiums for cost-cap insurance ($0.28 million) and property value insurance ($0.6 million), as:

$$NPV = \frac{\$12\,m}{(1.1)^2} - \frac{\$2.5\,m}{1.1} - \frac{\$2.5\,m}{(1.1)^2} - \frac{\$1.0\,m}{(1.1)^2} - \$1.5m - \$0.28m - \$0.6m = \$2.4\,m$$

The expected value of the property is discounted using the risk-free interest rate because of all the risks have been hedged. The expected cost of cleanup is discounted using the risk-free interest rate because the uncertainty in clean cost is accounted for by the cost overrun variable. The expected cost overrun is now discounted using the risk-free rate because Land4Sale bought a cost-cap insurance policy. The cost of the insurance premium and buying the put option are included in the NPV calculation. Based on a NPV valuation considering insurance premiums, Land4Sale would accept this investment opportunity.

5   Closing

The value of risk hedging techniques and using proper project valuation techniques were presented in a very simple framework in this paper to emphasize the importance of the concepts. The main conclusion of this paper is that brownfield redevelopment may be an attractive investment if proper risk management and valuation techniques are used.

The importance of using the proper "hurdle" rate and valuation techniques (i.e., ROV) that account for managerial flexibility was emphasized because it is important in management decision-making during the "go/no go" project decision, but also because it provides managers with key management rule throughout the life of the project.

Finally, as environmental risk becomes traded (e.g., emissions market), and market prices for these type of risks become available, the advantage of using valuation techniques (i.e., ROV) that are similar to techniques that are commonly used in the financial markets will become increasingly apparent.

6   References

[1] Majd, S., and Pindyck, R. S. Time to Build and Investment Decisions, Journal of Financial Economics, Vol. 18, pp 7-27, 1987.

[2] Copeland, T and Antikarov, V. "Real Options: A Practitioner's guide", Texere, New York, 2001.

[3] Black, F. and Scholes, M., The Pricing of options and corporate liabilities, Journal of Political Economy 81, pp. 637-659, 1973.

[4] Hull, J. Options, Futures and Other Derivative Securities, 4th Edition, Prentice-Hall, 2000.

[5] Cox, J., Ross, S., and Rubinstein, M. Option Pricing: A simplified approach, Journal of Financial Economics 7, No. 3, pp. 229-263, 1979.

Fixed price cleanups as useful tools to eliminate risks in brownfields redevelopment

R. Greenberg & D. Cervino
Environmental Waste Management Associates, LLC,
Parsippany, New Jersey, USA

Abstract

Stringent environmental regulations that impose significant and costly requirements for the cleanup of contamination from historical operations previously created an atmosphere where developers, inventors and lenders are hesitant to involve themselves with environmentally impaired Brownfields properties. Traditional risk shifting mechanisms such as contractual indemnity and escrow provisions are being replaced or supplemented by new innovative environmental liability and Brownfields restoration insurance products. Environmental cleanup cost cap and pollution liability policies can protect developers, sellers, investors and lenders from third party claims, costs associated with addressing areas of unknown contamination discovered after closing and during development, legal defense costs, as well as cleanup costs that exceed a guaranteed cost estimate. Environmental Waste Management Associates, LLC (EWMA) has pioneered a fixed-price cleanup program known as SECUR-IT^{SM}, backed by a combination cleanup cost cap and pollution liability insurance policy. The program is used by developers to effectively eliminate the environmental risks and unexpected cost of developing Brownfields because the insurance policy covers all risks both known and unknown. Under this program, the fixed price guarantee and policy can be issued prior to regulatory approval of a cleanup plan and with minimal paperwork or initial investigation. This program is attractive for developers because it caps their environmental liabilities at a site and addresses any unexpected cleanup costs which may have otherwise delayed or jeopardized the development.

The discussion will outline case studies of Brownfields redevelopment projects in which the SECUR-IT^{SM} program worked to successfully allocate environmental risks associated with redevelopment of contaminated properties. EWMA assisted a national developer in the rehabilitation of former cosmetics manufacturing facility regulated by state and federal authorities into a mixed-use commercial facility. EWMA used a variety of cleanup technologies to remediate both soil and groundwater contamination, including the ISOTEC patented remediation technology for extensive chlorinated solvent soil and groundwater contamination. EWMA's remediation system replaced a pump and treat system. EWMA completed the work for 50% of the projected cleanup cost in a fraction of the time proposed by a competitor. A second case study will cover EWMA's remediation that converted a million square foot paper cup manufacturing facility on 130 acres into an adult residential and assisted living community. Over 100 areas of concern were addressed in two years. The developer will be reimbursed approximately 75% of its cost through public funding for Brownfields.

1 Introduction

The ability to quantify environmental risks can make or break a real estate or commercial transaction and promote settlement of cost recovery litigation. Clean properties are becoming more and more difficult to locate in desirable urban and suburban areas across the United States since the mid 1990's. Environmentalists and open space advocates are pressuring governmental authorities and developers to target urban center for redevelopment. Local government has the power to condemn underutilized properties and to designate developers to replan and rebuild them to revitalize the area and the tax base. Former abandoned or obsolete manufacturing facilities are continually being converted to mixed-use residential/retail or commercial office and industrial parks. Parties on both sides of the transaction and their lenders are now hesitant to give or rely solely on the traditional contractual protections such as defense and indemnification provisions. This is so because many assets and properties are being sold by single asset entities, government entities and financial institutions through foreclosure or condemnation.

2 Innovative environmental insurance products

A new insurance product, environmental liability insurance policies, known as Pollution Legal Liability or PLL policies, provide coverage for onsite third party bodily injury, first party property damage and offsite third party bodily injury and property damage resulting from pollution conditions. Onsite coverage also includes loss of rental value in the event contamination causes operations to be reduced or closed. These products are relatively new and have become commercially available in the 1990's. There was a need for this product because traditional general liability policies have contained absolute pollution exclusion clauses since the late 1980's.

In most commercial settings, buyers require contractual defense and indemnification through sale contract or insurance if any contamination is discovered after closing. Sellers are generally not willing to offer an indemnity and are single asset entities that cannot guarantee coverage. Developer's concerns in acquiring and developing contaminated properties include availability of financing, delays in construction and unexpected construction costs, including costs of addressing existing or newly discovered contamination. Contamination is usually discovered during trenching for utilities and foundations and results in difficulty in renting the project due to the stigma of the contamination or default under loan terms.

Known liabilities and costs covered by a remedial action plan to address existing contamination are typically excluded from a PLL policy coverage. Therefore, insurers created the Cleanup Cost Cap or CCC policy to cover cost overruns or additional cleanup costs incurred after a governmental clearance due to a change in government action levels. For cost effective coverage insurers typically require a government approved remedial plan prior to providing CCC coverage and that the remedial costs at least $1 million.

The benefits of combined Pollution Legal Liability/Cleanup Cost Cap (PLL/CCC) insurance is that it virtually eliminates the risks and uncertainty involved with redevelopment projects requiring extensive environmental remediation. Developers can enhance and secure their investments by purchasing land at a lower cost and capping the cleanup cost and mitigating environmental risks. The insurance provides protection to lenders, tenants and subsequent owners who can be named as additional insurers on the policy. Insurance coverage secures the developer's financial position as well as the lender's collateral. Coverage terms can extend up to 20 years with limits up to $100 million. Typical coverage periods for PLL/CCC policies run three year, five year and ten year terms with limits of liability at $2 million, $5 million or $10 million with deductibles of $25,000, $50,000 and $100,000. Premiums for PLL/CCC policies range from $20,000 to $100,000, depending upon the nature of the use of the property and the type and extent of contamination present, as well as the projected cleanup costs, costs of natural resource damages, loss of rent, delay in store openings or diminution in property values can be covered by the new environmental insurance products.

Environmental insurance can cover the unexpected cost of offsite disposal or capping of newly discovered contaminated soil encountered during development or delays in the completion of construction and tenant expected opening dates or business interruption due to contamination. Developer's risks include a change in cleanup standards by a governmental authority after an initial acceptance and sign off a cleanup. Once the developer has cleaned up the contamination at the property to the appropriate governmental standards and has been issued a No Further Action letter, the developer does not want to be responsible for further cleanup actions in the event the state or federal authority changes its cleanup standards and orders the property owner or developer to conduct further remediation. There is a type of insurance coverage known as regulatory reopener coverage that covers costs of conducting additional required remediation after a No Further Action Letter is issued. This coverage is included in most PLL/CCC policies. The policies are negotiated to protect buyers, sellers or third parties such as lenders or tenants to cover a variety of risks, including lost rent or business interruption. The large insurance companies in the market include American International Group, Kemper Environmental, ECS, Zurich Insurance and Chubb. The carriers are competitive. There are no standard policies or premiums, but specimen policies and products available are provided in the insurance companies' literature and websites. Policy language can be manuscripted to cover exceptional circumstances. The field is rapidly changing and the programs are flexible. Premium costs for cleanup cost cap policies typically range from 6 to 10 percent of the estimated cleanup. Coverage for cleanup cost cap is typically two times the projected cleanup cost. However, it is increasingly more difficult to secure insurance for cleanup under $1million and where a remedial action work plan has not yet been approved by governmental authorities due to a large volume of claims in these types of cases.

3 Fixed price cleanups with insurance

Innovative insurance policies alone do not always allow a complex transaction proceed. Environmental consulting and remediation firms need to provide services that are geared towards quantifying the risks and costs of a cleanup some times long before a remedial action plan is approved and insurance is commercially available. Fixed-priced guaranteed cleanups provide an exact figure on the developer's environmental costs. Only a few progressive consulting and remediation firms like EWMA provide cleanup guarantee and combined insurance CCC/PLL environmental services in the early stages of an investigation. These guarantees typically address the known conditions at the property. A PLL policy is needed to cover unknown conditions that arise after the closing or during construction and to cover third party claims for bodily injury and property damage arising out of the contamination. The CCC portion of the insurance policy is designed to cover cost overruns, typically at a level of two times the anticipated cleanup cost with the fixed price cost as the self insured retention (terminology for deductible in the CCC market.) Fixed price cleanup together with innovative insurance products are needed to allow complex transactions to proceed.

The remainder of this paper reviews case studies where fixed price environmental cleanups backed by a PLL/CCC insurance policy were instrumental in the redevelopment of contaminated properties.

3.1 Case Studies

3.1.1 Cosmetic manufacturer facility

A prominent developer, a joint venture of a New Jersey based developer and a Connecticut capital investment group was negotiating the purchase of a 63-acre former cosmetic-manufacturing complex. Redevelopment plans called for demolition of most of the vacant 750,000 square foot plant and replacing it with a new 390,000 square foot industrial space for the cosmetic manufacturer and a new 150,000 square foot retail and warehouse development on 18 acres of previously undeveloped land. The seller did not want to assume the liability for the discovery of contamination during excavation for the new building foundations.

The owner of the property, the manufacturer of cosmetics, was going through the transaction triggered Industrial Site Recovery Act (ISRA) investigation and remediation process for industrial property transfers, during which 23 areas of environmental concern were identified. While a majority of these areas of concern were addressed to the satisfaction of the NJDEP, some soil and ground water issues were outstanding. As such, the owner was concerned with potential liability associated with future discovery of unknown contamination, and was unwilling to sell the property without certain types of assurances.

EWMA was retained by the developer to provide those assurances, while investigating and remediating any of the remaining areas of concern. The file information reviewed by EWMA indicated that shallow, intermediate and deep ground water underlying the site was contaminated with chlorinated organic compounds. A pump and treat system was previously installed at the property to address this contamination issue, and the consultants for the owner projected the ground water remediation costs to be in excess of $3 million. Additionally, arsenic was found during previous sampling activities, but the extent of contamination had not been fully delineated.

After reviewing pertinent site data regarding the ground water contamination, and conducting a minimal amount of site investigation work to completely delineate the extent of arsenic contamination, EWMA was able to provide a guaranteed fixed-price to remediate the remaining areas of concern. The guaranteed, fixed-price for remediation was then used by EWMA to obtain a Cleanup Cost Cap insurance policy for the developer in an amount five times greater than the cleanup cost; something that is typically provided by insurance carriers only after receipt of governmental remedial action workplan approval. EWMA was also able to assist the owner of the property in the procurement of insurance coverage to protect against the future liability associated with unknown contamination.

EWMA's remediation proposal included the installation of enhancements to the ground water pump and treat system, along with the installation of additional wells to optimize the recovery rates in both the shallow and intermediate zones. To further accelerate the process, EWMA proposed the use of vacuum-enhanced total fluids recovery combined with a chemical oxidation process. To remediate volatile organic contaminants from the soil, EWMA would use the recovery wells for soil vapor extraction (SVE) purposes.

To complete the $49 million purchase and redevelopment transaction, the developer purchased an $8 million Pollution Legal Liability/ Cleanup Cost Cap environmental insurance policy to cover any unexpected cleanup costs. The seller, wanted to limit its exposure for cleanup cost overruns and the legal liabilities involved in the $1million cleanup of the property. The developers agreed to purchase the $8 million insurance policy for a premium from $100,000 to $200,000. The cleanup involved the treating of chlorinated solvent groundwater contamination. EWMA was able to reduce the overall cost and time for the remediation at a guaranteed price. Without the guarantee and insurance, the seller was hesitant to sell the property for redevelopment.

3.1.2 Airport plaza
Another experienced national developer used environmental insurance and a fixed price cleanup guarantee by EWMA to purchase a 62 acre portion of an existing 188 acre city owned airport in New Jersey. The parcel was under utilized, contaminated and formerly used by light industrial failing local businesses. The property was redeveloped into a 650,000 square foot mixed use center, comprising of a hotel, movie theatre and retail stores rented by prominent national retailers such as Home Depot and Wal-Mart. The city had been trying to redevelop the property for three decades.

The $60 million project had a 60-day due diligence period. EWMA estimated the cost of the cleanup of the petroleum and chlorinated solvent groundwater contamination (various jet fuel, gasoline and other petroleum storage facilities for airport and plane maintenance) at $1.2 million. The cleanup approach and cost needed to be determined prior to state environmental agency approval.

The partners paid a $100,000 premium for PLL/CCC environmental insurance policy that covers $1million of unexpected cleanup cost overruns or newly discovered contamination. The developer is also taking advantage of a state cleanup cost reimbursement program known as a Redevelopment Agreement to recoup up to 75% of its cleanup costs once the project generated sufficient state tax revenue to reimburse cleanup costs.

The cleanup and development are to be conducted simultaneously and was expected to take 18 months. The project remained on schedule and no cost overruns were incurred.

3.1.3 Paper cup manufacturer

A 1 million square foot plant in Holmdel, New Jersey was used for almost 40 years to manufacture paper cups. The plant closed in 1990 and the building remained vacant for nearly ten years. The blighted facility plagued the predominantly residential community and was contaminated with asbestos, underground storage tanks, oils, solvents and ash piles. An enterprising developer saw this project as a great opportunity and decided to assume the cleanup responsibilities.

EWMA's client, the new owner of a 1-million square foot manufacturing facility built in the 1960s, triggered New Jersey's Industrial Site Recovery Act (ISRA). The client planned to redevelop the industrial 121-acre site into a mixed-use residential and commercial property. This site was one of the first to be eligible for New Jersey's Brownfield and Contaminated Site Remediation Act, which allows developers to recoup up to 75% of the environmental cleanup costs through a Redevelopment Agreement. Before the site could be developed, however, both the town where the site is located, and many of the site developers required the client to obtain a "No Further Action" (NFA) letter. EWMA was retained to complete the ISRA process, investigate and remediate the areas of environmental concern, and obtain the required NFA letter for the contamination at the site from the NJDEP.

After a long history of agricultural use as an orchard and vineyard, the site was developed in 1961 for use as a waxed paper product manufacturing facility. More recently, the property has been used for the disassembly and reclamation of electronic computer components. During the ISRA process, a total of 70 areas of concern were identified, including a waste water treatment area consisting of two lined settling lagoons, one classifier/activated sludge/chlorination basin, one treatment shed, and three sludge drying beds. Additional areas of concern consisted of underground and above ground storage tanks, hazardous material storage areas, railroad spurs, floor drains, trenches, sumps, incinerators, incinerator ash disposal area, and a one-mile long tunnel system that piped electricity, compressed air, non-contact cooling water and process waste water. Numerous impacts to both soil and ground water were identified at the facility by EWMA.

A major portion of EWMA's investigation and remediation of the site included the closure of the wastewater treatment area. Items of concern within the waste water treatment area included ground water impacts associated with settling lagoons, discharge points associated with treated waste water, the process waste water settling lagoons and associated sludge drying beds, and a sanitary waste clarified/activated sludge/chlorination basin. Large piles of sludge generated during the operation of the wastewater treatment area were also identified within the wastewater treatment area.

EWMA created significant disposal cost savings for the client by pre-treating the remaining process wastewater within the wastewater treatment area. The use of a portable treatment system to pre-treat the remaining 30,000 gallons of wastewater reduced contaminant concentrations to levels that could be safely discharged to the municipal sanitary sewer system. Additionally, EWMA coordinated the removal and disposal of over 1,000 tons of residual sludge material within the wastewater treatment area. Following wastewater treatment area closure and demolition, EWMA performed the required sediment and surface water sampling to demonstrate that the treated wastewater had not adversely impacted the stream that received the waste.

EWMA also investigated and remediated floating product and soil impacts associated with the fuel oil underground and above ground storage tanks, soil and ground water impacts associated with chlorinated solvents used during the manufacturing activities, and impacts associated with over 2,000 tons of residual incinerator ash contaminated with metals.

Since EWMA began working on the project, additional investigation and delineation has been completed, the entire 1-million square foot building has been demolished, work on the areas of concern identified at the site has been completed, and the property has been redeveloped.   By developing and implementing creative solutions to contamination identified at the site, EWMA has been able to cost-effectively investigate and remediate the areas of concern in a timely manner, allowing the client to proceed with the redevelopment and reclamation of the site. In fact, the project has progressed so well that it has been included in the NJ State Brownfields Redevelopment Task Force Resource Guide as well as other documents featuring brownfield redevelopment success stories.

The developer converted the factory into a mix of commercial and residential use. The new site will consist of an office building, a retail center with 20 stores and restaurants, 158 units of adult community housing; 110 assisted living (nursing) units and a 130-bed nursing home.

The cleanup will cost $2 million and up to 75% of the costs will be reimbursed by a state cleanup cost Redevelopment Agreement when the project generates sufficient state tax revenue that exceeds the cost of the cleanup.

4 Conclusion

There are multiple risks that need to be carefully managed and preferably mitigated when developers purchase and redevelop contaminated property. Property owners and developers have to finance the development and manage the risks of known and unknown contamination, they prefer to place the "liability" on the asset side of the balance sheet.   The innovative new environmental insurance products, together with guaranteed fixed price cleanups discussed in the article are now available to protect developers, owners, subsequent owners and their tenants from environmental liabilities at a property and that emanate from the property, whether these conditions are known or unknown at the time of acquisition.

Section 7:
Multimedia modelling and assessment

Integrated decision making for brownfields properties

J. R. Rocco[1], L. H. Wilson[1] & R. B. Gilbert[2]
[1]Sage Risk Solutions, LLC
[2]The University of Texas at Austin, Department of Civil Engineering

Abstract

Successful decision-making for the reuse and redevelopment of a brownfields property must incorporate complex social, economic, and environmental issues to achieve the critical outcomes of brownfields redevelopment – the environmental cleanup and beneficial reuse of the property. The decision making process must be open to and representative of the diverse concerns and issues of the many stakeholders that are impacted or may be impacted by a brownfields property. It must be well defined, agreed to by the stakeholders, and applied in a consistent manner. Further, the decision making process must ensure that information and data are effectively managed, appropriately linked to models and the decision making process, and effectively communicated to diverse audiences.

This paper presents a framework for effectively implementing the complex decision-making process necessary to address the reuse of a brownfields property based on clearly defined decision attributes, risk profile values and endpoint utilities. A major component of this framework is the use of a GIS application linked with tabular databases and models for analysis of environmental and other pertinent data, risk assessment, reuse planning, decision analysis for evaluating environmental and reuse alternatives, and visualization of results. These applications not only provide an excellent visualization mechanism, but also when linked with environmental and other information databases, provide an engine for calculating environmental and economic risks, evaluating uncertainty, and evaluating reuse alternatives.

Introduction

A brownfields property by definition is an abandoned, idled or under-used industrial or commercial property where expansion or reuse is complicated by a real or perceived environmental condition [1]. However, the environmental condition is not the only impediment to the reuse of most brownfields properties. Reuse for many properties considered to be brownfields, possibly the largest portion, is further complicated by a real or perceived economic or financial condition of a property, or the area surrounding the property, that limits the marketability or commercial reuse of a property. In addition, even for properties with economic driver for redevelopment, community and political conditions may be contrary to economically driven reuse alternatives. These economic and social conditions can also have a significant impact on the reuse planning for a property. Therefore, a variety of complex environmental, social, and economic issues must be identified and integrated into the decision making process in order to achieve a successful and sustainable reuse of the property.

Reuse planning for brownfields properties

Figure 1 depicts the reuse planning process for a brownfields property. Reuse planning for brownfields properties is most importantly a stakeholder driven process where stakeholders are individuals, organizations or other entities that directly affect or are directly affected by the brownfields property or its reuse. Stakeholders include responsible parties, landowners or developers, environmental and other regulatory agencies, local zoning, building and economic development officials, as well as those who live and/or work around the brownfields site [2]. This diversity in stakeholders can result in a corresponding diversity of concerns and objectives that may conflict or at least

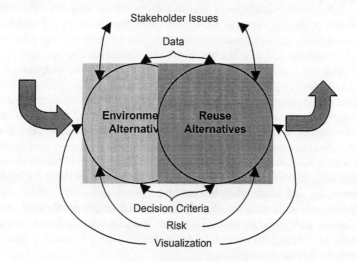

Figure 1: The reuse planning process.

impede the reuse effort. For this reason, a successful and sustainable brownfields reuse is one in which the concerns and objectives of all stakeholders are incorporated into the decision making process.

The reuse planning process consists of two key components that must be addressed in order to achieve a successful reuse plan. First, alternatives to address the environmental condition of the property must be identified and implemented in a manner that is protective of human health and the environment and in compliance with regulatory requirements. Second, alternatives to address the reuse of the property must be identified and implemented in a manner that provides for a beneficial and sustainable reuse of the property. In some cases, the environmental condition is the driver or major impediment for the reuse to address community concerns for health and safety or regulatory compliance. In other cases, the idled and abandoned condition is the driver or major impediment to placing the property back into a beneficial use to address the economic and social needs of the community. In either case, the need to address one condition requires the resolution of the other condition.

The decision on the approach taken to address the environmental condition of a property can not only affect the economics of reuse (e.g., prohibitively high cost for the implementation of an environmental alternative) but also may place physical limitations on the use of, or access to, the property (e.g., engineering controls and activity and use limitations to prevent human exposure to impacted media). Likewise, the decision on the reuse of the property can not only have an effect, positive (e.g., increased jobs) or negative (e.g., increased traffic), on the surrounding community, but also may place physical and technical limitations on the environmental alternatives that may be implemented on the property (e.g., unrestricted land use). In addition, a beneficial reuse may not always be for commercial purposes. Green space, recreational facilities, and other community needs may also be addressed through brownfields redevelopment. For this reason, the evaluation and selection of environmental and reuse alternatives are inter-related and one must consider the other. In other words, the selection and design of environmental alternatives should reflect and accommodate likely reuses for the property while the selection of reuse alternatives should reflect the economics and practicality of environmental alternatives. Both should be focused on providing a beneficial reuse for the property that is economically viable and sustainable into the future.

The reuse planning process requires an understanding of both the environmental and economic risks associated with the property. Environmental risks address the potential for adverse health or ecological effects for current and future occupants of the property and are driven by the activity and land use of the property. Economic risks relate to both short-term and long-term liabilities including the current and potential future investments, returns on those investments, and costs to implement, maintain, or modify in the future, the environmental alternative.

The reuse planning process requires the collection and management of the data and information needed to support the reuse planning decisions. Reuse planning activities may require large amounts of data to characterize and evaluate the environmental condition of the property and to understand the reuse potential for the property. These data may be classified in six general categories:

[1] Environmental site characterization – data and information about the potential sources, source areas, presence and distribution of chemicals of concern.

[2] Reuse site characterization – data and information related to the value of the real estate, surrounding land use and economic issues, community and neighborhood needs, market information, and infrastructure.

[3] Environmental risk assessment – data and information to evaluate the potential land uses and potential impacts on human health and the environment.

[4] Reuse risk assessment – data and information to evaluate potential options for redevelopment including economics, acceptability, and sustainability.

[5] Environmental limitations – data and information to identify limiting factors to redevelopment based on the environmental condition of the property including institutional and engineering controls.

[6] Reuse limitations – data and information to identify limitations on reuse of the property including environmental condition, economic conditions, neighborhood issues, and infrastructure issues.

The reuse planning process requires an effective means to communicate and visualize the data and information and the results of evaluations and decisions and to educate the stakeholders in order to facilitate their participation in the decision making process.

Brownfields decision analysis framework

It can be seen then, that the decision making process for reuse planning for a brownfields property must integrate the concerns and objectives of the stakeholders, alternatives to address the environmental condition of the property, and alternatives for reuse of the property. Further, the decision making process must be based on appropriate and sufficient information and data, and decision objectives and attributes that are clearly defined and agreed to by all stakeholders. Therefore, the foundation of effective decisions is comprised of proper management of data and information, sufficient quantity and quality of data and information, and clearly defined decision attributes.

Data management

To provide a better understanding of the available information and a mechanism for organizing and managing currently available and future data and information, a spatial database using a Geographic Information System (GIS) and a tabular database, using a relational database, provides an effective tool. Using the spatial database, the information can be linked to property specific features (e.g., chemical of concern distribution, property boundaries, physical features) as well as area or regional features (e.g., surface waters, transportation, utilities). The GIS links the maps or spatial database and the tabular database. The tabular database manages existing and future data and information and provides an easily manageable format for storing, adding, and retrieving different types of information. The spatial database contains geographically referenced site and regional features. Using the spatial data and the tabular data, evaluations can be conducted to support various reuse decisions (e.g., environmental fate and transport, uncertainty analyses, transportation patterns, demographics). Another significant benefit of a spatial database and GIS is its use as a tool to visualize the results of analyses, the potential alternatives considered, the outcomes of various decisions, and to provide information to allow stakeholders to understand and participate in the process. Figure 2 shows an example GIS application to evaluate the environmental risk associated with multiple sources at a site, where the probability of being below a risk based concentration and the coefficient of variation are displayed on a map. [3].

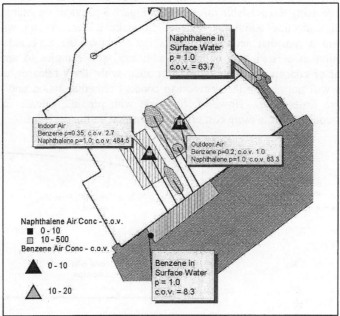

Figure 2 GIS Environmental Risk Example (Source: [3]).

Data quantity and quality

A critical component of any decision making process is having a sufficient quantity of data at a level of quality that supports the decision. The quantity and quality of data at brownfields properties can vary significantly. For example, some properties may have extensive environmental assessment data focused on site characterization, but very little data to support a risk assessment or a remedy alternative evaluation. Insufficient data or data that is questionable will generally result in a decision that is conservative and likely more costly. However, data collection can also be a costly portion of reuse planning if it is not focused on the decisions to be made. It is important to ensure that the value of the data or information collected is consistent with the decisions that are to be made. The collection of data that do not add value to the decision can not only be costly, but also obscure the decision. It is, therefore, not beneficial to collect data without understanding the potential impacts of those data on the decision.

To address this, a value of information analysis can be conducted to identify the type and quantity of data that would be of most benefit in the decision-making. This analysis provides a mechanism to quantitatively evaluate the impact of data collection on a decision [4]. The value of information analysis is conducted before gathering data on the premise that if gathering additional data would change a decision based on currently available data, the collection of additional data would be of value. Considering the value as a cost, the value of collecting information would essentially be the difference between the expected cost of the decision if data collection is initiated and the expected cost of the alternative using the available information. Figure 3 provides an example of the value of collecting additional samples to make a decision on whether to implement a remedial action [4]. This figure shows the expected value of information associated with collecting additional soil samples to analyze for chemical of concern concentrations. For areas with likely releases, additional samples will not change the decision to conduct remedial action and therefore they have little value. However, for areas with possible releases additional samples could result in more cost effective remedial action decision.

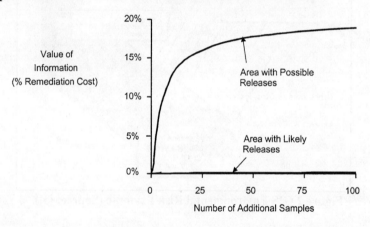

Figure 3 Value of Information Example (After: [4]).

Decision objectives and attributes

A decision is a choice between two or more alternatives to achieve defined objectives. In order to make a decision, the alternatives, the desired objective or objectives and the factors or attributes needed to arrive at a decision that achieves the objective(s) must be defined. A decision can be focused on a single objective (e.g., least costly alternative) or multiple objectives (e.g., least costly alternative that meets regulatory requirements, makes the property available for redevelopment in less than one year, and provides job opportunities for the community).

The decision attributes incorporate the data and information needed to evaluate the alternatives (e.g., cost of each alternative, regulatory requirements, time to implement alternative, jobs opportunities). As discussed previously, reuse planning for brownfields properties is a multi-decision process that involves diverse stakeholders with differing opinions and objectives. Although most decisions inherently consider a variety of attributes, explicitly and quantitatively identifying the attributes, particularly in a multi-stakeholder process, can be very difficult. If only one individual is to make the decision with no other interests to consider, then the attributes used to make the decision are limited to what is important to that individual and the definition of the attributes may not be significant in the decision process. However, for a decision that will require the consensus of a group of stakeholders in the reuse planning of a brownfields property, the attributes on which the decision will be based need to be clearly and consistently defined. Further, in order to make decisions that are defensible to the stakeholders, it is necessary to identify and agree on the decision attributes prior to making the decision.

Brownfields decisions

The decision analysis framework is only valuable if the specific objectives and clear decision attributes are established prior to making the decision. One approach is the use of multi-attribute utility theory to identify and incorporate the decision attributes into the framework [5]. The first step in this approach is to define the objectives. The second step is to establish the decision attributes. The final step is to develop a method or model for comparison of alternatives.

Defining the objectives

Defining the objectives requires developing a clear understanding of the decision to be made, the objectives to be achieved by the decision and the alternatives to evaluate. For example, when addressing the environmental condition, a basic decision may be whether a remedy must be implemented. This decision may be a simple yes or no decision, with an objective to meet general regulatory standards. It may also be a decision between no remedy and several remedy alternatives and different land use alternatives. However, as discussed previously, for a brownfields property the environmental alternatives and the reuse alternatives are interrelated and the environmental remedy may be dependent on the reuse alternatives proposed for a property. For example, where a risk-based approach is taken for determining the environmental condition and the need for remedy alternatives, the specific reuse plan will be an important consideration in identifying the objectives and alternatives for an environmental remedy. Therefore, defining the objectives also requires a prioritization of decisions to be made and an understanding and incorporation of other decisions that are critical to the specific decision being addressed.

As an illustration, an environmental remedy must be selected for a brownfields property. A specific reuse plan has been identified for the property that includes mixed recreational and commercial developments. The cost of the environmental remedy will be a significant determining factor to the implementation of the reuse plan. In addition, it is important that the environmental remedy not significantly interfere (e.g., minimize the amount of space needed for treatment equipment) with the planned recreational and commercial developments. Community stakeholders support the reuse plan but are concerned that the environmental remedy ensures the health and safety of the surrounding neighborhood. Several environmental remedy alternatives have been developed considering the planned reuse of the property. Table 1 provides example objectives that may be identified for the environmental remedy selection decision.

Table 1 – Example decision objectives and attributes

Objectives	Decision Attribute	Risk Profile	Endpoint Utility
Comply with regulatory requirements	Yes or No	Neutral	0.25
Minimize interference with planned reuse	Range of Cost	Averse	0.1
Integrate environmental remedy into redevelopment activities	Range of Economic Benefit	Affinitive	0.1
Minimize cost	Range of Cost	Affinitive	0.15
Minimize time to implement and complete	Range of Time	Affinitive	0.1
Achieve an endpoint that is protective of health and safety	Range of Risk	Averse	0.25

Establishing decision attributes

The second step is to establish the criteria or attributes that must be considered when evaluating each decision objective. Decision attributes generally include both quantitative and qualitative aspects, such as cost and regulatory compliance. Table **Error! Reference source not found.**1 provides example decision attributes that may be associated with each objective.

Develop a method for comparison of alternatives

The third step is to develop a common scale of valuing the objectives and to apply the objectives to the decision alternatives. A useful tool to use here is multi-attribute utility theory [6]. Utility provides a common scale of valuing objectives, with 0 being the value of the worst possible outcome and 1 being the value of the best possible outcome. Each attribute has a relationship or function that shows how the utility varies with the attribute. The attribute is typically normalized so that the worst possible value for the attribute is 0 (e.g., the maximum possible cost) and the best possible value is 1 (e.g., the minimum possible time). The shape of the utility function indicates the risk profile for that attribute (Figure 3). If the stakeholder is risk neutral, then the utility function is linear and the change in utility associated with a change in the attribute is constant for all values of the attribute (Figure 3). If the stakeholder is risk averse, then the change in utility is much greater for small values of the attribute versus large values (Figure 3); this means that the stakeholder will be very inclined to avoid small values of the attribute. The opposite is true for a stakeholder that is risk affinitive (Figure 3), meaning that the stakeholder will be inclined to take chances to obtain large values of the attribute. Example risk profiles for the individual attributes are listed in Table 1. This example corresponds to a stakeholder with a large reserve of capital because they are willing and able to take chances to maximize benefits and minimize costs.

The individual utility functions for multiple attributes are then combined by weighting or valuing the importance of achieving each individual objective (maximizing that attribute) with respect to the other objectives. This relative weighting is achieved by assigning endpoint utilities for each of the objectives (Table 1). The endpoint utility indicates on a scale of 0 to 1 the utility associated

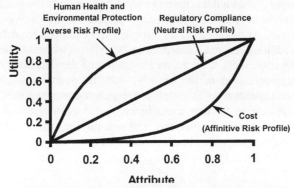

Figure 3: Individual Utility Functions (After: [4])

with achieving the maximum possible value for that attribute and the minimum possible values for all other attributes. In this scale, 0 is the utility associated with the worst possible outcome (the minimum attribute for all objectives) and 1 is the utility associated with the best possible outcome (the maximum attribute for all objectives). These endpoint utilities indicate how much each individual objective is valued, with respect to the other objectives, by the stakeholders. For example, utilizing extensive activity and land use controls might satisfy a cost objective, but may be considered less favorable by the community stakeholders, whereas an alternative that requires extensive removal and treatment of impacted media, may satisfy the community health and safety concerns but not the cost and time objectives. In this case, the community stakeholders value the "protect health and safety" objective more than the "minimize cost" objective. Determining the endpoint utilities for the objectives is an important component of the decision framework and may be the most difficult to resolve, since it incorporates the value judgments of the stakeholders. However, this framework provides a very helpful guide to incorporating the input of multiple and diverse stakeholders in a consistent and clear manner.

For each alternative, the decision objectives are reviewed and a risk profile value assigned that represents the appropriate criterion for each objective. The selected risk profile value is essentially "adjusted" by the endpoint utility [6] and a total utility value is obtained for a particular combination of attributes.

Finally, each alternative is evaluated to estimate the probability of achieving different attribute values for each objective. The expected total utility is then obtained by summing up each possible total utility multiplied by the probability of achieving it for each alternative. The alternative that has the highest expected total utility is the alternative that best meets all of the objectives of the stakeholders for the decision.

Summary

Reuse planning for brownfields properties must address complex social, political, environmental, and economic factors. Further, the reuse planning process is a complex, stakeholder-driven process that requires the concerns and objectives of diverse stakeholders to be incorporated in the decision making process, and the various stakeholders be informed and equipped to participate in the process. In addition, to make decisions in the context of reuse planning for brownfields properties, there must be an integration of the decisions related to the environmental condition of the property and the potential reuse for the property. The brownfields redevelopment process can be accomplished using an integrated GIS application for data management, value of information analyses for data quality and quantity evaluations, and risk profile values and endpoint utilities in a quantitative, reproducible decision process.

References

[1] United States Environmental Protection Agency (USEPA). Office of Solid Waste and Emergency Response Web Site, Washington, DC, www.epa.gov/swerosps/bf

[2] American Society for Testing and Materials (ASTM). *E-1984-98 Standard Guide for Sustainable Brownfields Redevelopment*, West Conshohocken, PA, 1998.

[3] Hay Wilson, L. "A Spatial Risk Assessment Methodology for Environmental Risk-Based Decision Making at Large, Complex Facilities," Ph.D. Dissertation, The University of Texas at Austin, Department of Civil Engineering, Austin, Texas, 2000.

[4] Gilbert, R. B. "Questions for the Future of Risk Assessment in Environmental Geotechnics," *Proceedings*, Fourth International Congress on Environmental Geotechnics, Rio de Janeiro, Brazil, 2002.

[5] Ang, A. H-S. and Tang W. H. *Probability Concepts in Engineering Planning and Design*. John Wiley & Sons, New York, 1984.

[6] Keeney, R. L. and Raiffa, H. *Decisions with Multiple Objectives*, John Wiley & Sons, New York, 1976.

Multimedia-modeling integration development environment

M. A. Pelton & B. L. Hoopes
Battelle, Pacific Northwest Division, USA

Abstract

There are many framework systems available; however, the purpose of the framework presented here is to capitalize on the successes of the Framework for Risk Analysis in Multimedia Environmental Systems (FRAMES) and Multimedia Multi-pathway Multi-receptor Risk Assessment (3MRA) methodology as applied to the Hazardous Waste Identification Rule (HWIR) while focusing on the development of software tools to simplify the module developer's effort of integrating a module into the framework. A module in this plug and play framework can be described as one or more codes, models, or databases that meet the framework communication protocol and can be placed in the visual conceptualization as a discrete part of an analysis. A framework such as this can be used to conceptualize and model the unique scenarios brought about by a Brownfields assessment. In a plug and play system users choose modules without having to worry whether the modules can communicate with each other allowing the user to focus on conceptualization.

An Application Programming Interface (API) has been developed for this framework and is implemented as a Dynamic Link Library (DLL), or shared library. The protocol developed for linking modules together is in the form of data dictionaries, which are designed for flexibility. The approach is to focus on developing these protocols (i.e., boundary conditions) between modules using a distributive environment, which allows multiple developers to collaborate on the same boundary conditions between modules in real time. The API is also used for boundary condition input and output (I/O). This enables the developer to consistently access data needed by other modules without the burden of educating their module on multiple file formats, making population of shared-data sources efficient and consistent. System editors are provided to set up the shared-data sources and the information needed to communicate the module's role in the plug and play system, thus easing the developers work load.

1 Introduction

Environmental assessments have grown from a single media being examined to multimedia multi-pathway environmental assessments. To that end, multimedia multi-model frameworks have been developed. These frameworks in the past, like MEPAS (Multimedia Environmental Pollutant Assessment System) and RESRAD (Residual Radioactivity), have always been hard wired together. Efforts such as FRAMES and 3MRA have helped to change the way in which model integration is approached. The general concepts and specifications of FRAMES are provided in Whelan et al. (1997) [1]. The fact that legacy codes exist and are almost always preferable make an applications programmer's interface (API) an unwelcome but necessary solution if the desired outcome is to integrate models. An API means writing wrappers (pre and post processors) for legacy models, which inevitably can change the way in which a model performs. It also means greater flexibility and manageability in the long run.

What really matters in the development of an API is the audience of modelers and scientists that will use it. Experience thus far has shown this audience to use FORTRAN and C with a little Visual Basic of late. The use of data structures is not present in most FORTRAN legacy codes. So more complex ways of process communication, like DCOM (Data Common Object Model), and SOAP (Simple Object Access Protocol), seemed too complex for a scientist who wants to write or wrap a simple model written in FORTRAN. These issues and others have resulted in an API that links and compiles with FORTRAN, C, and Visual Basic. The API, written in C++, has very little overhead compared to the more complex systems and will compile on most platforms.

A host program, the Framework Development Environment (FDE), executes the API. The FDE is a collection of editors and tools that allow users to define, link, select, and interact with a confederation of environmental codes for environmental and human health analyses. The FDE is an environment where developers can easily integrate models and databases as modules and users can link and run those modules as multimedia environmental analysis.

2 Framework API

The API accomplishes several tasks that would normally be done each time one model is wired to another. In March of 2000, a group of environmental modelers and computer engineers met at the Nuclear Regulatory Commission (NRC) in Washington, D.C. to compile a list of requirements for a multimedia multi-model plug and play framework that facilitated model integration [2]. Most of the requirements from the meeting were used to develop the API. This section gives a very brief overview of the API. Some of the main requirements include but are not limited to:

- handle unit conversions in a consistent and convenient manner

- provide a mechanism that allows for data retrieval of offline as well as online data sources
- provide a mechanism that allows for accessing and running remote models (i.e., remote computing)
- provide a common input/output (I/O) mechanism
- provide error checking for I/O (e.g., variable range, type, units, and cardinality)
- support the latest versions of Borland C++ Builder, Microsoft Visual C++, Lahey FORTRAN-90, and Digital Visual FORTRAN-90 compilers

Whelan et al.[3] provides a complete list of requirements and design. To meet these requirements, a common way to describe data to the system needed to be developed. These meta-data descriptions are at the core of the API. Meta data is the information used to describe data content. The 3MRA system used a set of dictionaries to define the meta-data for module inputs and boundary conditions. The MRA example was used and generalized for the API. These dictionaries are used to describe the meta-data necessary to aid the framework in execution management, variable parameterization, and I/O. The framework uses five types of dictionaries to help define the system architecture.

- The "StartUp" dictionary defines variables for system management (e.g., module list, window settings).
- The "Conversion" dictionary defines the variables for the measurement-conversion library (e.g., measure, unit).
- The "Module" dictionary defines the variables for a module component (e.g., name, version, executable file locations).
- The "Simulation" dictionary defines the variables for a problem conceptualization (e.g. module connectivity)
- The "Data" dictionary defines the variables for module communication (e.g., boundary conditions, module specific inputs).

The API is divided into three categories: Conversion, Dataset, and the System. Each API has an appropriate module or header file for each supported version of compiler. The developer is expected to include the module or header files and compile with the module's processor codes to gain access to the Framework API.

2.1 Conversion API

The Conversion API provides the function calls that allow the developer to work with any linear conversion. The developer is given the capability to add, delete, and retrieve measure and unit labels as well as editing and retrieving the unit conversion. The Convert function is called with a measure (string), an initial value (double), an initial unit (string), and the unit the value is to be converted to (string) as input. The function returns the converted value.

2.2 Dataset API

The Dataset API provides function calls needed to manipulate dictionaries and datasets (see Section 3.2). All classes of modules that are defined are required to register with the framework a set of connection schemes. A connection scheme is a set of dictionaries that define a module's produced and consumed datasets during a simulation (see Section 3.2). When a module is executed during a simulation, one of the defined connecting schemes is passed on to the module through the API. This enables the module the capability to identify its function and access the appropriate datasets. The Dataset API will be the most heavily used and allows the module developer to:

- open and close a session with the API
- retrieve expected dataset names and dictionary types to be produced
- retrieve expected dataset names and dictionary types to be consumed
- read and write data datasets that have been defined by a dictionary.

2.3 System API

The System API provides function calls used mostly by the FDE, and other 'System' modules like sensitivity and calibration modules. These 'System' modules must be defined as such in the Module Editor by setting the module's class type to 'System' (see Section 3.2). Setting the module class flag to 'System' allows the module to gain access to the functions that can manipulate the entire framework at runtime.

3 Framework development environment

The FDE is a suite of editors designed to manage, view, and set up the underlying infrastructure of the framework. The FDE consists of four editors: Conversion Editor, Dictionary Editor, Module Editor, and a Domain Editor. The Conversion Editor creates, edits and manages the measures and units used by dictionaries and datasets to aid in the automatic conversion of units during data retrieval. The Dictionary Editor creates, edits and manages dictionaries that the API uses for defining and categorizing datasets. The Module Editor creates, edits and manages. The Domain Editor is used to define and organize a palette of modules used by the Framework Simulation Editor. The FDE simply displays current system configuration while giving access to the same. The editors have very little logic, treating the API like a black box expecting all verification and consistency checks to be done by the API. When the editors are loaded, all changes made are live, and once changes are saved, they cannot be undone.

Another tool provided but not addressed in this document is the Update Manager that ties to a Linkage Server via http. The Linkage Server is a central information store that facilitates collaboration. The Update Manager provides the capability to acquire and manage the latest updates to conversions, dictionaries, and modules available from the Linkage Server.

3.1 Conversion editor

The Conversion Editor provides an interface for creating, modifying, and deleting measures and their associated unit conversions. Conversions provide the framework with the capability to convert units to and from any desired unit as long as the unit's conversion to the measure's base unit is defined. This provides a modeler the ability to produce and consume data in their desired units without worrying about other modeler's unit expectations. The conversions maintained by the framework are also bound directly to the dictionaries in the framework. Therefore, all conversions must be defined before there can be a reference to a unit by a dictionary. Requested dataset values made with a unit reference that is not contained in the list of conversions can only be consumed or produced in that unit. Currently, only linear conversions are supported.

Each measure must have a base unit defined. Therefore, the Conversion Editor automatically prompts the user to define a base unit after adding a measure. When converting units, the base unit is used as a link between all other units within the measure. A measure categorizes a collection of units that inherit the same measuring properties. For example, Time is a measure with seconds, minutes, hours, and days being related units of Time. Each measure has a base unit that all other units must provide a conversion to that base unit. Using Time as a measure, seconds is a logical choice for the base unit. Once a unit has been added the unit is capable of converting to every related unit by reversing the respective unit conversions. For example, if millisecond were added to the collection of Time and seconds, minutes, hours, and days already in the collection, then milliseconds could be converted to or from any other unit in Time by first converting to the base unit then convert to the desired unit. This reduces the amount of complexity in defining relationships between units because not every combination of units needs to be defined. [4]

3.2 Dictionary editor

Dictionaries in the framework describe the layout of how information is stored in a dataset. All dictionaries in the framework must have a unique name and are stored as an ASCII comma-separated file. The file can reside anywhere on the user's system. Dictionaries can be created outside of the editor using any kind of ASCII text editor. However, dictionaries created outside of the Dictionary Editor must be registered and checked for consistency with the Dictionary Editor by simply opening the dictionary file in the editor. The properties for a dictionary are:

- Variable Count—number of variable declarations in the file
- Dictionary Description—self explanatory
- Dictionary Name—self explanatory
- Privilege—dictionary scope – visibility to other modules
- Version—version number of the file
- Updated—flag to indicate whether the file has changed
- Variable Declarations—set of fields that declare a variable.

Privilege, Version, and Update provide the API with the capability to develop the dictionaries in a distributive environment. This allows developers from different parts of the world to share dictionaries (a.k.a., boundary conditions). Once boundary conditions are shared, then modules can communicate.

Currently, there are four intrinsic data types available for variable declaration from the Dataset API: string, integer, float, and logical. Each variable declaration has these properties:

- Name—self explanatory
- Description—self explanatory
- Dimension—number of indices required for retrieval of a value
- Data Type—type of data stream (one of string, float, integer, logical)
- Primary Key—flag to indicate if variable is a data selection key
- Scalar—flag to indicate if variable is scalar for the defined indices
- Minimum— [minimum length of string][minimum value of float/integer][blank for logical]
- Maximum—[maximum length of string] [maximum value of float/integer][blank for logical]
- Measure—string indicating the measure
- Unit—string indicating the unit of measure
- Stochastic—flag to indicate if variable is stochastic
- Preposition—string used to help describe the who, what, when, where, why
- Index 1, Index 2, …—the list of index references defined for the variable

It is expected that module developers will define the appropriate variables for their respective modules. Special attention must be given to indices of variables. Variables are forward packed, that is, when you have an index of 3, then an index of 1 and 2 also exist. The values will be zero filled or blank filled. All indices must be defined with another variable. For example, the vadose-zone variable kd could be stored by location and chemical. Therefore, the variable declaration for location and chemical must exist. Multidimensional variables are allowed as an index provided their indices are provided by the remaining indices in the variable declaration. No circular dependencies are allowed.

3.3 Module editor

The Module Editor provides the developer the ability to define a module for execution in the framework. A module class determines the level at which modules communicate with the API and how the module execution handled.

There are four classes of modules:

- System—Read/Write/Execute privileges, dataset completeness expected
- Database—Read/Write privileges, dataset completeness not expected
- Model—Read/Write privileges, dataset completeness expected
- Viewer—Read Only privileges

System modules are expected to know a lot about the inner workings of the API and framework. System modules have the capability to do completely manipulate the framework of conversions, dictionaries, modules, and domains. Database and Model classes have the same privileges, except they are prevented from accessing system functions. The only real difference between Database and Model is the expectation on the boundary conditions. Boundary conditions must be complete when output by a Model and/or a System module. Viewers are just that; they provide read-only access to the data.

Module properties consist of five types of information: executable, reference, company, developer, requirements, and connection schemes, most of which is for documentation purposes. Critical for module integration is the executable and connection scheme information. Modules manipulate data and do so by specifying the dictionaries that define the datasets they will manipulate, this declaration is known as a connection scheme. Executable information gathered includes location of interface and model executable, version, and command-line options.

Connection-scheme information is the most critical part necessary for module connectivity and execution management. Connection schemes and the class of the module determine how the framework allows the module to connect and the level at which the module can access the API. Modules need to have connection schemes defined to describe the type of information a module expects to consume and produce. A scheme should be defined for every combination of dictionaries the module being developed can consume and produce. A developer could think of these schemes as function calls to the module.

3.4 Domain editor

The Domain Editor allows the user to create tiers of icons that have the same or similar physical meaning. The developer as well as the end user can choose where modules should appear in their constructed domains. Domains consist of four predefined module classes: Database, Model, System, and Viewer. Under each of these classes, the user can create a group of modules or a grouping of sub-groups of modules. Modules can be placed in more than one tier, but can only be assigned to one class of module. Each domain, class, group, and sub-group can be assigned a unique icon. The icon should convey the physical aspects of the respective tier they represent. By assigning meaningful icons to the tiers, users provide their own understanding and meaning to the visual conceptualization of the simulation displayed by the Simulation Editor.

4 Framework module tester / dataset editor

This tool is designed to assist module developers in testing their modules within and outside of the Simulation Editor, as it pertains to consuming and producing datasets defined by module connection schemes. It is significant in that this tool allows for modules to be developed and tested in parallel with other modules, alleviating the need to develop and test dependent modules sequentially. In other words, modules that consume data produced by another can be tested without having to wait for the producing module to be completed. The editor runs in three modes determined by command-line arguments. It can be run standalone outside of the Simulation Editor, or within the Simulation Editor as a model user interface (UI) or model executable when specified as such in the Module Editor.

In any case, the tool assumes responsibility for assessing the modules' mode and generating an environment complete with a selected connection scheme (if standalone) and to populate those datasets specific to the module scheme in question. Produced datasets are populated with random values. For every variable in the dataset dictionary, values are generated using the intrinsic random function with the specified range and data type; random lengths are used for strings.

In all modes, the layout of the editor includes a hierarchical display of the applicable datasets and variables accompanied by a tabular display of actual values and indices. The dataset display lists dataset names and identifies each dataset as model specific input, consumed, produced, or referenced. Each variable is identified by description and name. The user clicks on a variable description to view its value(s). The user can also change any value within the data type and range.

4.1 Standalone mode

In Standalone mode, the editor facilitates testing a module's functionality and behavior without the additional complexity of interacting with the Simulation Editor and generating an appropriate simulation. As a standalone component running outside the framework environment, the editor displays the Simulation Editor icon palette from which the user selects a domain, class, group, sub-group, module, and scheme of interest. Given the selected scheme, the editor interacts with the API to generate a simulation (in the background) consistent with that scheme so the module executables will "get the right information" when later they query the API about connections and datasets. The tool then populates all consumed and referenced datasets with random data and displays the dataset contents to the user for viewing and optional editing. To test the module executables, there are buttons on the editor to invoke the UI and/or model as defined in the module editor. When a model executable is invoked and it runs successfully, the user can choose to view the datasets produced by that model.

The testing advantage here is that an environment supportive of the module scheme is guaranteed to exist when the UI or model is executed, and any datasets consumed or referenced by that UI or model will exist and be fully populated. This provides for testing dictionaries and module executable interactions with those dictionaries and corresponding datasets.

4.2 Model input mode

The dataset editor can be assigned as a model user interface (MUI). In this mode the editor limits the functionality to populating the model specific input dataset and allowing the user to view/edit/save that dataset. This capability is useful if there is no MUI for a module that requires user input or the MUI is in development but the model is ready to integrate and test.

4.3 Model run mode

The dataset editor can be assigned as a model executable. In this mode, the editor limits the functionality to populating the datasets produced by the module connection scheme identified by the simulation connectivity. This capability is useful for testing modules that expect to consume datasets that are not available through other modules, for whatever reason.

5 Future tools

The following is a list of recommendations on future additions and updates to the set of tools used to integrate modules and access the API. This list is not meant to be all-inclusive, but has been suggested by users and developers. This list can be used to help prioritize modifications of the software system. Recommended additions and modifications are:

- Define operators that perform on variables defined by dictionaries. Could prove to be very powerful for future model development. Also would cut down on communication traffic with the API reading and writing of variables.
- Flat file string substitution tool to aid in user input through a generic interface. The Multimedia Integrated Modeling System (MIMS) uses this tool to aid in its legacy code integration.
- Create variable declaration mapping for legacy code, a very interesting idea. The object here is to have UI that allows the developer to generate a mapping of legacy code variables to dictionary variables. After the mapping is complete, the program would then generate a file to be included with the legacy code for compiling. Then the developer is able to remove or comment out the legacy variable declaration and start using the API directly with the original variable names.

6 Summary

The API and model integration tools presented here were designed to automate tasks normally done when hard wiring models together and provide engineers the tools needed to integrate legacy models required by regulations and/or regulators to conduct contaminant exposure and risk assessments at brownfield sites. The power of the API and tools provide developers the opportunity to seamlessly link their model with others and gives them access to a platform for comparison.

Acknowledgments

The U.S. Environmental Protection Agency (EPA), U.S. Department of Energy (DOE), U.S. Army Corps of Engineers, U.S. Nuclear Regulatory Commission, and Pacific Northwest National Laboratory (PNNL) under Contract DE-AC06-76RL01830 support this work. A special thank you to K.J. Castleton, K. E. Dorow, G. Whelan, and F.C. Rutz from PNNL for their guidance and support and W.C. Cosby of PNNL for editing the manuscript. PNNL is operated for DOE by Battelle.

References

[1] Whelan, G., Castleton, K.J., Buck, J.W., Gelston, G.M., Hoopes, B.L., Pelton, M.A., Strenge, D.L., & Kickert, R.N., *Concepts of a Framework for Risk Analysis in Multimedia Environmental Systems (FRAMES)*. PNNL-11748. Pacific Northwest National Laboratory, Richland, Washington. 1997.

[2] Whelan, G. & Nicholson, T., *Proceedings of the Environmental Software Systems Compatibility and Linkage Workshop*. PNNL-13654. Pacific Northwest National Laboratory, Richland, Washington, 2001.

[3] Whelan, G., Pelton, M. A., Dorow, K.E., Gelston, G.M., & Castleton, K.J., *Merger Between 3MRA and FRAMES-V1: Design*, PNWD-3145. Pacific Northwest National Laboratory, Richland Washington.

[4] Elder, M., & Pelton, M.A., *Framework Development Environment*, to be published. Pacific Northwest National Laboratory, Richland Washington.

Database integration in a multimedia-modeling environment

K. Dorow
Pacific Northwest National Laboratory, U.S. Department of Energy, USA

Abstract

Integration of data from disparate remote sources has direct applicability to modeling, which can support Brownfield assessments. To accomplish this task, a data integration framework needs to be established. A key element in this framework is the metadata that creates the relationship between the pieces of information that are important in the multimedia modeling environment and the information that is stored in the remote data source. The design philosophy is to allow modelers and database owners to collaborate by defining this metadata in such a way that allows interaction between their components. The main parts of this framework include tools to facilitate metadata definition, database extraction plan creation, automated extraction plan execution / data retrieval, and a central clearing house for metadata and modeling / database resources. Cross-platform compatibility (using Java) and standard communications protocols (http / https) allow these parts to run in a wide variety of computing environments (Local Area Networks, Internet, etc.), and, therefore, this framework provides many benefits. Because of the specific data relationships described in the metadata, the amount of data that have to be transferred is kept to a minimum (only the data that fulfill a specific request are provided as opposed to transferring the complete contents of a data source). This allows for real-time data extraction from the actual source. Also, the framework sets up collaborative responsibilities such that the different types of participants have control over the areas in which they have domain knowledge-the modelers are responsible for defining the data relevant to their models, while the database owners are responsible for mapping the contents of the database using the metadata definitions. Finally, the data extraction mechanism allows for the ability to control access to the data and what data are made available.

1 Introduction

Modeling and database integration have always been important parts of providing environmental assessments. In most cases, this integration has been performed in a very un-automated fashion (large amounts of human interaction required). Typically, a modeler / assessor will identify the source(s) of data that can be used for a particular assessment. He or she will then acquire a copy of the complete data source(s) and have to develop some method for extracting the relevant information for that particular assessment. This causes several complications. In most cases the modeler / assessor will not be completely familiar with the structure of the data source(s), making it much more difficult to come up with an efficient plan for extraction of the required data. Also, because the modeler / assessor is now dealing with a copy of the data source(s), it can quickly become stale—the original data source(s) can receive updates, while the copy remains unchanged (unless the modeler / assessor is diligent enough to constantly be checking for updates). Even in the later case, the modeler / assessor may still run into problems if the new update to the data source has structural changes, which may render all of the previous extraction plan work useless.

To overcome these problems in the general case, a system, known as the Framework for Risk Analysis in Multimedia Environmental Systems (FRAMES), has been developed at Pacific Northwest National Laboratory in cooperation with the Environmental Protection Agency (EPA) and the Department of Defense-Engineer Research and Development Center (DOD-ERDC). The rest of this paper discusses the details of the data integration components of the FRAMES system and is structured as follows: Section 2 will provide a high level architectural diagram and discuss some of the requirements and goals for the system. Section 3 will take a look at FRAMES dictionaries (modeling metadata) and the FRAMES Development Environment (FDE). Section 4 will cover the database server components—the Data Owners Tool (DOT) and the Data Extraction Tool (DET). Section 5 will discuss the client component--the Data Client Editor (DCE). Section 6 will focus on the Linkage Server, which is the central information clearing house. Finally Section 7 will offer a summary.

2 Goals / requirements and architecture

The following list of goals was created for the development of the FRAMES Data Integration tools:

1. Create a mechanism that will allow FRAMES users to access disparate databases across the Internet in real-time.
2. Place collaborative responsibilities such that the different types of participants have control over the areas in which they have domain knowledge.

3. Provide tools such that the incorporation of data is easy and automated as possible.
4. Minimize the amount of data that needs to be transferred when fulfilling data requests

Along with these goals the following list of requirements were placed on the implementation:

1. The server-side software must be cross-platform compatible (run on Windows, Solaris, and Linux systems at a minimum) and can only rely on open source off-the-shelf components (to allow for a minimal cost distribution).
2. The system should have the ability to interface and extract data from a wide variety of disparate data sources (JDBC, ODBC, etc.).
3. The system must be able to operate in a variety of networking environments (corporate LAN's, Internet, through firewalls, etc.).

Based on these goals and requirements, the architecture depicted in Figure 1 was developed.

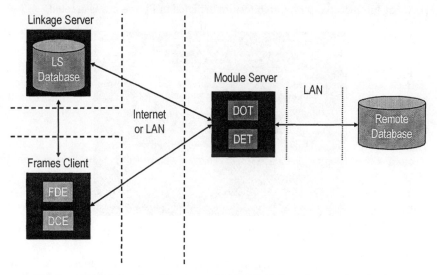

Figure 1: High level architecture of FRAMES Data Integration System

The sections that follow will explain each of these components in detail.

3 Dictionaries and the Framework Development Environment

One of the most important findings of the FRAMES project (which has been an on going effort to provide an environment where models and databases can be linked together seamlessly) is that the size of the data definitions for model-to-model and database-to-model interactions is small—typically on the order of 10 to 20 parameters. This small number of parameters allows them to be succinctly defined and packaged into small text based files that FRAMES refers to as dictionary files. Dictionary files contain the metadata definitions for the information that models / databases can consume or produce. With these dictionary file definitions created, model-to-model and database-to-model interactions can be validated based on the ability of each participant to supply or consume (depending on the role in the interaction) the data required by the other participants in the interaction.

FRAMES provides the user with tools to generate / edit dictionary file definitions for a model or database. In addition, these tools provide an interface to the central information clearing house (the Linkage Server component, which will be discussed in Section 6) to allow for the publishing and retrieval of shared dictionary files. These tools are an integrated part of the FRAMES client software and are known as the Framework Development Environment.

4 Module Server

To facilitate the sharing of database information across a network environment (LAN or Internet) the server architecture depicted in Figure 2 was created.

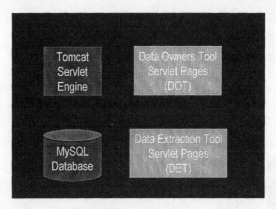

Figure 2: FRAMES module server architecture

This server acts as an information gateway and requires an Internet or LAN connection to FRAMES clients and a LAN connection to any databases in which it will provide information from. It contains two open source off-the-shelf components—the Jakarta-Tomcat servlet engine and a MySQL database. It also contains two custom-build applications for FRAMES—the Data Owners Tool (DOT) and the Data Extraction Tool (DET).

The DOT is a java servlet based tool that is designed to help data source owners create data extraction plans based on FRAMES dictionary files. It can connect to multiple databases (any JDBC compliant) and provides a wizard-type browser-based interface to create SQL queries that map data from the data to a specific dataset defined by a dictionary (also known as an extraction plan). Once an extraction plan has been created and tested, the data owner can store it in the local repository (the MySQL database) for later use by the DET. The DOT also provides the data owner with an interface for retrieving shared dictionary files from the central information clearing house (the Linkage Server component, which will be discussed in Section 6).

The DET is also a Java servlet based application that provides an automated interface for executing stored extraction plans (defined in the DOT) and returning the corresponding data back to the requesting FRAMES client. All communication between the DET and a FRAMES client takes place via HTTP / HTTPS which allows the DET to function through most corporate firewalls without reconfiguration. Username / password authentication is available to control access to the data.

5 Data Client Editor

On the FRAMES client side, a Data Client Editor (DCE) is provided to interface and extract data from FRAMES Module Servers and pass it on to other interconnected models. It communicates directly via HTTP / HTTPS to FRAMES Module Servers to request and process data from remote data sources. It allows for hierarchical data source combinations. This means that multiple data providers can be selected to fulfill a specific dataset, and the extracted data is loaded into the dataset by the order established in the hierarchy—the dataset is first populated by the primary source, then any holes in the dataset can be filled in by the alternative sources provided. The DCE also provides a mechanism for the user to review and make any edits to the data that has been retrieved (before passing it on to the interconnected model). The DCE also allows for local storage of extracted data in case the user would like to re-run a simulation with exactly the same data (normally the DCE would re-query a FRAMES Module Server to make sure that the most up-to-date information from the remote data source gets used).

6 Linkage server

As mentioned in earlier sections, the FRAMES Linkage Server component is the central clearing house for FRAMES information. The server resides at a well-known Internet address so that all FRAMES clients and Module Servers can communicate with it. It can store and distribute dictionary files, locations and content available on Module Servers, new models, and FRAMES software updates. The application is a java servlet based web application that supports uploading and downloading files (similar to FTP, but with HTTP / HTTPS). The architecture of the Linkage Server is similar to the Module Server—it contains two off-the-shelf components (Jakarta-Tomcat servlet engine and MySQL database) and a custom java servlet application to support the file transfers. Like the Module Server, it supports username / password authentication for security.

7 Summary

The intent of the FRAMES Data Integration tools project is to provide a means to easily integrate data from remote disparate sources into an interconnected modeling environment in a seamless manner. The methodology of the implementation allows for the participants (modeler, data owner, assessor) to focus on the areas where they have domain knowledge. The modeler can focus on defining the modeling metadata definitions for his or her model in FRAMES dictionary files. The data owner can focus on generating SQL extraction plans that map the information from the data source to datasets defined by FRAMES dictionaries. The assessor can focus on generating the required model / database interactions necessary to perform the assessment. Because of the mechanisms provided, data can be extracted in real time (via the Internet) in a secure manner and the amount of data transferred is kept to a minimum.

References

[1] Whelan, G., Castleton, K.J., Buck, J.W., Gelston, G.M., Hoopes, B.L., Pelton, M.A., Strenge, D.L., & Kickert, R.N., *Concepts of a Framework for Risk Analysis in Multimedia Environmental Systems (FRAMES)*. PNNL-11748. Pacific Northwest National Laboratory, Richland, Washington. 1997.

[2] Whelan, G. & Nicholson, T., *Proceedings of the Environmental Software Systems Compatibility and Linkage Workshop.* PNNL-13654. Pacific Northwest National Laboratory, Richland, Washington, 2001.

[3] Whelan, G., Pelton, M. A., Dorow, K.E., Gelston, G.M., Castleton, K.J., *Merger Between 3MRA and FRAMES-V1: Design*, PNNL-????. Pacific Northwest National Laboratory, Richland Washington.

[4] Elder, M., Pelton, M.A., *Framework Development Environment*, PNNL-????. Pacific Northwest National Laboratory, Richland Washington.

An overview of USEPA's integrated multi-media, multi-pathway, and multi-receptor exposure and risk assessment tool - The 3MRA Model

B. Johnson, S. Kroner, D. Cozzie, & Z. Saleem
U.S. Environmental Protection Agency, Office of Solid Waste
United States

Abstract

Risk based tools are often useful for making quick yet informed decisions regarding the threats posed by a waste, an emission, or a remedial setting. Such tools may prove useful in Brownfields settings where the primary objective is returning properties to productive use in the most efficient manner possible. This paper chronicles the development of a state-of-the-art risk based standard setting tool, USEPA's Multi-media, multi-pathway, multi-receptor exposure and risk assessment tool – the 3MRA model. The history of its development, nature of the data, the establishment of a site-based approach, the underlying Monte Carlo scheme, the unique software design, and the potential future applications are summarized.

1 Background

In the 1990's there was a substantial effort by Academia, Federal and State organizations and the private sector to develop risk assessment tools that could be used to help establish risk-based screening levels. These levels, depending on how they are developed, can be used to help inform a variety of decisions such as the need to initiate or complete remediation, where and how to manage wastes, or to help decide whether technology based standards are protective of human health or the environment. Similar decisions need to be made at Brownfield sites and the availability of sound and relevant risk based standards can help facilitate the efficient restoration of such sites and their return to productive uses. This paper chronicles and describes the development of one risk based standard setting tool that has the potential for a wide range of uses in waste programs.

In 1995 the Office of Solid Waste (OSW) in the Environmental Protection Agency (EPA) proposed to amend existing regulations for disposal of listed hazardous wastes under the Resource Conservation and Recovery Act (RCRA). The December 1995 proposal (60 FR 6634, December 21, 1995) outlined a rule that was designed to establish constituent-specific exit levels for low risk solid wastes. Under the proposal, waste generators of listed hazardous wastes that could meet the new concentration-based criteria would no longer be subject to the hazardous waste management system specified under Subtitle C of RCRA. Basically, this established a risk-based "floor" for low risk hazardous wastes that would encourage pollution prevention, waste minimization, and the development of innovative waste treatment technologies. The purpose of the rule making was to reduce possible over-regulation arising from the capture of wastes with low levels of hazardous constituents in the federal hazardous waste management system.

In 1995, a Subcommittee of EPA's Science Advisory Board (SAB) conducted a review of the proposed methodology for calculating exit concentrations and in May of 1996 published its findings in USEPA [1]. In addition to this review, EPA's Office of Research and Development (ORD), and numerous industrial and environmental stakeholders, also reviewed the proposed methodology. While there was a collective conclusion that the methodology lacked the scientific defensibility for its intended regulatory use, the SAB Subcommittee made a number of recommendations in their findings that, when addressed, should provide an adequate scientific basis for revising and validating a risk-based methodology for establishing national-level exit criteria. The SAB made the following general recommendations for revising and validating a national methodology:

1) develop a true multi-pathway risk assessment in which a receptor receives a contaminant from a source via all pathways concurrently, is exposed to the contaminant via different routes, and accounts for the dose corresponding to each route in an integrated way;
2) maintain mass balance;
3) conduct substantial validation of the methodology and its elements, against actual data derived from either the laboratory or field, prior to implementation of the model;
4) conduct a systematic examination of parameters to ensure a consistent and uniform application of the proposed approach, and further, the full suite of uncertainties needs to be addressed for the final methodology;
5) discard the proposed screening procedure for selecting the initial subset of chemicals for ecological analysis and instead require that a minimum data set be satisfied before ecologically based exit criteria are calculated;
6) seek the substantive participation, input, and peer review by Agency

scientists and outside peer review groups as necessary, to evaluate the individual components of the methodology in much greater detail; and, 7) reorganize and rewrite the documentation for both clarity and ease of use.

2 System

As a result of the methodology reviews, the Office of Solid Waste collaborated with the ORD to develop and document a sound science foundation, supporting data for the assessment, and related software technology for an integrated, multi-media model (entitled 3MRA) following the recommendations of the SAB and other reviewers. The Multi-media, Multi-pathway, and Multi-receptor Risk Analysis (3MRA) system represents a collection of science-based models and databases that have been integrated into a software infrastructure that is based on the FRAMES (Framework for Risk Analysis in Multimedia Environmental Systems) concept. The FRAMES-based technology provides a computer-based environment for linking environmental models and databases and managing the large amounts of information within the system, including the visualization of outputs. (Buck et al., [2]; Whelan et al, [3]).

The 3MRA system is a national-level risk assessment tool that estimates human and ecological chronic risks resulting from long-term chemical release from land-based waste management units. Alternatively, the 3MRA system is also capable of estimating the levels in waste that must be met in order to avoid exceeding risk thresholds of concern. The 3MRA system (Table 1) consists of: (1) 17 science-based modules that estimate chemical fate, transport, exposure, and risk (Figure 2); (2) 5 data processors, that select data for model execution; manage information transfer within the system; and, provide a visualization of the system outputs; and (3) multiple databases that contain the data for the assessment. The model implements a forward-calculating approach that begins with selected concentrations of a chemical in waste and estimates the associated hazards and risks to human and ecological receptors. The system evaluates simultaneous exposures from multiple exposure pathways that are defined for each receptor. A given receptor is subject to exposures from applicable pathways simultaneously. The aggregate risk to any individual receptor is defined as the sum of the risks from each pathway over a given time period. Given that the exposure in the different media can occur over significantly different times, aggregation of risk is performed for exposures that occur at the same time. Exposures and risks are estimated for all receptors at various distances within a 2km radius of the waste management unit giving this tool the unique feature of being able to evaluate the distribution of individual risks across a potentially and variably exposed population.

The 3MRA modeling system was designed specifically to facilitate Monte Carlo simulation methods and to address both uncertainty and variability in the risk outputs. Statistical distributions for many input modeling parameters were

Table 1. Summary of the 3MRA modules and their functions

SOURCE MODULES:	EXPOSURE AND RISK MODULES:
1) Landfill (LF)	14) Ecological Exposure (EE)
2) Surface Impoundment (SI)	15) Ecological Risk (ER)
3) Waste Pile (WP)	16) Human Exposure (HE)
4) Land Application Unit (LAU)	17) Human Risk (HR)
5) Tanks (AT)	DATA PROCESSORS:
MEDIA MODULES:	1) Site Definition Processor
6) Air (AIR)	2) Multimedia Site Processor
7) Surface Water (SW)	3) Exit Level Processor 1
8) Vadose Zone (VZ)	4) Exit Level Processor 2
9) Saturated Zone (AQ)	5) Risk Visualization Processor
10) Watershed (WS)	DATA BASES:
FOOD WEB MODULES:	1) Site
11) Aquatic Food Web (AF)	2) Regional
12) Farm Food Chain (FF)	3) National
13) Terrestrial Food Web (TF)	4) Chemical, Physical, Toxicological

developed and upon iterative implementation of the model yields outputs that allow evaluation of variability and uncertainty including the evaluation of the range and distribution of potential exposures and risks occurring at waste sites across the United States.

A total of 201 individual nonhazardous industrial waste management sites across the United States were randomly selected to represent the national variability in waste management scenarios and locations. The methodology for selecting these

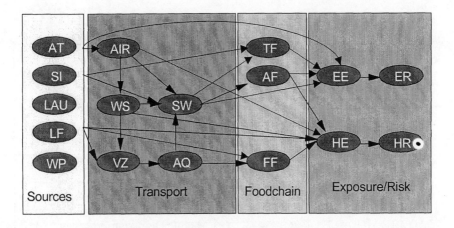

Figure 1. Illustration of the relationship among the 3MRA science modules

sites allows for measures of protection to be calculated probabilistically at the site level and aggregated over the 201 sites to estimate national statistics. The 3MRA system relies on a tiered approach for populating data files for each site evaluation. The approach is referred to as "site-based" in that the model does not require a complete site-specific characterization of the environmental conditions and human and ecological receptors at a specific site. Rather, the approach allows the assignment of data values according to a three- level protocol. Data values are filled first with data at a site level (Figure 2); when site data are not available, a statistically sampled value from a geographically relevant regional distribution of values are used; and lacking a representative regional distribution for the variable, a value from a national distribution are assigned. The site-based data are intended to provide a representative data set for a national assessment.

3 Model Integrity

The EPA focused significant efforts to ensure the scientific integrity of the 3MRA system and its results during system development and post-development. While complete validation of a modeling approach would be the ultimate proof for a multi-media system like the 3MRA, EPA did not find a multi-media data set to compare with the system's predictive outputs. Therefore, EPA implemented specific steps to build a level of confidence in the system to ensure that the system will present a reasonable estimate of nationwide risk for the application at the national scale to hazardous waste management problems. First, the overall technical approach and each individual science-based module included in 3MRA has been peer reviewed. Groups of 3-5 peer reviewers provided critical feedback

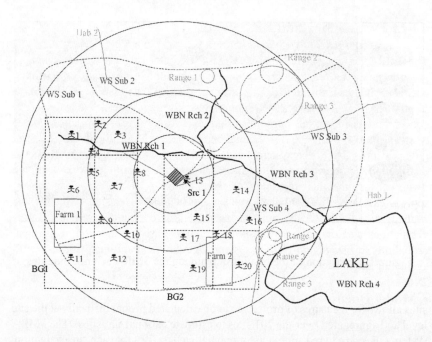

Figure 2. An idealized illustration of the site-level characterization data

about the science-based modules. All told, over 45 independent experts reviewed the science modules to ensure that the theoretical concepts describing the processes within release, fate, transport, uptake, exposure, and risk components were adequate representations of the processes to be evaluated. Second, all software components and databases underwent a series of tests to verify that the software and data were performing properly. These procedures, test plans, test packages, and test results are fully documented and available to the public. Third, the 3MRA system has undergone a comparison analysis with EPA's Total Risk Integrated Methodology that is currently under development. The objective of the model comparison effort was to increase confidence that the 3MRA modeling system produces estimates consistent with other multi-media models. The model comparison study was conducted using an actual industrial site where environmental monitoring data for mercury were available that included information on the source concentrations and some environmental concentrations across different media. Also, a number of the codes used in the model are legacy codes that have undergone considerable validation, testing and verification as part of their development. Finally, a substantial research effort is underway by EPA's ORD to explore the response and sensitivity of the model outputs and trends using state-of-the-art statistical methods.

4 Framework Flexibility

The 3MRA system is a national-level risk-based assessment tool that evaluates potential human and ecological health exposures resulting from long-term (chronic) chemical releases from land-based waste management units . The overall objective of the 3MRA system is to support establishing safe chemical-specific concentrations for wastes currently listed as hazardous below which the listed wastes would be eligible for exemption from hazardous waste management requirements. The 3MRA system was designed to provide flexibility in producing distributions of hazards or risks at sites that could potentially manage exempted waste because the final regulatory decision framework for defining chemical-specific exit levels has not been formulated. The model is designed to allow the evaluation of human health impacts to the general population or selected sub-populations and the impact of varying the measures of protection at different probability levels (Figure 3). The system has similar capabilities with respect to evaluating the impacts on ecological systems.

5 Model Documentation

The EPA has prepared a significant number of reports and documents at various levels of technical complexity that describe the 3MRA modeling system. Our intent is to provide different levels of presentation depending on the intended audience. The documents present: 1) the goals of the 3MRA model in terms of the regulatory needs; 2) the assessment strategy and conceptual basis for the model; 3) the science supporting the assessment; 4) the data used within the model and the their sources; 5) the quality control measures practiced during the system development; and 6) system sensitivity and component verification efforts. These documents are intended to be the primary means by which the general public would become familiar with the 3MRA system. A set of reports that describe the detailed technical descriptions of the assessment methodology, integrating software technology, science-based modules, and data are available at:
http://www.epa.gov/epaoswer/hazwaste/id/hwirwste/risk.htm.

Also, an early version of the model was presented for public comment (65 FR 138, July 18,2000). The information presented in this notice illustrates one application of the model in a national regulatory setting.

6 Future Application of the System Technology

The EPA has invested significant time and resources in the development of the FRAMES-based 3MRA system. This effort was in direct response to earlier concerns about the conduct of multi-media modeling for national assessments and was designed to be an environmental information processing system. The EPA has research efforts underway to build upon and expand the application of the

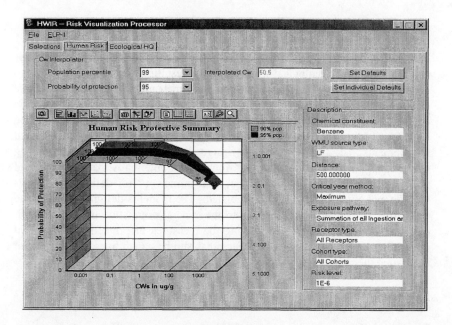

Figure 3. An example of the risk visualization tool illustrating probabilistic outputs

technology beyond the current national-level effort. Examples of these system expansions include: assimilation of different models with more spatially/temporally resolved simulation capabilities to be used for site-specific assessments; a framework to facilitate the sharing and adapting of components, databases, and data analysis tools to answer differing programmatic questions; a user interface to facilitate data entry and visually construct a conceptual site model; and the capability to have world wide web access

Disclaimer: While the research and development efforts associated with 3MRA have been funded by EPA this paper reflects the views of the authors and not those of the US Environmental Protection Agency. This article has not been reviewed for its consistency with present or anticipated EPA policies. Mention of trade names or commercial products does not constitute endorsement or recommendation for use.

7 References

[1] US Environmental Protection Agency. Review of a Methodology for Establishing Human Health and Ecologically Based Exit Criteria for the Hazardous Waste Identification Rule (HWIR), EPA-SAB-EC-96-002, 1996.

[2]Buck, J.W., Nicholson, T., & Whelan G. 4.0 Integrated Multimedia Models and Systems Presented at the Workshop. *Proc. of the Environmental Software Systems Compatibility and Linkage Workshop,* eds.G. Whelan & T.J. Nicholson, Pacific Northwest Laboratories, Richland, WA and U.S. Nuclear Regulatory Commission, Washington, D.C., pp.4.1-4.55, 2000.

[3]Whelan, G.,Pelton, M.,Laniak, G., & Beam P., Multi Agency Modeling Platform Supporting Long-Term Site Issues. *Proc. of the 2001 International Containment and Remediation Technology Conf. and Exhib.,* eds. Chamberlain, S., Gatchett, A. ,Quinn, J., Everett, L.,Lefebvre, P., & Shoemaker, S., Florida State University, Tallahassee, Florida, pp. 334-336.

Army risk assessment modeling system for evaluating health impacts associated with exposure to chemicals

M. S. Dortch & J. A. Gerald
U.S. Army Engineer Research and Development Center, United States

Abstract

The Army Risk Assessment Modeling System (ARAMS) is a computer-based, knowledge delivery, and decision support system that integrates multimedia fate/transport, exposure, uptake, and effects of chemicals to assess human and ecological health impacts and risks. ARAMS is being developed to: reduce the time and cost for conducting site-specific health risk assessments; provide more uniform methods for conducting risk assessments with more reliable risk estimates; and reduce the cost of remediation by establishing more reasonable cleanup targets. ARAMS is based on the risk assessment paradigm of combining exposure and effects assessment to characterize risk. ARAMS uses the Framework for Risk Analysis in Multimedia Environmental Systems (FRAMES) for the system main chassis. FRAMES is a modular modeling framework that allows seamless linkage of disparate models and databases. ARAMS contains various tools needed for risk assessment, such as viewers, report generators, and a module for assessing uncertainty. ARAMS allows for both screening-level and focused (i.e., comprehensive) assessments and has linkages to Web-based databases to allow the use of up-to-date information for conducting risk assessments. ARAMS is being fielded in stages with new versions released approximately annually. Although the primary focus of ARAMS development is associated with models and methods for assessing contaminant exposure and effects on military ranges and bases, it is generally applicable to any setting with contaminated sources or media.

1 Introduction

U.S. Department of Defense (DoD) operations of munitions plants, bases, training ranges, and other facilities have resulted in the release of chemicals, military relevant compounds (MRCs), and other contaminants to soil, groundwater, surface water, and air. The risks to both human and ecological health associated with multimedia exposure to these compounds must be evaluated. The U.S Army utilizes risk assessment procedures to determine cleanup target levels and to evaluate remediation alternatives to provide the most cost-effective approach to reach these levels. Currently, risk assessment procedures are plagued by high levels of uncertainty in both the estimate of effects of the contaminants as well as the probability of exposure. This uncertainty results in excessively conservative risk estimates and levels of cleanup, driving the cost of the cleanup to prohibitive levels.

The U.S. Army Engineer Research and Development Center (ERDC) is developing a computer-based, knowledge delivery, and decision support system that integrates multimedia fate/transport, exposure, uptake, and effects of chemicals to assess human and ecological health impacts and risks. The Army Risk Assessment Modeling System (ARAMS) is based on the widely accepted risk paradigm, where exposure and effects assessments are integrated to characterize risk. The development of ARAMS is expected to significantly reduce the time required to conduct risk assessments, thus, significantly reducing cost. Additionally, substantial cost savings for cleanup are expected by reducing the influence of uncertainty in setting cleanup targets. ARAMS is an ongoing development that is being accomplished in phases. This paper discusses developments for version 1.0 and presents an example application.

2 Development description

The overall strategy for ARAMS development is to build the system in stages, with lower-order (i.e., screening-level) assessment features included first, then higher-order (i.e., comprehensive or more focused) assessment features included later. Screening-level methods involve the use of simpler, analytical models or models of limited dimensions for exposure assessment. Such models or methods make simplifying assumptions that reduce data requirements and time and effort to apply. The comprehensive assessments will involve the use of more complex, numerical exposure models that are often multi-dimensional and involve more detail, but require more data, time, and effort to apply while yielding more information. Higher-order methods may also involve more comprehensive methods for assessing effects, such projections of ecological populations, whereas hazard quotients are used now. ARAMS 1.0 features only the lower-order capabilities, whereas later versions will feature higher-order methods.

2.1 System framework

The development strategy also features the use of an object-oriented, system framework to construct environmental pathways and exposure routes and to link various models or databases for exposure and effects. The system framework is based on the Framework for Risk Analysis in Multi-media Environmental System (FRAMES) [1] developed by Battelle Memorial Institute, which operates the Pacific Northwest National Laboratory (PNNL) of the U.S. Department of Energy. FRAMES enables the user to specify, through objects, the pathways and risk scenarios and to choose which particular model or database to use for each object. New models, databases, or methods can be added to FRAMES objects (i.e., modules) as long as the new model/database/method consumes and produces the type of information characteristic for the module. During the course of ARAMS development, modifications were made to FRAMES to allow new types of modules, such as ecological exposure and effects.

2.2 Modules

ARAMS/FRAMES contains the following modules:
- database for chemical-specific physicochemical properties, bio-accumulation factors, and human health effects;
- source terms for describing initial contaminant concentrations or release characteristics;
- fate/transport modules for air, surface water, soil-vadose zone, and groundwater;
- human exposure pathways;
- human receptor routes;
- human health impacts assessment;
- ecological effects databases;
- ecological exposure pathways;
- ecological health impacts assessment;
- sensitivity and uncertainty analysis; and
- GIS.

Each module can host multiple models, methods, or databases to accomplish module objectives. For example, there are several models/methods hosted within the Source Term Module, such as computed source term release and known soil and water concentrations, from which the user can select the model most appropriate for their application.

The sensitivity and uncertainty module uses Monte Carlo analysis with Latin Hypercube sampling and user specified distributions for stochastic parameters. The GIS module is still under development. Microsoft Excel-based plotting packages are provided for each module. A report generator for the risk assessment guidance for Superfund (RAGS) is also available for reporting results in a form that is compatible with existing procedures.

2.3 Models and databases

All the models within ARAMS/FRAMES modules at this time can be classified as screening-level models. There is a suite of models within the system referred to as MEPAS (Multi-media Environmental Pollutant Assessment System) [2]. MEPAS provides models for: sources; fate/transport in air, streams, vadose zone, and groundwater; exposure pathways for water, air, soil, crops, fish, and farm animals; human uptake routes; and human health impacts. The MEPAS fate/transport models are typically analytical solutions to simplified transport equations, such as one-dimensional advection with three-dimensional diffusion. Other models have been added to the system, such as the surface water, contaminant fate model, RECOVERY [3], the Hydrologic Evaluation of Lechate Production and Quality (HELPQ) [4], a model for theoretical bio-accumulation potential (TBP) for estimating aquatic species uptake [5], and a Wildlife Ecological Assessment Program, WEAP [6], for computing ecological health impacts. A Terrestrial Wildlife Ecological Model (TWEM) is being added during 2002 to compute exposure doses for terrestrial species. There are plans to bring other screening-level models into the system, and work is underway to ling to comprehensive models, such as those in the Groundwater Modeling System (GMS) [7]. The GMS is a modeling environment for 3D, numerical subsurface/groundwater flow and reactive transport models. It should be noted that there are also input screens for entering measured data for contaminated media, rather than applying models to derive exposure concentrations.

The system contains a client-based, chemical-specific database for physicochemical properties required for fate/transport, bioaccumulation factors, and human health effects (i.e., reference doses/concentrations and cancer slope factors). Chemicals can be queried by name or CAS (Chemical Abstracts Service) number. The system links to an on-line (web-based) database for aquatic ecological health effects referred to as the Environmental Residue Effects Database (ERED) (http://www.wes.army.mil/el/ered/index.html). ERED is updated periodically, thus the reason for the on-line linkage, which is seamless allowing effects data queried by species and chemical to be pulled into the system automatically. On-line linkage to a terrestrial ecological effects database of toxicity reference values is being developed during 2002.

2 Example application

An example application is presented to demonstrate several of the software features. This example was kept relatively simple to fit within paper size constraints and represents risk assessment for a hypothetical brownfield site located near a residential area. An adult human is the target receptor for this example, but other receptors could be considered, such as a child. Soil at the brownfield site is contaminated with cadmium as a result of operation of a battery manufacturing plant. The current reasonable maximum exposure (RME) soil concentration of cadmium based upon field sampling and measurement is 1000 mg/kg. Community stakeholders and public health officials want to know the present health risks for a residential area located across the road from the brownfield site.

The exposure routes for this scenario are considered to be soil inhalation (from resuspension), soil ingestion, and soil dermal contact. Groundwater and surface water contamination are not concerns for this site. The following FRAMES modules are required for this application: source term, human exposure, human receptor, and human health impacts. The FRAMES object workspace representing this scenario is shown in Figure 1. A known or specified source model is used for the source module, and each of the other three modules contains MEPAS models that were used for this example. The model parameters used for each of the MEPAS models are shown in Table 1. The soil concentration is assumed to be constant, i.e., no decay or leaching. Since there are few toxicity reference values for dermal contact, the MEPAS health impacts model uses the oral reference dose and cancer slope factor along with the GI absorption fraction to calculate dermal reference values per EPA guidance. Dermal reference dose is obtained by multiplying the oral reference dose by the GI absorption fraction, whereas dermal cancer slope is obtained by dividing the oral slope by the GI absorption fraction. Health impacts for this example are shown in Table 2 in terms of noncarcinogenic hazard index (HI) and carcinogenic cancer risk. Health impacts are reported for each exposure route and combined routes. An example like this can be set up and run very quickly (on the order of minutes to about an hour, depending on the user's experience and familiarity with the system).

3 Conclusions

ARAMS is being developed as a versatile tool for conducting screening-level or more focused assessments for both human and ecological health impacts. Although a simple example was presented here, far more complex assessments can be conducted. For example, in the case presented here, one could model a time varying source released to the vadose zone with decay and leaching, thus producing time-varying exposure concentrations, doses, and health impacts, allowing assessment of future health risks. Other environmental pathways could have been included too, such as air, groundwater, surface water, and food. Other receptor exposure routes could have been included as well, such as: water contact, ingestion, and inhalation during showering; air inhalation; and ingestion of contaminated food.

The object workspace enables the user to conceptualize the problem making assessments more intuitive. Also, the system is modular, which facilitates the addition of new models, databases, and assessment methods. ARAMS has been recently adapted to provide the capability to conduct tier 1 (or level 1) ecological assessments, which will be the topic of another paper. The tier 1 ecological assessments consider hazard quotients (i.e., body concentrations or exposure doses divided by toxicity reference values) without considering spatial aspects, such as home range, habitat use factors, etc. Tier 2 assessments will consider spatial aspects. Additionally, there are plans for considering species population risks.

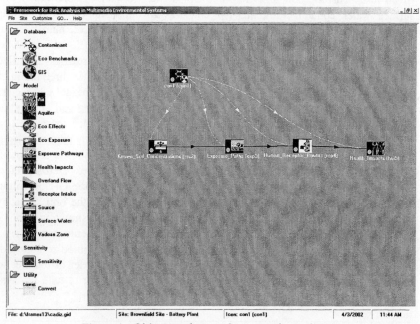

Figure 1. Object workspace for example application

Table 1. Input parameters for example case		
Model/Parameter	Units	Value
Exposure: soil leach rate	cm/yr	0
Exposure: airborne particulate mass loading factor (mass of soil that is suspended in the air above the contaminated soil)	kg/m^3	1×10^{-7}
Receptor: body weight	Kg	70
Receptor: exposure duration	Years	30

Table 1. Input parameters for example case, continued		
Model/Parameter	Units	Value
Receptor: skin area	cm^2	5800
Receptor: soil adherence factor	mg/cm^2	1.0
Receptor: frequency of soil dermal contact	events/day	1
Receptor: fraction of year that soil dermal contact occurs	unit-less	1
Receptor: soil ingestion rate	g/day	0.1
Receptor: fraction of year that soil ingestion occurs	unit-less	1
Receptor: inhalation rate for resuspended soil	m^3/day	20
Receptor: fraction of year that resuspended soil inhalation occurs	unit-less	1
Contaminant database: dermal absorption fraction for soil	unit-less	0.001
Contaminant database: Gastrointestinal absorption fraction, insoluble	unit-less	0.01
Contaminant database: inhalation reference dose	mg/kg/day	2.6×10^{-4}
Contaminant database: oral reference dose	mg/kg/day	0.001
Contaminant database: inhalation cancer slope factor	$(mg/kg/day)^{-1}$	6.1

Table 2. Health impacts for example case		
Exposure route	HI	Cancer risk
All routes	9.83	7.44×10^{-5}
Soil ingestion	1.43	NA
Soil inhalation	0.11	7.44×10^{-5}
Soil dermal contact	8.29	NA

NA – Not applicable, i.e., cadmium is considered by weight of evidence to be carcinogenic for inhalation only.

4 Acknowledgment

This project is funded by the Installation Restoration Research Program of the U.S. Army. The Chief of Engineers, U.S. Army Corps of Engineers, has granted permission to publish this paper. Assistance in interpreting system responses by Dr. Gene Whelan of Battelle, PNNL, is greatly appreciated. Dr. Whelan leads the development of FRAMES at PNNL. The contributions of Mr. Mitch Pelton of Battelle, PNNL, in developing new FRAMES features are recognized and appreciated.

References

[1] Whelan, G., Buck, J., Gelston, G. & Castleton, K., Framework for risk analysis in multimedia environmental systems - training manual, Pacific Northwest National Laboratory, Richland, WA, 1999.
[2] Buck, J., Castleton, K., Whelan, G., McDonald, J., Droppo, J., Sato, C., Strenge, D., & Streile, G., Multimedia environmental pollutant assessment system (MEPAS) application guidance, Pacific Northwest National Laboratory, Richland, WA, 1995.
[3] Boyer, J., Chapra, S., Ruiz, C., & Dortch, M., RECOVERY, a mathematical model to predict the temporal response of surface water to contaminated sediments, Technical Report W-94-4, U.S. Army Engineer Waterways Experiment Station, Vicksburg, MS, 1994.
[4] Aziz, N., & Schroeder, P., Documentation of the hydrologic evaluation of leachate production and quality (HELPQ) module, Technical Note EEPD-06-20, U.S. Army Engineer Waterways Experiment Station, Vicksburg, MS, 1999.

[5] Clarke, J., & McFarland, V., Assessing bioaccumulation in aquatic organisms exposed to contaminated sediments, Miscellaneous Paper D-91-2, U.S. Army Engineer Waterways Experiment Station, Vicksburg, MS, 1991.
[6] Whelan, G., Pelton, M., Taira, R., Rutz, F., & Gelston, G., Demonstration of the Wildlife Ecological Assessment Program (WEAP), Report number PNNL-13395, Pacific Northwest National Laboratory, Richland, WA, 2000.
[7] Environmental Modeling Research Laboratory (EMRL), Department of Defense groundwater modeling system - reference manual & tutorials, GMS V 3.0, Brigham Young University, Provo, UT, 1999.

A flexible object-oriented software framework for developing complex multimedia simulations

P.J. Sydelko, J. E. Dolph & J. H. Christiansen
Decision and Information Sciences Division, Argonne National Laboratory, USA

Abstract

Decision makers involved in brownfields redevelopment and long-term stewardship must consider environmental conditions, future-use potential, site ownership, area infrastructure, funding resources, cost recovery, regulations, risk and liability management, community relations, and expected return on investment in a comprehensive and integrated fashion to achieve desired results. Successful brownfields redevelopment requires the ability to assess the impacts of redevelopment options on multiple interrelated aspects of the ecosystem, both natural and societal. The Dynamic Information Architecture System (DIAS) is a flexible, extensible, object-oriented framework for developing and maintaining complex multidisciplinary simulations of a wide variety of application domains. The modeling domain of a specific DIAS-based simulation is determined by (1) software objects that represent the real-world entities that comprise the problem space (atmosphere, watershed, human), and (2) simulation models and other data processing applications that express the dynamic behaviors of the domain entities. Models and applications used to express dynamic behaviors can be either internal or external to DIAS, including existing legacy models written in various languages (FORTRAN, C, etc.). The ability to simulate the complex interplay of multimedia processes makes DIAS a promising tool for constructing applications for comprehensive community planning, including the assessment of multiple development and redevelopment scenarios.

Introduction

Brownfield lands can be characterized by the interplay of diverse natural and anthropogenic processes interacting across media and across a range of spatial and temporal scales. Effective risk management and long-term stewardship of brownfield lands typically involve gathering, integrating, and evaluating site-specific information regarding physical, chemical, and ecological processes and relationships. These data can be obtained by a variety of methods, including direct sampling and measurement of biological and environmental parameters, laboratory toxicity studies to develop dose-response relationships, extensive literature reviews, and mathematical modeling (EPA [1] and Campbell and Bartell [2]), to estimate contaminant- and species-specific doses and responses.

It is a major challenge to assemble a simulation system that can successfully capture the dynamics of complex ecological systems, such as brownfields, and an even more serious challenge to be able to adapt such a simulation to shifting and expanding analytical requirements and contexts.

Researchers have recognized the need for more integrated and comprehensive approaches to modeling and simulation that can assess several components of an ecological system simultaneously (Maxwell and Costanza [3]; Berry et al. [4]; Bennett et al. [5]; Frysinger et al. [6]; Fedra [7]; Zandbergen; [8], and Bartell et al., [9]).

Geographical information systems (GISs) have been widely used to visualize, integrate, and analyze spatial data pertinent to evaluating changes in ecological systems (for example, see Minns and Moore [10]; Akcakaya [11] (online); Band et al. [12]; DOE [13]; and Zandbergen [8]). Many of these efforts have resulted in the creation of larger, more comprehensive models that employ model-to-model or model-to-GIS linkages (Band et al. [12]; and Ortigosa et al. [14]).

The use of GIS software as an integration framework for ecological applications seems obvious because of the important role spatial dynamics has in evaluating complex ecological systems. Although these efforts have illustrated the potential of integrated modeling, they have created integration systems that are somewhat inflexible and that do not adequately reflect true interprocess dynamics. While a powerful tool for displaying and analyzing large data sets, GIS software packages do not provide an adaptive platform for integrating diverse models and simulations (Sydelko et al. [15]). For these reasons, the Dynamic Information Architecture System (DIAS) was developed. DIAS is a flexible, dynamic, and cost-effective object-oriented approach to integrating models.

Object-oriented architecture for integrated modeling

DIAS is a flexible and extensible object-oriented framework for developing integrated, multidisciplinary, dynamic simulations (Christiansen [16] and Sydelko et al. [17]). DIAS is domain-neutral, meaning that it is not designed for simulations specific to any one discipline or subject area; rather it supports the development of simulations for virtually any type of application. DIAS has been successfully utilized to build a wide range of simulations, including dynamic terrain- and weather-influenced military unit mobility assessment; integrated land management at military bases (Sydelko et al. [17,18]) a dynamic virtual oceanic environment; clinical, physiological, and logistical aspects of health care delivery; avian social behavior and population dynamics (Rewerts et al. [19]); and studies of agricultural sustainability under environmental stress in ancient Mesopotamia (Christiansen [16]).

An important design distinction of DIAS is that it is a framework – an environment and set of tools that developers utilize to build simulations. DIAS is not a model or a suite of models, but rather a flexible modeling environment within which developers build applications by either "wrapping" currently existing models and applications (including legacy models, database management systems, GISs, etc.), coding new in-line models, or building any combination of external and in-line models to create new multicomponent simulations. In this way, the DIAS framework allows new and/or existing legacy models and other applications to inter-operate in the same object environment.

The main components of a DIAS simulation are (1) software objects (Entity objects) that represent real-world entities such as atmosphere, soil, or river, and (2) simulation models (related to environmental fate and transport and ecological processes and responses) or other applications that express the dynamic behaviors of the real-world entities (such as air dispersion, stream flow, and soil microbial processes). The DIAS infrastructure makes it feasible to build, manipulate, and simulate complex ecological systems in which multiple objects interact via multiple dynamic environmental and ecological processes.

Many traditional model integration architectures create model-to-model links (Figure 1). However, as the number of models in the simulation suite grows, however, this approach becomes more cumbersome and more difficult to implement successfully. In addition, when new models are added or one model is replaced by another, the intermodel links in the system often have to undergo major revision to permit integration of the new models. To address this difficulty, DIAS extends the Object Paradigm by abstraction of the objects' dynamic behaviors, separating the "WHAT" from the "HOW." DIAS object class definitions contain an abstract description of the various aspects of the object's behavior (the WHAT), but no implementation details (the HOW); these details are addressed by other DIAS models or applications.

To illustrate the principle of behavior abstraction, assume a brownfields application includes a Brownfields Manager Entity object that exhibits a behavior that results in the removal of a contaminant from the soil. This behavior may be generically coded into the Brownfields Manager Entity object as "implement removal." This generic behavior "implement removal" represents the "WHAT." The implementation details for "HOW" the behavior "implement removal" specifically occurs would depend on the external simulation process(es) of interest. For example, depending on the objectives of the simulation, models that "implement removal" can include accelerated microbial degradation, phytoremediation, or direct soil removal.

The DIAS approach allows models to be linked to appropriate Entity objects at run time to meet the specific needs of a given simulation objective. This capability leads to even greater flexibility and extensibility of the simulation model. In DIAS, models communicate only with domain (Entity) objects, never directly with each other (Figure 1). From a software perspective, this makes it easy to add models, or swap alternative models in and out without major recoding.

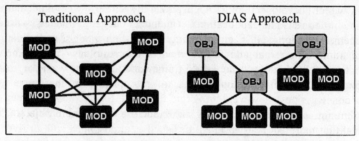

Figure 1: Traditional model-model interaction approach versus the
DIAS object-model interaction approach.

The Entity-centric design of DIAS also allows DIAS-based applications to more closely mimic the associations and relationships that exist in the real world, including dynamic feedback. In general, most models and applications have been designed to operate independently, even though effective decisions often call for assessing several components of a system simultaneously – in terms of their relationship to each other as well as how they affect broader management decisions. DIAS provides a framework for developing applications that address interprocess dynamics in a highly realistic way, in large part because the focus of the simulation is on the Entity objects and their interaction with one another. The DIAS design allows developers to articulate the dynamics of an ecosystem much more closely to the way we understand them, while at the same time not impose one-world view on the development of an application. DIAS interprocess flows are realistic because model processes affect the state of Entity objects, and thus reflect the true dynamics of the real-world system.

A schematic diagram of the DIAS architecture is shown in Figure 2. DIAS is a process-based, discrete event simulation framework. In complex DIAS simulations, both DIAS-internal and -external models and applications can interact with Entity objects to create dynamic simulations. External models or applications are made ready to participate in a simulation through a formalized process that "wraps" each model or application inside a software object for use in DIAS. This "wrapping" procedure enables the DIAS Entity objects to utilize external models to address behaviors. An important feature of DIAS is that the "wrapped" models and applications run in their native languages (e.g., FORTRAN, C, MODSIM, etc.) rather than requiring translation to a common or standard system language.

This ability to link external models and applications gives DIAS the ability to scale to increasingly complex problems. To adequately address the scientific domain of these new models, however, requires the use of domain (discipline) expertise in Entity object design. A future goal for DIAS is the development of Entity object libraries built by DIAS users for a wide variety of subject domains. The ability to reuse objects from colleagues and other researchers could speed application development. Each time a new application is developed using the DIAS framework, existing objects become more mature (new state and behavior are added) and new objects are added to the library. In this way, the library is continuously expanding, making future application development more efficient. This concept does, however, bring up the need for greater cross-discipline coordination or standardization regarding such issues as nomenclature, units, and data formats.

Entity objects are linked to the model (either internal or external to DIAS) through an object called the Process object. Process objects represent and formally define specific models that can implement specific abstract Entity object behaviors. The Process object is the only object with knowledge of both the Entity "world view" and the Model "world view." Therefore, the Process object is responsible for all data translation, unit conversions, data aggregation/disaggregation issues, etc. The Process object also controls the packaging of Entity data needed as input to models, as well as the unpackaging of model output data and its distribution to the Entities.

The data import/export utilities provide a mechanism to supply Entity objects with data for their state variables. These data-ingestion utilities have been developed to supply the object state variables from a variety of external data sources.

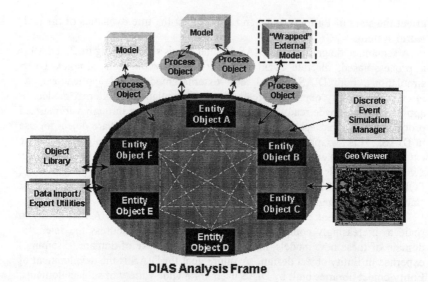

DIAS Analysis Frame

Figure 2: Schematic diagram of the DIAS architecture.

Another important feature of the DIAS framework is its ability to provide run-time feedback between models. Whereas many integration approaches require a static setup for simulation runs using a "hard-wired" sequencing of models and interactions, users of DIAS simulation applications have the freedom to choose various combinations of models and interactions for each new simulation run. Simulations are set up "on the fly," aided by an context-driven graphical user interface (GUI). The DIAS Simulation Analysis Frame establishes the connections between models and domain objects at run-time. The user need only indicate which combination of models should be included in the scenario, and this "context" drives simulation setup. An added benefit is the ability for users to track and visualize the simulation as it occurs.

The DIAS Framework includes Spatial Data Set (SDS) objects that allow Entity objects to express spatial variability in their parameters. SDSs are software objects carrying (1) a complete geometric specification for a 2-D or 3-D spatial partitioning scheme (grid, mesh, network, patchwork, etc.) that divides a region into 2-D or 3-D cells and (2) a collection of data elements for each cell. DIAS employs an object-oriented GIS module, called the GeoViewer (Figure 2), to provide real-time spatially oriented displays of an object's position and/or parameters. This GIS module is designed to navigate within a DIAS study area/frame to create, query, view, and manipulate objects. For each simulation implementation, model output parameters are generated at each time step of the simulation. In Figure 3, the GeoViewer is displaying four different aspects of a Integrated Dynamic Landscape and Analysis System (IDLAMS) simulation.

IDLAMS is a DIAS application for modeling military land management and land use impacts on vegetation and wildlife habitat on army installations (Sydelko et al. [17,18]).

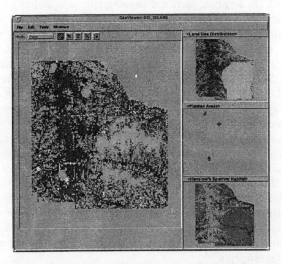

Figure 3: GeoViewer provides real-time spatially oriented displays for four different aspects of an IDLAMS simulation (vegetation, land use distribution, planted areas, and wildlife habitat).

DIAS can also link to external GISs that can perform spatial analysis and modeling functions. For instance, for the IDLAMS application, an external wildlife habitat model written as an Environmental Systems Research Institute (ESRI®) application, was linked into the overall IDLAMS model suite. In addition, DIAS can import and export in many spatial data formats through the data ingestion utilities.

Summary and Future Direction

DIAS is a process-based, discrete event simulation framework that can be used to develop and maintain a wide variety of multidisciplinary simulations. The framework is completely object-oriented and is domain neutral, making it useful for virtually any application domain. A key benefit of the DIAS framework for development of multidisciplinary models is the ability to allow new and/or existing legacy models as well as other applications to operate in the *same* object environment. The modular design of DIAS promotes this flexibility and extensibility, which enhances cost-effectiveness and the evolution of applications.

DIAS continues to be used to develop a wide variety of applications. Currently, efforts are underway to utilize DIAS for a three-compartment multimedia prototype model sponsored by the U.S. Environmental Protection Agency (EPA). In addition, there are ongoing efforts in building the Integrated Oceans application, a DIAS-based virtual maritime environment within which existing models are employed to simulate the transition of wind-generated waves in the deep water to waves in the near-shore environment, then to surf characteristics and currents. Another new application built using DIAS is CASCADE, the Complex Adaptive System Countermeasure Analysis Dynamic Environment. CASCADE is an object-oriented software system for building and running agent-based multidisciplinary simulations that concurrently address socioeconomic, psychological, environmental, etc., factors to support countermeasures analysis.

DIAS is an integration framework capable of integrating the wide variety of models, simulation and data necessary to assess the complexities of brownfields assessment. Entity objects can be created that represent the multiple interrelated aspects of the ecosystem, both natural and societal. The Entity objects can exhibit specific behaviors, either directly coded within DIAS or provided by linked external models and applications. DIAS can address both the spatial and temporal scale issues required for complex dynamic modeling necessary in brownfields assessments.

While the generic design of DIAS makes it a robust framework for multidisciplinary modeling, this benefit necessitates a more complex development environment than that typically seen in more domain-specific simulation development software. Currently, DIAS users must have advanced object-oriented software engineering skills to readily implement applications in the framework. The DIAS Application Programming Interface (API) provides instruction and examples that help guide application developers through creation of simulations, however, there is presently no GUI to assist this process, and developers must extend the framework objects directly by adding new source code. As a first step in assisting programmers, an API has been developed to help application developers utilize the framework.

Currently, DIAS users are typically Java programmers. However, the DIAS paradigm is beginning to be adopted by other research groups who are building and enhancing the application development interfaces for specific areas. The Army Corps of Engineers Engineering Research and Development Center has acquired DIAS to build modeling and simulation applications related to military land management. The EPA is developing the Multimedia Integrated Modeling System (MIMS) framework, a software infrastructure or environment that will support constructing, composing, executing, and evaluating complex modeling studies. MIMS is being developed to support complex cross-media modeling. The MIMS framework uses DIAS for its model coupling paradigm and execution management software. The DIAS software library provides basic templates for domain objects and models and capabilities for constructing interacting sets of models and executing those models in the proper order. The MIMS project is supporting the addition of new capabilities to the DIAS framework. An interface on top of DIAS, the MIMS framework layers generic user interfaces, well-defined parameter types, functionality that minimizes the programming effort required to incorporate new models into a MIMS simulation, and additional tools to support modelers.

References

[1] U.S. Environmental Protection Agency, *Wildlife Exposure Factors Handbook*, Vol. I, EPA/600/R-93/187a, Office of Research and Development, Washington, DC, 1993.

[2] Campbell, K.R. and Bartell, S.M., Ecological models and ecological risk assessment, *Risk Assessment: Logic and Measurement*, eds. M.C. Newman and C.L. Strojan, Ann Arbor Press: Chelsea, MI, pp. 69-100, 1998.

[3] Maxwell, T. and Costanza, R., Distributed modular spatial ecosystem modeling, *International Journal of Computer Simulation, Special Issue on Advanced Simulation Methodologies*, 5(3): 247-262, 1995.

[4] Berry, M.W., Flamm, R.O., Hazen,B.C., and MacIntyre, R.L., A system for modeling land-use change. *IEEE Computational Science & Engineering*, 3(1), pp. 24-35, 1996.

[5] Bennett, D.A., Armstrong, M.P., and Weirich, F., An object-oriented model base management system for environmental simulation, *GIS and Environmental Modeling: Progress and Research Issues*, eds. Goodchild et al., GIS World Books: Fort Collins, CO, pp. 439-444, 1996.

[6] Frysinger, S.P., Copperman, D.A., and Levantino, J.P., Environmental Decision Support Systems (EDSS): an open architecture integrating modeling and GIS, *GIS and Environmental Modeling: Progress and Research Issues*, eds. Goodchild et al., GIS World Books: Fort Collins, CO, pp. 357-362, 1996.

[7] Fedra, K., Distributed models and embedded GIS: integration strategies and case studies. *GIS and Environmental Modeling: Progress and Research Issues*, eds. Goodchild et al., GIS World Books: Fort Collins, CO, pp. 413-418,. 1996.

[8] Zandbergen, P.A., Urban watershed ecological risk assessment using GIS: a case study of the Brunette River watershed in British Columbia, Canada. *Journal of Hazardous Materials*, 61, pp. 163-173, 1998.

[9] Bartell, S.M., Lefebvre, G., Kaminski, G., Carreau, M., and Campbell, K.R., An ecosystem model for assessing ecological risks in Quebec rivers, lakes and reservoirs, *Ecological Modeling*, 124, pp. 43-67, 1999.

[10] Minns, C.K. and Moore, J.E., Predicting the impact of climate change on the spatial patterns of freshwater fish yield capability in eastern Canada, *Climate Change*, **22**, pp. 327-346, 1992.

[11]Akcakaya, H.R., Linking GIS with models of ecological risk assessment for endangered species, *Proc. of the 3rd Third Int. Conf./Workshop on Integrating Geographic Information Systems and Environmental Modeling CD-ROM*, 21-25 January, Santa Fe, NM, 1996.

[12] Band, L.E., Mackay, D.S., Creed, I.F., Semkin, R., and Jeffries, D., Ecosystem processes at the watershed scale: Sensitivity to potential climate change, *Limnology and Oceanography*, **41(5)**, pp. 928-938, 1996.

[13] U.S. Department of Energy, *Salt Lake City Area Integrated Projects Electric Power Marketing, Final Environmental Impact Statement*, DOE/EIS-0150, Western Area Power Administration, Salt Lake City, UT, 1996.

[14] Ortigosa, G.R., De Leo, G.A., and Gatto, M., VVF: integrating modeling and GIS in a software tool for habitat suitability assessment, *Environmental Modeling Software*, **15**, pp. 1-12, 2000.

[15] Sydelko, P.J., Majerus, K.A., Dolph, J.E., and Taxon, T.N., A dynamic object-oriented architecture approach to ecosystem modeling and simulation. *Proc. of the 1999 American Society of Photogrammetry and Remote Sensing (ASPRS) Annual Conf.*, Portland, OR, pp. 410-421, 1999.

[16] Christiansen, J.H., A flexible object-based software framework for modeling complex systems with interacting natural and societal processes, *Proc. of the 4th Int. Conf. on Integrating GIS and Environmental Modeling (GIS/EM4): Problems, Prospects and Research Needs*, Banff, Alberta, Canada, 2000.

[17] Sydelko, P.J., Hlohowskyj, I., Majerus, K., Christiansen, J.H., and Dolph, J., An object-oriented framework for dynamic ecosystem modeling: applications for integrated risk assessment, *The Science of the Total Environment*, **274**, pp. 271-281, 2001.

[18] Sydelko, P.J., Majerus, K.A., Dolph, J.E., and Taxon, T.N., An advanced object-based software framework for complex ecosystem modeling and simulation, *Proc. of the 4th Int. Conf. on Integrating GIS and Environmental Modeling (GIS/EM4): Problems, Prospects and Research Needs*, Banff, Alberta, Canada, 2000.

[19] Rewerts, C.C., Sydelko, P.J., Dolph, J.E., Shipiro, A.M., and Taxon, T.N., An object-oriented, individual-based, spatially explicit environmental model: A discussion of the approach to implementing the system, *Proc. of the 4th Int. Conf. on Integrating GIS and Environmental Modeling (GIS/EM4): Problems, Prospects and Research Needs*, Banff, Alberta, Canada, 2000.

Fugacity framework: web access and implementation for site assessment and rehabilitation

R.C. Sims, J.L. Sims, A.S. Gibbons, M.R. Baugh, M. McKonkie, J.K. Nieman, and W.J. Grenney
Utah State University, Logan, Utah, United States of America

Abstract

The fugacity-framework addresses multiple-media and multiple contaminant aspects of environmental site assessment. A U.S. EPA database of priority chemicals has been linked to the fugacity model for assessment of contaminant sources, transport and transformation, and exposure. Site-specific data are provided by the user. This tool provides a way for managers to visualize the behavior of toxic chemicals at a contaminated site, the effect of site-specific characteristics on contaminant distribution, the behavior of daughter products of degradation, and the associated risks to humans and the environment. The framework can be used to make decisions regarding protection of public health and the environment, site rehabilitation, and the sustainable development and economic recovery of impacted sites. The fugacity framework can be accessed at http://www.engineering.usu.edu/uwrl/ , Utah Water Research Laboratory.

1 Introduction

Contaminated site rehabilitation involves an assessment of contaminant distribution from source(s) to receptor(s) that may include humans and the environment. Contaminants within sources generally spread through release and migration that expands to larger volumes of soil, water, air, and/or oil phases that result in large-scale exposure and receptor scenarios (Figure 1) that must be controlled through costly risk management technologies of containment, removal, and/or destruction. The technologies selected often are not focused on the correct physical phase(s), resulting in inefficient or ineffective risk management and site rehabilitation efforts.

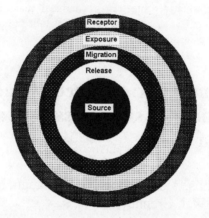

Figure 1. Linking sources to receptors including humans and the environment.

To address these issues of site rehabilitation, a fugacity framework for assessment of multi-media distribution (air, water, soil, and oil) of chemical contaminants in subsurface environments has been developed. The framework is utilized in an interactive mode and pedagogically designed for visual presentation to allow managers and decision makers to focus on making decisions regarding identification of information needs, site assessment, and site rehabilitation. This fugacity-based management system has been implemented to allow easy web access to users worldwide. This paper describes the three components of the fugacity-based management system, including: (1) fugacity framework and linked database, (2) instructional technology-based theory and tools for learning and user interaction, and (3) web-accessibility.

2 Objectives

The overall objective of this paper is to describe a web-accessible tool for managers, senior personnel, and other decision makers who may or may not have a technical background in mathematical modeling, multimedia contaminant behavior, and engineering treatment technologies.

The specific objectives of this paper include presentations of: (1) the fugacity-based framework as a useful tool for multi-media contaminant evaluation; (2) chemical database and site characterization requirements; (3) examples of applications of the fugacity framework; (4) instructional technology theory and application for learning and using the system; (5) web accessibility; and (6) future enhancements to the fugacity framework tool.

3 Fugacity-based Framework

The fugacity-framework is based on the conceptual site model (CSM) that consists of: (1) an Evaluative Site, (2) a chemical mass balance indicating distribution of contaminant(s) among physical phases at the evaluative site, and (3) a flow diagram to relate source(s) to receptor(s) pathways.

3.1 Evaluative Site.

The evaluator designs a "site," then explores the likely behavior of chemicals in that site. A sensitivity analysis of the variables can be performed to focus the types of information needed to evaluate fate and transport, risk, and control options. The concept of "evaluative environments" was introduced by G.L. Baughman and R.R. Lassiter [1]. Evaluative aquatic environments were used to develop the U.S. Environmental Protection Agency (USEPA) EXAMS model of chemical fate in rivers and lakes.

Physical phases of a subsurface evaluative site (Figure 2) include a solid phase and three fluid phases. The solid phase is composed of soil organic matter (SOM) and inorganic components including sand, silt, and clay that comprise the subsurface texture. Fluid phases include the water or leachate phase, the gas phase, and the oil or NAPL (non-aqueous phase liquid) phase.

3.2 Mass Balance Approach.

A chemical mass balance approach is directed at evaluating how contaminant chemicals are distributed among the physical phases at a site. Chemicals at

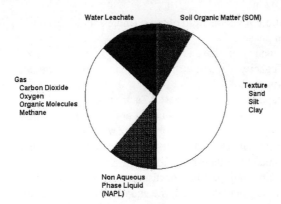

Figure 2. Evaluative site physical phases for fugacity framework.

Brownfield sites are contained within the physical phases shown in Figure 2. One or more physical phases link contaminant sources to receptors. Therefore, a mass balance approach is used to address: (1) chemical distribution among phases, (2) phase distribution at a site, (2) phase release or transport, (3) phase exposure/risk assessment, and (4) which phase(s) to manage with short-term or long-term technologies to prevent exposure. Chemical distribution among site phases can be determined through the application of fugacity calculations.

3.2.1 Fugacity

The word fugacity is derived from Latin meaning "to escape or to flee," and refers to the escaping tendency of a chemical from a particular subsurface phase. Fugacity was introduced by C.N. Lewis in 1901 and has been widely used in chemical process calculations. Fugacity is applicable to organic chemicals, but not to metals and other inorganic chemicals. Its convenience in environmental partitioning calculations became apparent after 1980. Donald Mackay, Professor of Chemical Engineering at the University of Toronto, Canada, introduced the concept of fugacity in the 1970's to express the distribution of organic pesticides among phases of the environment. His books "Multimedia Environmental Models – The Fugacity Approach" [2, 3] describe the basis and applications of fugacity for handling chemical distribution, reaction, advective flow, and diffusive and nondiffusive transport in multimedia environments.

Fugacity modeling provides a way for conducting a chemical mass-balance analysis. Fugacity uses partition coefficients for a chemical contaminant between soil and water (Kd), air and water (Kh), and oil and water (Ko) to describe the distribution of chemicals among physical phases comprising a multi-phase evaluative site. Physical phases can be evaluated with regard to: (1) sources, (2) transport and transformation, and (3) exposure.

Fugacity has the units of pressure (Pa). Relevant equations for Level I fugacity (equilibrium conditions among phases) are described below.

$$Ci \quad = \quad fi \; Zi \qquad (1)$$

Ci = concentration of chemical in phase i, (moles/cubic meter)
fi = fugacity in phase i (Pa)
Zi = fugacity capacity of phase i (mole/Pa-cubic meter)

The fugacity capacity of each phase can be determined using the following equations.

Za = fugacity capacity of the air phase, $Za = 1/RT = 4(10^{-4})$ (2)
    where R = university gas constant and T = temperature (K)

Zw = fugacity capacity of the water phase, $Zw = 1/H$ (3)
    where H = Henry's Law constant (Pa-cubic meter/mole)

Zs = fugacity capacity of the soil-solid phase, $Zs = Kp \; \rho_b \; Zw$ (4)
    where Kp = soil partition or distribution coefficient
    Kp can be calculated using the equation $Kp = (Koc)(foc)$ (5)
    Where Koc is the organic cabon partition coefficient and
    Foc is the fraction of organic carbon in the solid phase
    ρ_b = soil dry bulk density

Zo = fugacity capacity of the oil phase, $Zo = Zw \; Kow$ (6)
    where Kow = octanol-water partition coefficient

The amount (moles, m) in each phase is: $mi = Ci \; Vi = fi \; Zi \; Vi$ (7)

where Vi = volume of each phase

At equilibrium, since f1 = f2 = f3 ... = f, then f = M/ Σ (Zi Vi)   (8)

Thus, the equilibrium fugacity can be calculated from knowledge of the volumes and fugacity capacities of the subsurface phases and the total mass of chemical in the evaluative site. Then for each phase it is possible to calculate: (1) chemical concentration using equation (1), and (2) mass of chemical using equation (7). Fugacity has been used by Sims [4,5,6] for subsurface assessment.

When the amount of chemical released to the subsurface is not known, then a percentage, proportion, or relative amount, of chemical mass in each phase can be determined by substituting the value one (1) for the variable 'M' in equation (8). Volumes can also be estimated as percentages (%) such that the some of the four phases equals 100%. Such a determination is useful to indicate the phase(s) that would likely contain the most mass (%) of chemical. Field determination of chemical concentration could be used with an estimate of phase volume to determine the mass of chemical in the target phase. An estimate of the total mass of chemical at a site is possible since the mass percentage of chemical in each phase is known through the use of the fugacity calculations.

3.3 Flow diagram

A flow diagram incorporates site-specific data into an evaluative site and site specific mass balance information to assist in organizing the information to identify priority source phases with regard to release, secondary sources of contamination, exposure and the need for phase control to accomplish risk management and site rehabilitation (Figure 3). A flow diagram also serves as a common communication tool for site owners and operators, regulators, and consultants.

Site Conceptual Exposure Model

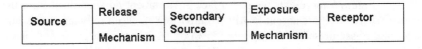

Figure 3. Flow diagram identifies priority physical phases that connect source to receptor through migration and exposure. Risk management technology selection is based on priority phase management goals.

4 Chemical database and site characteristics required

Fugacity calculations require numerical values for properties of chemical substances for the calculation of fugacity capacities, as shown in equations (3) through (6). Calculation of Henry's Law constant (eq. 3) involves water solubility and vapor pressure; calculation of Kp (eq. 4) can be accomplished through a knowledge of Koc and the fraction of soil or aquifer solid phase organic carbon (foc). Calculation of Zo (eq. 5) requires a value for the octanol-water partition coefficient (Kow). The chemical properties required are provided in an embedded chemical database as part of the fugacity framework.

The chemical database utilized in the fugacity framework was developed by the U.S. Environmental Protection Agency (U.S.EPA) [7] to provide information pertaining to fate and transport properties for contaminants commonly found at Superfund sites. The U.S. EPA database of approximately 70 chemicals therefore provides the values used for chemical properties required for fugacity calculations for chemicals relevant to Brownfield and other sites, and also serves as a peer-reviewed reference.

Site characteristics required for fugacity analysis include the volume (cubic meters) of each physical phase (air, water, soil or aquifer, and NAPL). If the volumes are not known, then relative percentages of the volumes can be estimated and used in the fugacity calculations so the total percent of all phases equals 100. In addition, the soil organic carbon, as percent by weight of the soil phase, and the soil dry bulk density (kg/cubic meter) are required.

5 Examples of applications of the fugacity framework

Examples of the applications of the fugacity framework include: (1) distribution of trichloroethylene (TCE) in the subsurface as influenced by the NAPL phase, (2) distribution of TCE parent compound and daughter products dichloro-ethylene (DCE) and vinyl chloride (VC) within the same subsurface site conditions, and (3) Provo City Ironton Economic Redevelopment Project, Utah.

Figure 4 illustrates the distribution of TCE in a subsurface environment composed of 2% NAPL, 25% air, 23% water, and 50% soil by volume, and represents a source area of TCE contamination. The majority of TCE (87%) is located within the NAPL that comprises only 2% of the volume of the subsurface, while only 7% is present in the soil-solid phase.

Figure 5 represents the subsurface environment downgradient from the source where NAPL is absent (NAPL volume = 0%). With only air, water, and soil-solid phase present, the majority (55%) of TCE is present in the soil phase that represents a large increase in the %TCE in soil phase from 7% in Figure 4.

Using the same site characteristics as identified in Figure 4 for TCE, the distributions of daughter products of anaerobic biodegradation DCE and VC are shown in Figures 6 and 7, respectively. Figure 6 shows that DCE is present in air (34%) compared to 2% for TCE, and Figure 7 shows that VC is present almost exclusively in the air phase (96%). Therefore in the subsurface, anaerobic transformation from TCE to DCE to VC results in a different

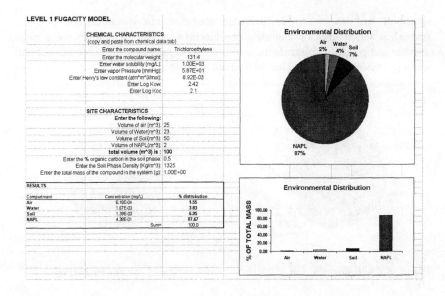

Figure 4. Fugacity Analysis of TCE in a subsurface environment with 2% (volume) NAPL.

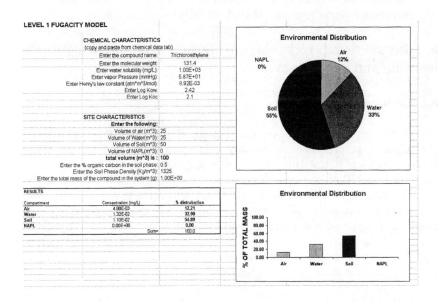

Figure 5. Fugacity Analysis of TCE in a subsurface environment with 0% (volume) NAPL.

distribution of daughter products. These assessments have implications with respect to which phases to monitor for which chemical, as well as the appropriate phase for control and treatment to prevent exposure to human health and the environment for each contaminant at a site.

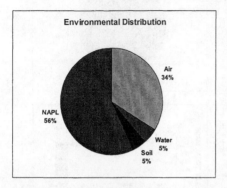
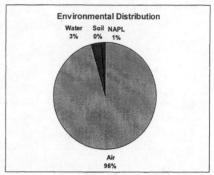

Figure 6. DCE Distribution   Figure 7. VC Distribution

The Provo City Ironton Economic Redevelopment Project was initiated in 1996 with a grant from the US EPA. The project involves the redevelopment of land contaminated with polycylic aromatic hydrocarbons and metals, and includes citizens, local businesses, and local industry. The fugacity framework is applicable to the evaluation of PAH source assessment and risk management with respect to human exposure and land rehabilitation.

6   Instructional technology - theory and application for learning and using the system.

6.1 The instructional problem.

Given the complex nature of the fugacity-framework, two instructional challenges were presented: 1) quickly help users understand the important concepts of fugacity and 2) teach them how to use the web-accessible fugacity-framework tool. Both novice and experienced users will need to understand and use the system. While novices may need considerable instructional aid, experienced users prefer quick access to the desired information with as few mouse and keyboard operations as possible. Whatever the case, a thoughtful design will help individuals and groups effectively make informed decisions regarding risk management and associated costs for control and treatment of toxic chemicals in subsurface environments.

6.2 Criteria.

There are several different potential use cases. Novices might want to simply explore the concepts and interface features or have direct instructional assistance.

Experienced users may want to seriously explore different "what if scenarios," the how and why fugacity works, or evaluate specific sites. The framework could also be used in educational settings by classroom teachers or individual students.

The criteria for this instructional design required a quick direct interaction between users and the model for these different purposes. Therefore a variety of features are required to meet the range of needs.

- Easy to use
- Intuitive
- Process to directly evaluate sites
- Method to track, record, compares, and saves past experiments
- Explanations
- Attention focusing to possibilities offered by the framework

6.3 Principles of solution

A theory of instructional design that humans learn from interacting with models of environments, cause-effect systems, and expert behavior was selected. Model-centered instruction as characterized by A.S. Gibbons [8] prescribes methods for making human/model interactions effective by promoting model observation, interaction, and problem solving. MCI's central premise is "that the most effective and efficient instruction takes place through experiencing realia or models in the presence of a variety of instructional augmentations designed to facilitate learning from the experience." The interface was designed to be intuitive to incorporate the criteria that users need to quickly learn how to use the fugacity framework. Features were utilized that meet the needs of both novice and experienced users. The instructional interface incorporates a visually simple design with minimal text. The user control options are clear with quick access to required information.

Instructional features were provided for both the novice and experienced user (see Figure 8). Features provided for the novice user include: sequenced demonstrations based on problem solving that range from easy to complex, embedded didactics that provide explanations of important functions, instructional road signs that point out key events, and coherence between visual elements of the interface and the model. Features that are included for the experienced users are: quickly accessible data and results, a multi-problem representation trace that allows the user to visualize inputs and results from different initial conditions over a series of "what if scenarios", and downloadable results in the desired format.

Figure 8 shows subsurface display elements that correspond to the novice and expert user features described previously.

1. **Chemical**. Selects the chemical and displays the properties.
2. **Environmental Distributions**. Displays how contaminant chemicals are distributed among physical phases.
3. **Evaluative Site**. Shows a subsurface representation of the contaminant distribution for the selected chemical.

4. **Site Characteristics**. Selects and displays the percent of physical phases in the subsurface environment.
5. **Representation Trace**. Records and displays from a series of "experiments" or "what-if scenarios."
6. **Demonstration**. Provides a series of instructional demonstrations.
7. **Practice**. Presents the user with various problems or case studies.

Not shown in the diagram is the export feature that enables users to save a session's data into portable file format.

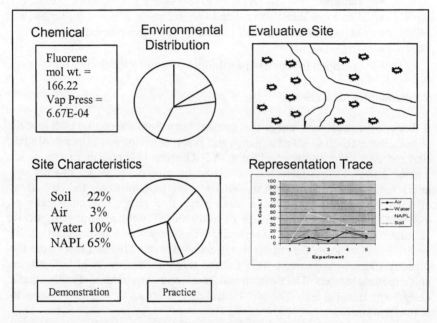

Figure 8. User Interface for Fugacity framework for site assessment.

6.4 Conclusion

The primary challenge of the instructional design was to enable users at all levels to quickly understand the concepts of fugacity and teach them how to use the web accessible Fugacity based management tool. Model-centered instruction with an intuitive interface and relevant instructional augmentations was selected to provide users with instructional experiences of the fugacity model. In the end, meeting these challenges will enable users at all levels to characterize and evaluate their own specific sites, engage in fruitful dialogue with other stakeholders, and make informed decisions regarding risk management and associated costs for control and treatment of toxic chemicals in subsurface environments.

7 Web accessibility and downloading to the local computer

There is a continuing and rapid change in internet technology and development tools. The number and types of development products, and the uncertainty associated with the longevity of each introduces major complexity into the selection process. We considered the three implementation paradigms for delivering the model over the internet: 1) running the model on the server, 2) down loading the model to the local computer, and 3) a combination of both where the database resides on the server and an executable module is downloaded to the local computer. Each method has advantages and disadvantages in terms of accessibility, standardization, support, and extendibility.

Accessibility to applications running on the server' "server apps'" is independent of the local platform (i.e. the model can be accessed by a variety of operating systems found on MacIntosh and PC computers); the only requirement is a conventional web browser and online connection to the internet. Applications designed for downloading to the local computer, "local apps", are specific for each class of operating systems (e.g. Mac and PC), and separate executable modules must be coded and compiled specifically for each.

Standardization in this case means that at any time all users are operating with the same model version and dataset. If strict consistency is an issue then standardization is essential.

Continuing support is a major issue for all models. Rarely is a model released and not modified shortly thereafter. Each update to a local app has to be distributed to the entire user group; only one update implementation is needed for a server app. This issue applies to both the model itself and to the training materials for the operation of the model.

Extendibility refers to the architecture of the model - can it be enhanced with major new features while maintaining the basic features and look-and-feel of the original.

Comparing the server app with the local app, the server app has a clear advantage in standardization and support. In terms of accessibility the server app can be operated from any local computer having a conventional browser while the local app is specific to a particular platform. On the other hand, the user must be connected to the internet in order to use the server app. Extendibility weights heavily in favor of the local app because server apps are constrained by the functionality and speed of the browser.

The fugacity model was developed as a local app because extendibility was a major issue. Borland Delphi® (object Pascal) was used as the development tool. Delphi provides a tight compiled module that can be easily downloaded and installed onto the local computer from the internet. Delphi is a powerful computer language with excellent interface components, and we have found that Delphi is good compromise between MS Visual Basic® and C++. Delphi combines much of the ease-of-use of Visual Basic with the programming power of C++.

8 Future enhancements to the fugacity framework tool.

A web-accessible database of risk management technologies and associated costs for control and treatment of toxic chemicals to prevent human and environmental exposure will be linked to the fugacity modeling framework in future enhancements. In addition, the number of chemicals and chemical properties contained in the chemical database will be increased to more comprehensively represent Brownfield chemicals.

References

[1] Baughman, G.L. and R.R. Lassiter. Estimating the hazard of chemical substances to aquatic life. In: J. Cairns, Jr., K.G. Dickson, A.W. Maki, eds. American Society of Testing and Materials Technical Publication 657, Philadelphia, PA, 1978.

[2] Mackay, D. Multimedia Environmental Models. The Fugacity Approach. 2nd. Ed. CRC Press, Boca Raton, FL, 2001.

[3] Mackay, D. Multimedia Environmental Models. The Fugacity Approach. Lewis Publishers. Chelsea, MI, 1991.

[4] Sims, R.C. U.S. EPA RCRA Corrective Action Workshop. Chapter 8, Conceptual Site Model. U.S.EPA Washington, D.C. 1999-2000.

[5] Sims, R.C. Champion International Superfund Site, Libby, Montana. Bioremediation Field Performance Evaluation. EPA/600/R-95/156. 1996.

[6] Sims, R.C., J.S. Ginn, W.J. Doucette. Chemical mass balance approach for estimating fate and transport of PAH metabolites in the subsurface environment. *Polycyclic Aromatic Compounds* 5:225-234, 1994.

[7] U.S. Environmental Protection Agency.Subsurface Contamination Reference Guide. EPA/540/2-90/011, October 1990. Office of Emergency and Remedial response, U.S. Environmental Protection Agency, Washington, D.C.,1990.

[8] Gibbons, A.S. Model-centered instruction. (J. Scandura, J. Durnin, and J.M. Spector, Eds). J. Structural Instruction and Intelligent Systems, 14(4), pp. 511-540. New York Taylor & Francis (Pub.), 2001.

Acknowledgement: Support for this project was provided by a grant from the Huntsman Environmental Research Center (HERC) at Utah State University, Utah, U.S.A. (Dr. Maurice Thomas, Director), and by funds provided by the Utah Water Research Laboratory and the U.S. Environmental Protection Agency, Washington, D.C., U.S.A.

Concepts associated with a unified life cycle analysis

G. Whelan[1,2], M.S. Peffers[1] & D.A. Tolle[3]
[1]Battelle-Pacific Northwest Division, USA
[2]Washington State University_Tri_Cities, USA
[3]Battelle Columbus Organization, USA

Abstract

There is a risk associated with most things in the world, and all things have a life cycle unto themselves, even brownfields. Many components can be described by a "cycle of life." For example, five such components are life form, chemical, process, activity, and idea, although many more may exist. Brownfields may touch upon several of these life cycles. Each life cycle can be represented as independent software; therefore, a software technology structure is being formulated to allow for the seamless linkage of software products, representing various life-cycle aspects. Because classes of these life cycles tend to be independent of each other, the current research programs and efforts do not have to be revamped; therefore, this unified life-cycle paradigm builds upon current technology and is backward compatible while embracing future technology. Only when two of these life cycles coincide and one impacts the other is there connectivity and a transfer of information at the interface. The current framework approaches (e.g., FRAMES, 3MRA, etc.) have a design that is amenable to capturing 1) many of these underlying philosophical concepts to assure backward compatibility of diverse independent assessment frameworks and 2) linkage communication to help transfer the needed information at the points of intersection. The key effort will be to identify 1) linkage points (i.e., portals) between life cycles, 2) the type and form of data passing between life cycles, and 3) conditions when life cycles interact and communicate. This paper discusses design aspects associated with a unified life-cycle analysis, which can support not only brownfields but also other types of assessments.

1 Introduction

The conceptualization of the problem is the Conceptual Site Model (CSM), which represents the analyst's understanding of the problem, problem components, spatial relationships, and flow of information between components. The real world is very complicated and the traditional way to approximate the real world is to simplify and compartmentalize it into more manageable "pieces." These generally represent what we understand, tending to group all of the things that we do not know or understand into a selected group of parameters; hence, our conceptualization of the real world tends to be a function of what we know and understand. The intent is to design a framework that allows this conceptualization to change and grow more sophisticated; as our understanding of the real world grows, so we may more accurately estimate the impacts associated with our anthropogenic activities.

Because the world is traditionally compartmentalized and the flow of information is from compartment to compartment, the basic conceptualization of the problem begins with a flow of information from beginning to end. Over the life of a

- life-form, which follows the life of human or other organism from its inception through growth and eventually to death. These models account for gender, race, economic standing, location, lifetime financial development, mobility, etc., and can be used to determine what factors create the greatest risk to the individual. [e.g., micro_environmental modeling (SHEDs, LifeLine, Mentor)] [1].
- process, which tracks the life of a manufacturing process and its products [e.g., production life cycle (LC_Advantage software)].
- activity, which tracks the life of an activity [e.g., waste site clean_up, certain job type (CEO, policeman, fireman)].
- idea, which follows the ramifications of an idea from its inception to closure [e.g., risk from implementing an idea (cultural revolution in China, Taliban concept of Islam, prohibition in the United States, etc.)].
- chemical, which follows the life of the chemical from its inception into the environment, through multiple media transport and fate, to exposure and risk to sensitive receptors [e.g., traditional risk assessment paradigm (TRIM, RESRAD, MEPAS, MMSOILS, 3MRA, ARAMS, MIMS)] [2],

a beginning and end can be defined, and, hence, a flow diagram can be constructed to link the individual components, whether they be models, databases, or other frameworks. Each of the "life cycles" listed above represents, in effect, types of frameworks. To capture the structural relationship between the principle components required in estimating human exposure to chemicals, multiple frameworks, representing existing legacy software, may need to be linked to provide the most scientifically defensible picture of the impacts to chemical exposure. Because the real world is compartmentalized, each of these life cycles is also compartmentalized, meaning that the various life cycles can be linked to address the demanding questions associated with the identification, facilitation, and communication of generic research that will characterize exposure to chemicals and raise the confidence and lower the uncertainty for quantitative estimates of exposure associated with potential effects to chemicals.

Unified Life-Cycle Analysis

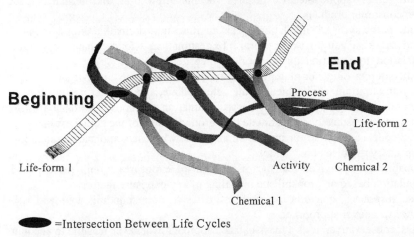

=Intersection Between Life Cycles

Figure 1: Abstract example of the linkage between different types of framework life-cycle ribbons

An abstract view of these various life-cycle "ribbons" is illustrated in Figure 1. Each of these framework ribbons represents a flow of information from the beginning to the end of an assessment. The Life form ribbon could represent micro-environmental modeling. The Chemical ribbons could represent the standard fate and transport risk assessment paradigm for two different chemicals, and where the chemical ribbons cross the lifeform ribbon (i.e., black dots), data are transferred. When there is no intersection and no data are transferred between ribbons, then black dots do not appear as the ribbons are actually skew. A similar case can be made for the Activity and Process framework ribbons, if applicable.

2 Example design requirements

Currently, there is great interest in designing an overarching software framework that is comprehensive, logical, and useful for industrial needs [3]. In particular, the framework needs to satisfy an objectively derived set of specifications, rather than merely duplicate existing modeling approaches. In particular, the chemistry industry wants the ability to adequately assess the risks associated with chemicals released into the environment. Researchers require tools for accurately estimating human exposure under a variety of exposure scenarios. As an example of linking multiple life-cycles, the chemistry industry is interested in the linkage of the standard risk-assessment paradigm (i.e, chemical life cycle) with that of the life-cycle of a human (i.e., micro-environmental modeling), containing the following requirements [3]:
• Comprehensive: applicable to exposure scenarios of interest
• Modular: consisting of modules (algorithms and databases) that can be easily updated and exchanged without affecting other parts of the framework

- User_friendly: PC application with a menu_driven interface
- Multi_route: applicable to exposures via inhalation, oral, and dermal contact with consumer products
- Multi_pathway: inhalation   (air_to_lungs);   dermal   (liquid_to_skin, solid_to_skin,   air_to_skin);   oral   (ingestion   in   food,   hand_to_mouth, inhalation_to_ingestion, air_to_food_to_ingestion)
- Multi_source: single or multiple compounds with the same target organ
- Varying duration: applicable to acute, intermediate, and long_term exposures
- Accurate: integrates state_of_the_art estimation methods and databases to estimate or reasonably overestimate the "ground_truth" of the actual exposure
- Open code: accessible for inspection and review by users and stakeholders (no proprietary or "black box" code)
- Probabilistic: provides realistic distribution of exposures within the exposed population based on probabilistic modeling of key exposure factors
- Dose_response: converts exposure estimates to corresponding dose and risk values whenever appropriate
- Mass_conservative: uses a mass-balance approach whenever feasible to account for fate and transport of pollutant mass.
- System User Interface – The System User Interface (SUI) represents the forum for visually describing the Conceptual Site Model (CSM). The CSM is a mechanism to convey the problem to the user in graphical form and directly interacts with the user, important to be user_friendly and relatively intuitive.
- Data Transfer Protocols – The data transfer protocols describe the foundation upon which data are universally transferred throughout the system and represent the heart of the framework.
- Sensitivity/Uncertainty Considerations – Utilizing the data transfer protocols, so sensitivity/uncertainty and parameter estimation models can be incorporated into the system.
- Model Space and Time Considerations – Protocol for linking disparate models in both space and time (e.g., disparate numerical models).
- System Integration Tools – To help facilitate the ability of linking disparate models, databases, and frameworks, a series of system software helpers (i.e., "editors") are required to step the model or database developer through the integration (and application) process.
- Server Side – Software tools, which allow the user to link to remote models and databases, are presented.
- Lock and Key – Software tools that allow a user to fix the CSM by locking icon types and connections between icons and lock the models that are available under each icon.

3 Illustrative example

The American Chemistry Council formulated a series of problems and exposure scenarios, which do not directly translate to a brownfields analysis but have many similar component types that one may see implemented [3]. Using one of these scenarios (i.e., indoor exposure), an illustrative example is presented that demonstrates the linkage of the standard risk assessment paradigm of fate, transport, exposure, and risk analyses (i.e., chemical life-cycle associated with human exposure) with that of the standard micro-environmental modeling paradigm, which follows the life stages of a human with its associated exposures and subsequent risks. By allowing these two disparate paradigms to seamlessly communicate, more detailed and accurate assessments can be performed to holistically view the impacts of contaminants that are released into the environment.

Table 1 presents a summary of the illustrative example outlining the exposure of Semi-Volatile Organic Compounds (SVOC) contained in plastics (e.g., furniture, wall coverings, food warp, pacifiers, nipples, toys, etc.) through the life-stages of a child from 0 to 18 years of age. The example addresses diffusion of SVOC in semi-porous or non-porous media, volatilization or vaporization of semi-volatile organic from surfaces, transport of airborne contaminants in the indoor environment, adsorption or absorption of SVOCs onto particulates, child activity/potential exposure (e.g., hand-to-mouth, object-to-mouth, and fate in child digestive and respiratory systems), and dermal and oral absorptions (fate in circulatory system, transport of plastic packaging and goods into home) [3].

Figure 2 presents a flow of information diagram, describing the general steps for implementing the exposure scenario as the analysis sequences through the various modeling systems. As this figure suggests, there are elements of the traditional risk paradigm (i.e., source releases, transport and fate, deposition, etc.) as well as elements of micro-environmental modeling (i.e., activity patterns and life stages).

Figure 3 expands the life-stage from fetus (conception) through newborn, infant, toddler, child, preteen, and teanager. Activities and exposure types vary by life stage, yet their aggregate and health effects vary and accumulate over the 0 to 18 year life cycle. Activity patterns, uptate rates, and exposure pathways and routes are potentially unique to the fetus/child from years 0 to 18. This situation is very similar to the different activity patterns associated with various cohorts, which can be impacted by brownfield sites. As people age, their activities change. As an infant, inadvertent soil ingestion and inhalation of vapors may be primary routes of concern. Older children may experience negative impacts from exposures to contaminants or solid waste. Activity patterns are also different, since older children are more mobile and tend to have access to more dangerous situations and constituents.

Some micro-environmental models are designed to explore and account for these life-stages and activity patterns. These models, though, were never designed to accurately account for transport and fate of contaminants in the environment or to address detailed mechanistically based impacts from exposure as pharmokinetic models might provide. The standard risk paradigm with their detailed fate and transport modeling tools can provide time and spatially varying contaminant concentrations, which then could feed a micro-environmental model at the appropriate life stage, corresponding to the appropriate activity pattern. Linking this exposure with a pharmokinetic model could provide the body concentrations to determine health impacts in a more accurate manner. In effect, the health effects can be more accurately assessed by combining the best attributes of the standard risk paradigm (e.g., transport and fate) with those of life-stage analyses.

Table 1. Example exposure scenario of a child (0 to 18 years of age) exposure at home to semi-volatile organic compounds [3]

Scenario Designed to Demonstrate: • Child exposure • Multimedia, multi-pathway, and multi-route exposure • Multi-chemical exposure • Intermittent source • Multiple receptor behavior patterns • Changing receptor characteristics over time (i.e., exposure patterns will change as child grows) **Residence Description:** • 1_story Townhouse with 1700 ft^2 in MA • Electric heat pump, forced air and AC • 3 bedrooms, 1.5 baths, kitchen, living and dining rooms **Exposure in Home to SVOC:** • Chemicals: phthalate esters (e.g., Diethylhexylphthalate [DEHP] and butlybenzylphthalate [BBP]) in plastic • Exposure to phthalates in plastic furniture covers, carpet backing, and shower curtains • Exposure to phthalates in plastic food wrap and containers • Exposure to phthalates in plastic toys **Child Age Time Lines:** Newborn (0 to 0.25 yrs); Infant (0.25 to 2 yrs); Toddler (2 to 3 yrs); Child (3 to 6 yrs); Preteen (6 to 12); Teenager (12 to 18)

Figure 4 provides a more detailed view of the modeling structure, trying to maintain descriptive differences between the transport and fate, and life-stage analyses, which are highlighted in gray. The parallelograms represent data supplied by user-input or databases (e.g., EPA Exposure Factors Handbook, EPA Child Exposure Handbook, IRIS, HEAST). The rectangles represent models with the arrows showing the quality and quantity of information flowing from one component to the next. When linked with the Physiologically Based PharmacoKinetic or Pharmakodynamic (PBPK or PBPD, respectively) models, the micro-environmental modeling can provide a more accurate view of impacts to organs and body burdens for calculating risk to the individual. TherdbASE is EPA's Total Human Exposure Risk database and Advanced Simulation Environment, which is an integrated database and analytical modeling system used in exposure-assessment calculations.

Figure 5 provides the most detailed flow-of-information diagram, associated with the modeling sequences for this problem. Figure 5 shows the sequential linkages between source, environmental transport, exposure scenario, pharmokinetics, effects and impacts modeling. The volumetric cylinders represent user-supplied input or databases, rectangles represent models, and ellipses represent information transferred between components. This figure focuses less

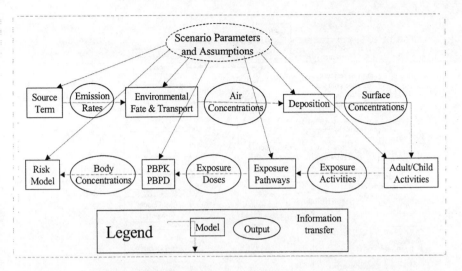

Figure 2: Information flow diagram linking fate and transport modeling with child/adult activity relationships [3]

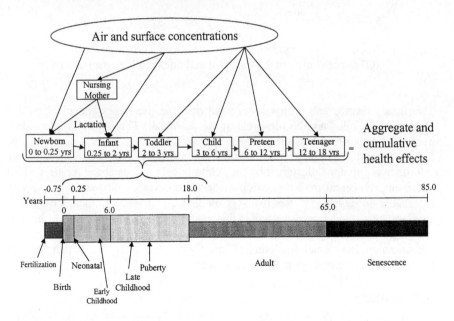

Figure 3: Expanded life-stages from fetus through newborn, infant, toddler, child, preteen, and teanager [3]

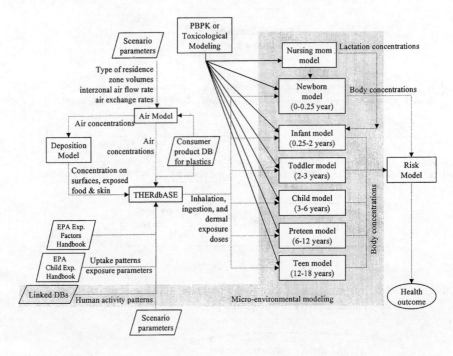

Figure 4: View of the modeling structure, trying to maintain descriptive differences between the transport and fate, and life_stage analyses

This figure focuses less on the activity patterns and life-stages and more on the mechanics of following the life-cycle of the chemical. This figure is interesting because it demonstrates that the linkages between independent modeling systems (i.e., fate and transport versus micro-environmental) can occur through an information portal, as illustrated by the "child 0 to 18 yr" database (or file). The micro-environmental modeling can provide the necessary information to support the PBPK modeling. If the focus is on fate and transport, then the micro-environmental modeling can support this assessment. If the focus is on human life-stages and activity patterns, then the fate and transport can be used to support the micro-environmental modeling. This figure shows that each independent modeling system can support the other system.

4 Summary

A decade ago, researchers were touting the gains of linking individual models together to conceptualize and capture a more holistic understanding of how anthropogenic activities have impacted the environment. Like the individual models that preceded them, no one framework can adequately address all situations; hence, frameworks are now being linked in a manner that models used to be linked [2].

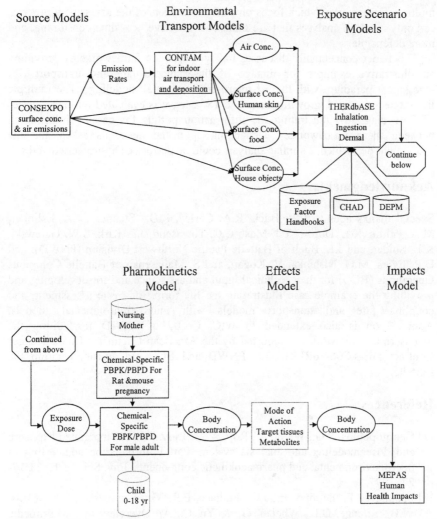

Figure 5: Flow Diagram showing the sequential linkages between source, environmental transport, exposure scenario, pharmokinetics, effects and impacts modeling [3]

A brownfield site analysis may require many tools to complete an accurate assessment, as to the potential disposition of the site. Many of the sites that sit empty or idle today were initially developed because of their prime location and easy accessibility to transportation systems and utilities. The potential value of these properties is a function of whether they can be reintroduced as commercially viable entities. Evaluating the ramifications of the no-action alternatives and comparing them to the impacts of remediating these sites provides a clearer understanding of the decisions that are made. Having the ability to couple multiple frameworks, which focus on different aspects of the assessment process, can only result in analyses that are more cost effective, less time consuming, and more defensible.

This paper conceptually discusses linkages between frameworks by providing an illustrative example for linkage of the standard fate and transport risk assessment paradigm with that of micro-environmental modeling. The multiple life stages of a child, growing from 0 to 18 years, was evaluated through a series of flow diagrams, illustrating the information portals for passing information between these frameworks. Depending upon the questions asked and the assessment performed, a similar analysis could be performed at brownfield sites.

Acknowledgments

Special thanks goes to J.W. Buck, R.A. Corley, R.D. Stenner, M.A. Eslinger, M.A. Pelton, K.E. Dorow, T.J. Mast, C.C. Townsend, J.L. Kirk, S.W. Gajewski, K.L. Soldat, and J.L. Buelt of Battelle-Pacific Northwest Division (PNWD), and D.P. Evers, M.G. Nishioka, V. Kogan, and S. Mahasenan of Battelle Columbus Operations (BCO) for their technical input and assistance and for developing and providing the example case illustrating the life-form, life-cycle assessment, and coupling fate and transport models with micro-environmental models. Appreciation is also extended to W.C. Cosby of PNWD for editing the manuscript. This work is supported by the American Chemistry Council under Contract DE-AC06-76RL01831. PNWD and BCO are divisions within the Battelle.

References

[1] Georgopoulos, P.G., Walia, A., Roy, A., & Lioy, P.J., An integrated exposure and dose modeling and analysis system. Part I: formulation and testing of microenvironmental and pharmacokinetic components, Env. Sci. Tech., 31:17-27, 1997.
[2] Laniak, G. F., Droppo, Jr., J.G., Faillace, E.R., Gnanapragasam, E.K., Mills, W.B., Strenge, D.L., Whelan, G., & Yu, C., An Overview of a Multimedia Benchmarking Analysis for Three Risk Assessment Models: RESRAD, MMSOILS, and MEPAS, *Risk Analysis*, **17(2)**:203_214, 1997.
[3] Buck, J.W., Tolle, D.A., Whelan, G., Peffers, M.S., Evers, D.P., Corley, R.A., Eslinger, M.A., Pelton, M.A., Townsend, C.C., Stenner, R.D., Mast, T.J., Nishioka, M.G., Kogan, V., Dorow, K.E., Kirk, J.L., & Mahasenan, S., *Design of the Comprehensive Chemical Exposure Framework for American Chemistry Council*, Battelle-Pacific Northwest Division, Richland, Washington, 2002.

Section 8:
Lessons from the field

Public policies that foster contaminated land recycling– expanding the horizon

C. M. Morgan & P. A. Brown
Marasco Newton Group, Ltd.

Abstract

This paper identifies the public policy levers that have been developed to address brownfields cleanup and reuse and asks whether the best practices identified to date are sufficient to address the problems posed by individual brownfield properties as well as the larger local and regional challenges implicated by these properties. More specifically, this paper suggests that the current, limited linear approach to brownfields cleanup and reuse (i.e., site identification, assessment, cleanup, and redevelopment) is insufficient to meet the complex problems associated with brownfields cleanup and sustainable reuse. Despite the great advances seen in many local, regional, and national reuse programs, the creation of new brownfields is still exceeding the number of sites reclaimed. The purpose of this paper is to propose and explore a more comprehensive, institutionalized approach to brownfields that addresses prevention and regeneration to ensure that the number, size, and complexity of "new" brownfields created is minimized as much as possible.

Introduction

Virtually every country—regardless of size, population, geographic characteristics, political structure, legal system, industrial maturity, economic base—has or will face the issues associated with contaminated properties that are underused or have been abandoned. While the terms used to describe these properties vary from country to country (e.g., brownfields, derelict lands, vacant lands, contaminated properties, *friches*), the environmental, social, cultural, and economic challenges posed by these properties vary only in detail and scope. Many countries with mature or declining industrialized economies have spent the last several decades addressing the legacy of contaminated properties left behind after the manufacturing plants and warehouses have closed down or moved on. Others, still in the midst of developing or expanding their industrial economies, have yet to face these issues. The industrialized countries have an opportunity to develop a holistic brownfields prevention and sustainable regeneration model that can apply to industrialized and developing countries alike.

Practically every country has its own way to define a brownfield site. In fact, individual countries may have multiple definitions, depending on the national or local agency addressing the issue. For purposes of this paper, we will rely on the baseline definition recently created as part of the new American brownfields law which defines a brownfield site as "real property, the expansion, redevelopment, or reuse of which may be complicated by the presence or potential presence of a hazardous substance, pollutant, or contaminant." [1] These sites can range from small parcels housing former dry-cleaning facilities or gas stations to large, multi-hundred acre sites that once served as mining operations, port facilities, landfills, or steel manufacturing plants. Although there are no reliable quantitative data on the magnitude of the brownfields problems faced by industrialized nations, most experts agree that the problem is pervasive. For example, some estimates of the number of brownfield properties include 450,000 in the U.S., 200,000 in Germany, and 100,000 each in France, the Netherlands, and the United Kingdom. [2]

The magnitude of the brownfields problem is matched only by its complexity. The reasons brownfields are "created" are as numerous as there are sites. While there are typically specific issues that force the abandonment of any given site, there are a several causes common to brownfields creation, including: market forces, including shifts toward more service-based operations, movement of industrialized operations to other regions or countries, individual bad business decisions, information asymmetry; public policies, such as increased environmental regulations, incentives to develop in other areas, impacts from transportation or other infrastructure decisions; and other societal or cultural factors, including demographic shifts or movement of workforce populations.

Similar forces drive the continuation of brownfield sites in abandoned or idled states. Both the creation and continuation of brownfields contribute to decreased property and tax values, foster vandalism and crime, lead to blight and slum conditions, and promote urban sprawl. On the positive side, brownfield sites typically benefit from excellent urban core and/or waterfront locations, provide easy access to transportation and other public infrastructure, frequently provide historical or cultural connections for the community, and offer proximity to readily available workforce pools.

Under some circumstances in market-driven economies (e.g., robust economic growth, sites in prime locations, sites with little or no actual contamination), market forces will return a brownfield back to productive use. Because this paper focuses primarily on the programmatic responses by local, regional, and national governments, however, we do not address those sites that are reclaimed by Adam Smith's so-called "invisible hand." Rather, this paper addresses the programs and policy levers that have been developed to reclaim those sites that require public intervention. Because the specific causes vary by location and the associated legal, economic, environmental, and other social factors, the appropriate public policy solutions required to remedy the creation of brownfields also vary. To a large extent this paper focuses on the environmental issues since they are the issues that distinguish a brownfield site from other abandoned or vacant properties.

Public policy responses to date: A limited linear model

Many local, regional, and national programs have been modified or created to respond to the multi dimensional problems posed by brownfields. The programs and specific policy levers are as varied as the issues they seek to address. Most of the programs specifically targeted to redevelop brownfields have only arisen during the last five to ten years. Other beneficial, but non-targeted programs have been in place for decades. Perhaps the longest running land reclamation program is the United Kingdom's which clearly identified the need for public policy to address the vacant land problem beginning as early as the 1960's with the London Docklands redevelopment project. [3]

Although there are as many differences as commonalities across these programs, there are several developing trends that have proven application across many brownfields redevelopment programs. Several studies have been undertaken to identify best practices and tools being used among the brownfields redevelopment programs. [4] These programs offer a range of policy tools to promote brownfields reuse, including:

- *"Facilitation" services*–from public land acquisition and assembly, to fostering dialog among all the interested stakeholders. For example, in the Netherlands has used a high degree of public intervention to reclaim brownfields including acquiring the land and being heavily involved in reuse planning.
- *Financial incentives*–from direct funding for environmental assessments and market analyses, to tax credits and other incentives that promote land regeneration. For example, the U.S. Environmental Protection Agency's (US EPA) Brownfields Initiative provides grants for site assessment and cleanup.
- *Policy changes*–from clarifying environmental liability issues, to encouraging sustainable transportation and other infrastructure decisions. For example, Germany has established national uniform criteria for brownfield risk assessment and cleanup.
- *Information resources and technical assistance*–from innovative cleanup technologies, to sector-based reuse guides. For example, the United Kingdom has conducted several "derelict land" surveys to identify the scope of the problem and to ensure resources are targeted to needed areas. [5]

In addition, several "best practices" with broad transferability have been identified, including:

- *Utilizing Program or Project Champions.* Frequently, brownfields programs and projects fall or rise because of the energy and excitement created by one or more public or private sector leaders.

- *Integrating the Cleanup with the Reuse Planning.* Perhaps the most important best practice has been a greater consideration of the next use of the property (e.g., residential, commercial, recreational, industrial) when making cleanup decisions. Additionally, the actual design and implementation of the remedy can provide great time and resources savings when dovetailed with the reuse plans.
- *Taking a Coordinated, Multi-Disciplinary Approach.* Most projects require resources to address economic, environmental, and other social issues such as transportation and workforce training. This frequently requires horizontal approaches among a variety of public agencies (e.g., environmental, public health, economic development, planning, transportation) as well as vertical coordination (e.g., local, regional, and national agencies).
- *Forming Partnerships.* Because there are frequently so many issues and stakeholders impacted by a typical site cleanup and reuse project, public-private partnerships (e.g., government officials, lenders, property owners, developers, environmental professionals) are an integral part to almost every brownfield transaction. In addition, in most locales ensuring some type of community participation in cleanup and redevelopment decisions will frequently save time and money.
- *Using Public Resources to Leverage Private Investments.* Although sometimes a property will be returned to productive reuse solely through the use of public resources, it is often beneficial to use the public resources to stimulate the market to take action and carry the site through to reuse. [6]

Typically, these program and policy levers address the brownfields redevelopment challenge through a limited linear model, beginning with the identification of a site and ending with its redevelopment.Figure 1 illustrates this model. While each individual brownfield transaction is much more iterative than depicted by this model, the model is intended to reflect the institutionalized, public sector programmatic approaches that have been developed to address brownfields reclamation. For the most part, these programs have fairly defined start and end points–site discovery and reuse, respectively.

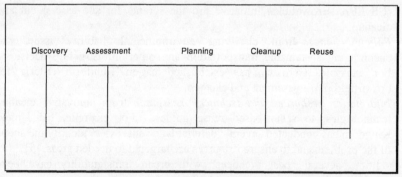

Figure 1. The Linear Brownfields Redevelopment Model

Given the fairly nascent state of most local, regional, and national brownfields redevelopment programs, this model has been very successful at initially framing the problem and accomplishing some impressive early successes.

Limitations of the linear redevelopment model

Despite these successes in reclaiming some of the world's brownfields, these early works and this limited linear model just scratch the surface in recycling these lands. Although there are no quantitative data to demonstrate the number of sites that are in various stages of this linear pipeline (e.g., how many have been identified, how many of those have been assessed, how many of those have been cleaned up, how many of those have been reused) our experience and research indicate that, at least in the US, the vast majority of sites are just entering the cleanup phase and that only a small minority have actually been cleaned up and put back into productive reuse.

While this early linear model has helped serve as a preliminary framework by which to initiate brownfields redevelopment, there is much more work to be done. Other brownfields scholars have lamented the rate of brownfields creation versus the rate of brownfields reclamation. [7] We concur and suggest that the narrow focus of the linear pipeline fails to account for two key points that continue to feed the creation of new brownfields. First, the beginning of the linear pipeline currently focuses on sites that have been contaminated and have been left abandoned, idle, or underused. This focus excludes a large source of potential brownfields that are poised to enter the pipeline without some sort of public policy focus. By focusing more on these sites and working to prevent the brownfield conditions, the volume in the pipeline will be greatly reduced.

Similarly, the end of the current linear approach aborts with the initial redevelopment. Apart from certain targeted, but random projects–the continuing sustainability of the site's reuse is not a key factor being systemically addressed by public policy. While this approach promotes temporary redevelopment and gets the sites off the brownfields lists, there are no institutionalized policy levers to minimize the chance of these sites reentering the brownfields pipeline. The market failures and public policy decisions that have lead to the world's current inventory of brownfield sites can rise again to flip these properties back into their current unproductive state unless more concentrated policy efforts are undertaken.

A holistic perspective: The brownfields prevention and regeneration model

As the previous discussion suggests, the prevailing perspective in the field of brownfields is one that is focused predominately on a narrow segment in the life cycle of contaminated properties redevelopment–from the point that a brownfield property is discovered to the point of its initial reuse. This perspective, while appropriate for addressing the immediate needs of the brownfields problem, often fails to see the larger context in which that problem exists. The key to this problem when viewed in a more holistic manner is centered not on such issues as environmental liability, cleanup, and land transfer, but rather on two key elements often not considered in the contemporary brownfields life cycle: prevention and sustainable regeneration

We do not propose that brownfields can be totally eliminated. Instead, we assert that the number, size, and complexity of brownfields being created and entering into the linear pipeline can be greatly reduced by applying some of the same focus, resources, and best practices described above to undertake more holistic, institutionalized approaches to preventing new brownfield sites and ensuring that the redevelopment of a site is conducted in such a manner to promote its sustained, continued use.

The prevention of brownfields is a concept that receives little attention in the field in large part due to the complexities of the root causes inherent in their formation. However, when controlling for the larger economic and demographic cycles that commonly occur in market-based societies, what separates a brownfield property from any other vacant, formerly used property is defined in large part by the real or perceived threat of contamination on that property. The formation of brownfield properties can be prevented, in large part, by the manner in which those properties are managed leading up to and during their "closure" (i.e., the cessation of operations requiring compliance with waste management regulations) and whether measures are taken to prevent, minimize, or properly manage environmental contamination. The emphasis on pollution prevention is important because it focuses on the proper closure of waste-generating facilities, the primary source of potential new brownfield properties.

Sustainability in the brownfields context, on the other hand, is a topic for which a lot has been written, although much of this body of work tends to be abstract in nature and addresses sustainability on a global or societal level. A review of the literature suggests that brownfields sustainability is typically thought of in terms of either process or outcome. The sustainable brownfields prevention and regeneration model is one that integrates ecological remediation, economic development, and social equity into an overall strategy that emphasizes land use within the boundaries set by the local communities' resources, and its current and future goals. [8] The goal is often described as a sustainable outcome whereby the reuse of the property is maintained in an ecologically sound manner over the long term to meet the needs and capacity of the community. The emphasis on ecologically sustainable reuse outcomes is important because it ensures that the property does not become a brownfield once again, the second source of potential new brownfield properties.

By including the concepts of prevention and sustainable regeneration into the life cycle of the brownfields model, a new holistic perspective is possible which expands upon the widely held view that brownfields development begins with site discovery and ends with reuse. As Figure 2 illustrates, the current linear brownfields redevelopment model falls within a larger life cycle model. This broader brownfields prevention and regeneration model takes into account the multi-generational life of a parcel of land. It readjusts the conceptual framework that currently applies public policy resources only after a parcel of land has been contaminated and abandoned (i.e., identified as a brownfield) and ceases those public policy levers upon the initial reuse. The implication of this new model is important in the brownfields arena because it shifts attention to those areas that are central to the creation of brownfields, yet typically are not well understood and rarely systemically addressed in a methodical or programmatic manner.

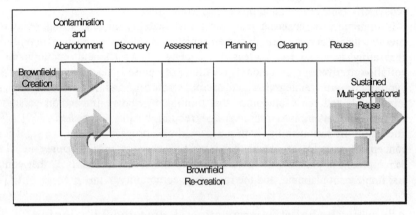

Figure 2. The Brownfields Prevention and Regeneration Model

Key elements of the brownfields prevention and regeneration model

The conceptual expansion of the brownfields life cycle to include considerations of brownfields prevention and sustainable reuse outcomes has several implications for public-sector brownfields programs. Specifically, adopting a more expansive view towards brownfields life cycle management suggests that government programs need to develop an institutionalized approach to brownfields prevention and regeneration. Fortunately, many of the best practices that have arisen from the more traditional brownfields programs (e.g., utilizing champions, taking a multi-disciplinary approach, forming partnerships, using public resources to leverage private investments) can be applied under this broader model. The key is to systemically address prevention and regeneration with the same focus, creativity, and flexibility that has been applied in developing the original brownfields redevelopment-focus approach. Other more targeted elements to focus efforts along the entire continuum of brownfields life cycle management might include:

- Pollution prevention measures that are taken at active facilities prior to and during closure to preempt the formation of new brownfields.
- Improved methods of implementing, monitoring, and enforcing mechanisms for managing contamination remaining on property and the use of the land.
- Recognition that the reuse of a brownfield extends beyond the initial reuse of the property and should be viewed as multi-generational.
- Incorporating ecologically sustainable practices into the reuse to prevent the

re-creation of brownfields.

From an ecological standpoint, brownfields can be prevented if the reuse of the property is sustainable over the long term and only if it is not degrading its environment or using up finite resources. Environmental concerns include protecting human and environmental health; having healthy ecosystems and habitat; reducing and/or eliminating pollution in water, air, and land; providing green spaces and parks for wildlife, recreation, and other uses; pursuing ecosystem management; and protecting biodiversity.

Strategies for ensuring ecological sustainability include using renewable energy sources at a rate less than their replacement, less dependence on pollution generating technologies, increased recycling of wastes and employment of pollution prevention methods, creation of closed resource loops where byproducts are reintegrated as useful resources and waste products are minimized, and reuse outcomes that minimize negative impact on ecological systems and attempt to preserve and rebuild healthy ecosystems. [9] The integration of sustainability with the brownfields process represents a significant conceptual shift to the extent that it reflects changes in the manner in which development is planned, the organization of the social mechanisms that control and implement planning, and the role of the community in that process. [10]

Adopting the holistic perspective: A piecemeal beginning

This holistic perspective toward brownfield prevention and regeneration appears to be gaining a foothold in several countries with established brownfields redevelopment-based programs.   There are many small-scale initiatives and projects that envision a public policy role in preventing and sustainably regenerating brownfields. Unfortunately, these projects are fairly random and piecemeal.   Together with the established best practices under the linear brownfields redevelopment model, however, these projects can serve as models upon which to develop a more systematic, institutionalized approach under the broader brownfields prevention and regeneration model.   Two examples of projects that have adopted elements of the holistic perspective to brownfields prevention and regeneration include:

- Emscher Park regeneration projects in the Ruhr valley of northwestern Germany, which are fostering the ecological and economic regeneration of hundreds of sites in the heavily industrialized region. [11]
- Waterfront Regeneration Trust projects in Toronto, Canada, which are integrating the management and conservation of natural resources with the redevelopment of contaminated properties, including a design template for building green infrastructure throughout Toronto's former industrial waterfront. [12]

Examples of how several organizations are beginning to adopt the emerging tenets of a holistic brownfields perspective include:

- The Groundwork Trust project in the United Kingdom has adopted an approach that embraces the concept of brownfields prevention by helping business parks upgrade their properties and physical surroundings as a way

to decrease vandalism and illegal activities, increase business, and bring in new tenants before such sites become vacant. [13]
- The American Society for Testing and Materials (ASTM) has published a standard guide that provides a flexible approach to facilitating the redevelopment of brownfields that embraces, among other things, an exit strategy at the site that ensures that new brownfields sites are not re-created; in other words, sustainable brownfields redevelopment. [14]
- The US EPA is implementing the RCRA-Brownfields Prevention Initiative with the goal of preventing brownfields and ensuring the successful cleanup and long-term, sustainable reuse of manufacturing facilities.
- US EPA Brownfields Initiative provides grants to communities to promote ecologically, economically, and socially sustainable reuse outcomes. When selecting projects for funding, US EPA specifically seeks proposals that link brownfields activities with other community empowerment, sustainable development and community livability efforts, and provides additional funding when these properties are targeted for green space uses.

Conclusion

The current, narrow perspective to brownfields redevelopment that is predominant in most established brownfields programs is limited in its ability to address the formation of new brownfields. By adopting a more holistic view of the brownfields life cycle, public programs can better respond to the challenges posed by the continued creation of brownfields despite great advances seen in many local, regional, and national reuse programs. The beginnings of such a holistic perspective are starting to find their way into individual redevelopment efforts, non-governmental initiatives, and even several government programs. However, the key to the more sustainable land management, including preventing brownfields—whether from the recent closure of contaminated properties or the re-creation of brownfields on properties once cleaned up and reused—rests on expanded public-sector involvement aimed at pollution prevention and sustainable reuse. To this end, scholars, public policy makers, and all brownfields practitioners must seize upon the best practices and tools developed under the current programs and begin to develop a systematic approach to institutionalize and implement the more holistic perspective envisioned under the brownfields prevention and regeneration model.

References

[1] See 42 U.S.C. Section 9601(39)(A).
[2] Organization for Economic Cooperation and Development (OECD), *Draft Interim Urban Brownfields Report*, May1998; and United States General Accounting Office, *Community Development–Reuse of Urban Industrial Sites*, GAO/RCED-95-172, 1995.
[3] Barry Woods, *Vacant Land in Europe*, Lincoln Institute of Land Policy

Working Paper. 1998.

[4] OECD, 1998; Woods, 1998; Detlef Grimski and Uwe Ferber, *Urban Brownfields in Europe*, Journal of Land Contamination and Reclamation, Volume 9, Number 1.   2001; Colin C Ferguson, *Assessing Risks from Contaminated Sites: Policy and Practice in 16 European Countries*, Journal of Land Contamination and Reclamation, Volume 7, Number 2.   1999; Joseph Shilling, *Sustainable Brownfields Redevelopment*, Proceedings from Mayors' Asia Pacific Environmental Summit,1999; Charles W. Powers, Frances E. Hoffman, Deborah E. Brown, Catherine Conner, *A Great Experiment: Brownfields Pilots Catalyze Revitalization*, The Institute for Responsible Management.   2000; International City/County Management Association (ICMA), *Brownfields Blueprints: A Study of the Showcase Communities Initiative*, June 2001.

[5] Ibid.

[6] Ibid.

[7] OECD, 1998; Woods, 1998.

[8]   US EPA, A *Sustainable Brownfields Model Framework*,   Publication No. EPA 500-R-99-001, January 1999.

[9]   US EPA, *Characteristics of Sustainable Brownfields Projects*, Publication No. EPA 500-R-98-001, July 1998.

[10]   Lachman, B.E., The RAND Corporation, *Linking Sustainable Community Activities to Pollution Prevention: A Source Book*, Publication No. MR-855-OSTP, 1997.

[11]   US EPA and the ICMA, *The International Building Exhibition (IBA), Preserving Open Space and Our Industrial Heritage Through Regional Brownfields Redevelopment: An International Brownfields Case Study*.

[12]   US EPA and the ICMA, *Waterfront Regeneration Trust, Integrating Eco-System Management with Brownfields Redevelopment and Local Land Use Planning: An International Brownfields Case Study*.

[13]   Groundwork UK, *Groundwork 2002: Building Sustainable Communities Through Joint Environmental Action*.

[14]   American Society for Testing and Materials, *Standard Guide for Process of Sustainable Brownfields Redevelopment*, Guide E 1984-98, January 1999.

More alike than different: local collaboration on brownfields issues in border communities

M. Singer
The International City County Management Association
United States of America

Abstract

In the Unites States (US) and Mexico there are many well-publicized national borders initiatives. Ultimately, land use and economic development decisions are made on a local level. For these reasons a regional approach to border brownfields redevelopment makes sense, the question is whether or not the geopolitical differences between the countries would be assets or challenges. This presentation looks historically at the dynamics, changes in populations and demographics over the past quarter of a century and the impact that they are having on environmental, social and economic factors today along the US Mexico Border.

The issues of simultaneous growth and decline, the changing economic base, outward growth on the cities' boundaries, increased transportation demands are the very characteristics that mark the regional brownfields issues. In the Juarez-El Paso region, even the periphery properties where new industries are moving have many brownfields issues. The lack of infrastructure in the barrios has created unregulated businesses such as auto-recyclers and illegal dumping. These environmental issues are pasted on top of significant social issues that are a part of many of the root problems.

Another challenge to the regional perspective is that of jurisdictional authorities and the dynamic between the federal roles of maintaining and enforcing policies and the state and local approach of solving problems through creative measures that depend on personal acquaintance, agreements and collaboration. As the regional perspective suggests and as the statistics support, there are more elements of life, economy and the environment that are shared between the border communities than not. For these reasons it makes sense to share strategies and resources in brownfields redevelopment on efforts, rather than work in isolation, ignoring how co-dependent the border communities are on each other.

Introduction and overview

On January 30-31, 2002, the International City County Management Association (ICMA) convened a group of local, state and federal officials as well as representatives from the private sector, non-profit and community organizations from the United States (US) and Mexico. The gathering was a two-day Forum to discuss local approaches to brownfields redevelopment along the US Mexico border, specifically looking at the region that includes El Paso, Texas in the US and Juarez, Chihuahua in Mexico. The goal of the Forum was to begin local dialogues among stakeholders to talk about the numerous and intersecting issues around brownfields and the redevelopment of industrial sites.

Why did we do it?
This Forum was one part of a number of ongoing projects being conducted by a range of organizations that include: studies, programs, working groups, technical assistance and conferences as both the United States and Mexico federal governments are paying closer attention to border issues. In spite of the numerous border initiatives, ICMA's marked the first that was focused on local governments and stakeholders as the agents for engaging in cooperative redevelopment. Why Have a Local Focus? Ultimately, land use and economic development decisions are made on a local level. The nuts and bolts of building better communities does not happen through national legislation alone. It happens by colleagues sharing ideas and resources to effect a change.

A regional perspective
Along the US Mexico border, the two countries function as one region, including shared and co-dependent economies, cultures and challenges. Over the last decade, with the passage of the North American Free Trade Agreement (NAFTA) which expedites the passage of raw materials and manufactured goods across the border, opening trading routes among North American countries, particularly US, Mexico and Canada much of the industry and strength of the economy has moved from the US to Mexico. Likewise, while the population of the El Paso-Juarez region has grown 35% in a decade, the city of Jaurez has increased by 52%. The population of the two cities is over two million. These changes are having an impact on environmental, social and economic factors today.

Geopolitical borders, created as lines of division, separation and protection also have crossing points which become shared spaces, creating a sense of connection and community. Along borders these points become blurred both in their geographical sense and in their intention. Is the crossing point a separation place or a connecting place? Borders as an area are continually changing functions and cultures, and interests, struggling between a zone and a line.

Border as a line   As a line, the border is a defense point, the US military guards against drugs, immigrants and other security threats. Mexico guards against firearms and commercial contraband.. Under NAFTA, the border, as a line, punctuated by crossing points, has been asked, in effect to fulfill two completely contradictory functions. On the one hand, it is to demarcate, separate, and prevent the passage of drugs and unauthorized people. On the other hand, it is designed to facilitate the passage of legal money and merchandise for the benefit of both countries. That is, to be closed and open simultaneously.

Border as a zone   Zone issues that are prevalent on both sides include: rapid population growth, a young population, low levels of education and high levels of poverty accompanied by high mortality rates and an abundance of health issues. The rapid population increase, the economic base (shifting from mining, agriculture and textiles on the US side to *maquiladoras* in Mexico) and social and cultural changes have left a population with varying abilities to attain basic employment, health care and food.

Closely related to the demographic data that lend themselves to thinking of the region as a zone, is economic data. The economy of the border zone has also grown enormously, especially in areas related to the *maquiladora* industry. The number of plants in Juárez alone has surpassed 300, generating approximately 250,000 jobs. Because almost all *maquiladora* products are destined for further assembly or sale in the US, cross-border traffic has mushroomed, for both goods and people. The total value of *maquiladora* exports has been put at approximately $19.3 billion dollars, including materials and value-added labor. While these numbers demonstrate that the economic worth of the region has grown, in reality the population has not prospered as one might expect. El Paso has become poorer, education levels have dropped, and the city is further behind other national indicator levels than it has ever been. Cd. Juárez is in many ways much better off than Mexican national averages, but is often said to have achieved this at a cost that is yet to be reckoned in infrastructure and quality of life issues.

Regional brownfields issues   These very issues, simultaneous growth and decline, the changing economic base, outward growth on the cities' boundaries, increased transportation demands are the very characteristics that mark the regional brownfields issues. The regional brownfields issues include: inadequate infrastructure to address changing needs, urban-core sites that have been abandoned or underutilized as industry has left or moved to the periphery of the city. The spread from the urban core to the edges of the city are bumping into and displacing pockets of poverty. Unlike other urban settings, *colonias* or *barrios clandestinos*, the unplanned, unregulated, residential areas created by the extreme need for low cost housing for the many immigrants arriving in the area, are built on the fringes of the city, not in the urban core.

Individual land recycling efforts

To understand and explore border collaboration on brownfields issues it is important to see what each community is doing in this effort. The point of this discussion is not to fully document each side's efforts, rather to point out the noteworthy aspects of them.

El Paso, Texas

In El Paso, there are two institutions that are formally addressing brownfields the City of El Paso and the Rio Grande Council of Governments (COG). Each organization has a formalized brownfields process, advisory boards and tries to attract investment in abandoned or underutilized properties and tries to work with current property owners to remediate sites. Both programs in El Paso sit squarely in the economic development and real estate arenas. While the City's program is focused in the poverty zone, the programming is not specifically targeted at poverty abatement, housing, public health or other issues. Rather, it is focused on site reuse, and secondarily on job creation. The COG program seeks to identify the properties with the highest potential for economic redevelopment, and assist property owners in assessing environmental contamination and plan for reuse. Both institutions, the City of El Paso and COG, typify US programs in their partnership building on the local level and in their connections with state and federal agencies. In these ways, the local programs have a network for technical assistance, fiscal incentives, liability relief, federal tax deductions, grants for assessment and outreach and other. Both the City of El Paso and COG have received US EPA Brownfields Assessment Pilots.

The tools used in El Paso uses are fiscal incentives, supplied through bonds, or through grants and loans from state and federal levels. The City of El Paso is seeking assistance from federal and state agencies to assist with its infrastructure needs. Both El Paso groups have an informal network for information sharing and feedback on brownfields issues, but there are not shared resources among stakeholders at the local level. Representatives stress their need for and use of federal and state funds to increase the city's capacity to lay the ground work for a long-term brownfields strategy that includes site inventories, widespread stakeholder involvement and prioritization for assessment and cleanup.

Juarez, Chihuahua

Juarez does not have any local, state or federal brownfields programs. The term brownfields does not even exist in Mexico. Nonetheless, Juarez works with state and federal agencies and integrates cleanup and land recycling efforts into its land use planning, infrastructure development and economic development. Most recently Juarez has been particularly focused on greenspace creation from contaminated parcels. Mexico in general, has a new industrial history, relative to the US, and is taking steps to prevent future instances of brownfields. There are programs on the local and national levels that address brownfields prevention. Locally, Jaurez is targeting its efforts to stop development from moving to greenfields. In 1999, the City of Juarez established an incentive program for landowners of greenfields to give their land "protection certification" wherein it will not be developed for industrial or manufacturing uses. This program encourages land recycling by limiting the amount of land available for development.

The Federal government is also working to address abandoned and under utilized sites before their histories and owners become lost and forgotten. As a part of that *Procuraduria Federal de Proteccion de Ambiente* (PROFEPA), the enforcement arm of Mexico's environmental agency, *Secretario de Medio Ambiente y Recursos Naturales* (SEMARNAT) in conducting a survey which documents abandoned maquiladora sites, contamination, existing infrastructure and other issues. Additionally, the law in Mexico states that before property is abandoned, the owners must go to PROFEPA to get approval to leave the site. If there is no contamination, property owners can leave the site. If there is contamination, owners must clean the site before departing, leaving its original state. In this program, PROFEPA prescribes the remediation. As with any environmental programs, there are flaws in this one. The laws are not very precise in what types of assessments are requires. And, like in the US, it is difficult to track down absentee landowners.

Urban redevelopment specialists in Juarez have been very creative in their use of tools and technologies to examine a number of cross-cutting topics such as population density, infrastructure, growth rates, and health and environmental issues. In particular, practitioners have taken advantage of geographic information systems (GIS) and remote sensing to overlay factors. Use of such tools demonstrates the ability to use existing data to look at land use, the environment, and economics in new and better-integrated ways. By using the remote sensing data, planners can literally step back and look at the big picture to gain a broader perspective of the scope of the issues facing the border region. This perspective facilitates a regional approach to redevelopment.

Juarez's interest in brownfields prevention and its use of technology to look broadly at existing issues as well as to model for future changes demonstrates the advantage of integrating brownfields approaches across a number of programs and agencies. Practitioners working in one department do not feel compelled to look at the inter-related social, environmental and economic issues through a single lens which could happen if a single agency is tasked to address the issues.

Shared local issues
In spite of all of the differences between brownfields cleanup in the United States and land recycling efforts in Mexico there is one very strong similarity— cleanup happens at the local level. Land use decisions are made locally and

programs to encourage appropriate land uses are developed by the local government and local stakeholders. Both cities have gone through varying economic turns that have created brownfields settings. Both cities understand that one of the underlying issues for any sustainable redevelopment is a greater investment in infrastructure development and repair. These shared approaches: local land use decision-making, concerns over economic development, infrastructure development and efforts to curb sprawl, coupled with individual strategies and tools serve as a basis for more directed collaboration and strategic planning in a way that takes advantage of the many qualities that both sides of the border share, rather than pitting them against each other. Collaborative approaches to land recycling are also beneficial to the border communities because that approach complements the mutually dependent aspects of life on the border, including shared economies, markets, job-bases, housing, social and cultural life. Since so many of the qualities that are embedded in brownfields issues are inter-dependent at the local level across the border, a comprehensive cross-border approach to clean up at the local level should be considered. But, what are the drawbacks?

Towards collaborative approaches

Individual challenges in brownfields work
These issues limit the success of individuals and programs in brownfields redevelopment.

Information and technology There are a number of information and technology issues raised that include establishing and maintaining networks, developing institutional knowledge, establishing inventories of sites, conducting education and outreach and increasing civic engagement. The area also includes technology issues such as knowledge about remediation techniques, risk communication, access to and use of technologies such as GIS and health information.

Social and cultural issues Social and cultural issues includes addressing environmental justice and equity issues like access to health care and standardized housing. Also included in this category are community based information issues, such as developing youth leadership, community advisory panels, attending to public safety issues. The challenges discussed here also address the social and cultural histories

Infrastructure Among the local governments the most resonant challenge is infrastructure needs. Infrastructure refers to the physical development of cities, such as roads, sewers and other components that help a city operate like a machine, moving people, good and wastes to their appropriate places. Other types of infrastructure are long term commitments to creating sustainable practices, land use planning, jurisdictional authority and regional cooperation.

Resources come in the form of financial, human and technical—and all of them are needed by all of the participants in the Forum. Finally, more human resources to work on brownfields and land recycling issues would facilitate collaboration, help to address the other challenges.

Institutional challenges to collaborative coordination
Day to day challenges to collaboration among local stakeholders are compounded when there are local stakeholders, state agencies from two different countries, and two federal governments with differing laws, languages and cultures.

National authorities It is difficult even for citizens of a country to understand the depth of its laws, the organization of its government and how programs are implemented in practical terms. This difficulty is compounded when trying to understand, another country's different laws and enforcement authorities. As frustrating as a country's laws can be for its citizens, there is a strong sense of nationalism that enters into discussions about the "rightness and propriety" of the laws and systems. These attitudes are hard to avoid in discussions about border programs. Whether or not these authorities would be a practical challenge on the local level is not certain, but they at least present a perceived challenge.

Matters of scale and past collaborative efforts One challenge to international collaboration that local stakeholders identified was that of scale. It is difficult for local practitioners to think of collaborating with their neighbors across the border as anything less than an international affair, which it is. In addition to the necessary levels of engagement and increased complexity of larger scale programs, they also attract different skill sets and knowledge because they are essentially policy efforts. Local stakeholders who may feel perfectly comfortable working with their neighbors and talking about the auto scrap yards in their communities do not relate to bi-national policy and program collaboration—even if in effect they are one in the same.

Leadership For any initiative to be successful it needs the backing of leadership. Staff members act under direction from their leaders. Leaders of various interests and levels need to work together to realize success. In political settings where there are changes in leadership positions, elections, leaders vying for political advantage, authority can become a hindrance to collaboration.

Managing work loads Nearly all stakeholders appreciate the idea of collaboration and understand why it is a good idea in terms of integrated approaches. But the size of the idea and the amount of work that is seen as needed to get the program launched, is intimidating. And the nature of brownfields issues is one that is never ending—there's always more than any staff member can be doing. Because of the unknown impact that it would have on their existing programs, stakeholders were reluctant to engage in collaboration.

Coordination The skills and knowledge required to corral diverse interests and personalities is greatly under-appreciated. One challenge that stakeholders pointed out to effective collaboration—both within communities and across the border—is the need for better coordination of interests, resources and people.

Scope of issues   There are border programs on other environmental media (such as urban air and watershed issues) that are easier to quantify, easier to gather stakeholders for and easier to measure progress on. In brownfields projects, there are diverse stakeholders who have an interest in the redevelopment, including community members, private sector representatives, health and environmental officials, finance interests and any of a number of local, state and federal officials working on land use, environmental or economic development issues.

Priorities and issues   Brownfields scenarios on both sides of the border are enmeshed in a host of social, economic and health issues so that it is difficult to know where to begin. Even though the setting on both sides of the border is filled with equally compelling challenges, by most indicators, the United States is a "developed country" with a stable economy, strong infrastructure and established social programs, whereas Mexico is considered a "developing" country with less stable economic and social structures. For these reasons it is difficult to develop parallel priorities. In addition, Mexican and American brownfields and brownfields-related approaches are coming from different perspectives.

Nationalism can be a challenge to international cooperation. Nationalism in and of itself is a good thing, it brings people together, it fosters civic engagement, and encourages understanding of national policies. However, as stakeholders participate and gain advantages of their own programs, they adopt the perspective that their way is the best way. In this attitude, stakeholders suggest that collaboration is not feasible because one country does not do what the other does—as if there are no opportunities for change or flexibility.

Developing a local collaborative program
Many of the challenges that exist for US Mexico border cooperation are institutional in nature and make formal work challenging. However, there is tremendous overlap in interests and opportunity, roles and priorities among border stakeholders. Understanding that peers from other interest groups and from across the border are facing similar challenges in their day-to-day work makes the possibility for local collaboration an even smarter idea.

Underneath the exciting discussion about collaboration lie two realities: comparing priorities and keeping any borders initiative extremely focused. A number of current and past initiatives have been large and wieldy and very slow to produce results. Since brownfields is a local issue and depends on local participation, a pilot program with very specific goals, clearly marked across time would have the highest potential for maintaining interest and participation and demonstrating the value of redeveloping industrial properties. This approach would create a visible success story that citizens could easily understand and participate in. This idea was widely supported and participants began to elaborate on how such a program could be established in a bi-national effort. As Forum participants elaborated on the idea – they discussed the importance that the full spectrum of issues be explored and that widespread participation be garnered.

Specific steps and ideas to a borders brownfields project.
In developing a model project, participants decided that doing a small redevelopment project would yield the highest opportunity for collaboration, success, tangible outcomes and lessons learned. Stakeholders suggested either two joint redevelopment projects, one on each side of the border, or one project that stakeholders from both sides would work on would be best. In either case the steps of the project would include:

Develop a razor sharp focus Participants could not over-emphasize the need to create a razor sharp focus that pursues only the goals of the project and does work to garner larger recognition, have political ambitions, or be anything other than a locally based project.

Get leadership This sort of project requires support from formal local leaders (e.g. elected officials and city managers) as well as support from other sectors including private sector, non profit, educations and faith-based leaders.

Create an advisory board that includes stakeholders from all brownfields interests, even if they may not have a direct role in this project. It would be important to compile a team and class to go through the model process and learn about all of the aspects of brownfields redevelopment.

Create a timeline. It is important that in a model project, there are specific deliverables that can be measured over time. The small successes clearly marked along the way create the energy and motivation to work through the setbacks and arduous aspects of the project.

Publicize the project. Like the timeline and list of deliverables, keeping a project high profile over a long period of time, continues to garner interest and support.

Open doors and transparent policies. Seeking and welcoming all stakeholders and maintaining open transactions increases understanding about how decisions are made, and diminished distrust among participants, even those with opposing points of view.

Community participation
One issues that participants discussed was that of community participation in land use, economic development and social programming issues. While the group did not have any concrete answers for the challenges. The following inter-connected issues were discussed as topics to be addressed and overcome in

developing better collaboration. Topics include: participation, information access and exchange, respect;

Additional issues

The brownfields border Forum was the first of its kind to successfully focus on local strategies and ideas for brownfields redevelopment along the border. Like all gatherings worth their mettle, there remain a number of issues to explor. They include:

Environmental justice – Environmental justice (EJ) issues in this region are not separated in a corner of a neighborhood along the border, on either side. They are not isolated to a few specific topics like lead paint, or job training needs. EJ issues riddle and underlie every aspect of the social, cultural, economic, environmental and health fabric along the border.

Infrastructure and transportation needs US and Mexico stakeholders repeatedly discussed the need to address both the decaying urban infrastructure, the need for modified infrastructure to include new technologies. In Mexico, the need also exists to unify its infrastructure efforts. In addition, the roads, bridges and cross points were established before NAFTA and the huge explosion of commerce and traffic along the border. Many of the *maquiladoras* owners are building the infrastructure to support their efforts including roads, waste water and housing for employees. However, given the rapid development and high turnover of sites, it is difficult to unify efforts, so that systems are inter-connected and it is difficult to enforce standards, assuring the short and long term viability of the land uses and infrastructure itself.

Border crossing – The business and governance of border crossing and the economic and environmental factors involved in it could become assets to brownfields prevention along the border. Materials that pass through the region on rail and truck shipments are inventoried and their destination in Mexico is noted. This information could be adapted for site inventories and inspections on hazardous materials inventories, storage and wastes. There is an inventory of raw materials crossing into Mexico and there are inventories of products leaving the country, so the calculation of wastes and by products and their disposal should be a relatively simple effort to track and could be crucial in monitoring materials and preventing brownfields, requiring coordination among federal agencies. This effort becomes a matter of data management, since the information is already gathered and available.

Private sector – Involvement and responsibility of private sector groups was broached in the Forum. Struggles that remain between US and Mexico stakeholders is some amount of finger-pointing about potentially responsible parties on either side of the border. There was not an active discussion of ways to pressure the private sector to adopt more responsible and sustainable practices. And there was not discussion about the possibility of US EPA pressuring companies with facilities in Mexico, but whose pollution is mobile to the US.

Conclusion

Because of the enormity and cross-cutting nature of the issues surrounding land recycling in the United States and Mexico approaches to local collaboration are necessarily difficult. Likewise, collaboration among local stakeholders is facilitated by support from regional, state and federal entities. While it is clear that widespread collaboration will not begin tomorrow, there are similarities and differences in approach, institutional and individuals challenges and incentives to working together and ideas about ways that effective collaboration might be realized.

Somerset County, New Jersey's center-based brownfields pilot

D. G. Roberts
Planning Department, Schoor DePalma Inc., USA

Abstract

As environmental and infrastructure constraints have made conventional suburban development more difficult and expensive, urban municipalities are turning progressively more toward brownfields redevelopment as a viable alternative to suburban sprawl. The Somerset County Center-Based Brownfields Pilot was initiated by Somerset County and the Business Partnership of Somerset County. Together, they are working with their consultant, Schoor DePalma Inc., under the guidance of the United States Environmental Protection Agency to establish a unified, cost-effective methodology for targeting brownfields for redevelopment in six of the county's more urban municipalities. The aim of the program is to select a specific brownfield site to benefit from environmental response analysis and end-use planning as a model by which all brownfields in the area can be redeveloped.

1 Introduction

Somerset County, New Jersey has been a growing industrial force since it was established in 1688 as an agricultural community, processing and transporting grain and other commodities. As Somerset became a more popular place to live, it became host to lumber mills, quarries, and other industry to satisfy the demand for homes and transportation. The Industrial Age brought textile mills to the region, which were replaced by chemical and pharmaceutical interests in the 1900's. All of this growth through the past three centuries in Somerset County has contributed to the creation of nearly 200,000 acres of brownfields sites throughout the county. In March 2001, Somerset County undertook an ambitious initiative to participate in the United States Environmental Protection

Agency's (USEPA) Brownfield Assessment Demonstration Pilot, a program that awards a competitive $200,000 grant to allow a county or municipal group to rank and inventory brownfields, target selected sites for assessment, and eventually propose plans for remediation and reuse of one or more sites. The pilot serves the long-term goal of its parent program, the EPA Brownfield Redevelopment Economic Initiative, which is to encourage the development of one proven, unified method of brownfields assessment, clean up and redevelopment.

2 The project

Somerset County's Center-Based Brownfields Pilot Project is now nearing the end of the site selection phase. The county's consultant for this initiative, Schoor DePalma Inc., one of the region's leading engineering and design firms, has been working over the past few months to complete the first phase of the project: updating the county's brownfield site inventory for Somerville, Manville, North Plainfield, Bound Brook, Raritan and South Bound Brook to include new sites from the New Jersey Department of Environmental Protection's (NJDEP) Known Contaminated Sites List. This process included interviews with leaders from each of the six municipalities and reviewing the lists of sites and the aerial photomaps, and prioritizing brownfield sites based on the likelihood of selection under EPA eligibility criteria.

   The Somerset County Brownfields Pilot Steering Committee, upon review of recommendations made by its Site Selection Subcommittee this past winter, has short listed five sites for consideration as pilots and they are currently under review by the EPA for eligibility under the program. These sites will undergo further review until three sites are recommended by the Steering Committee for selection as pilot sites. Once qualified, Schoor DePalma will conduct preliminary assessments and site investigations for each of the three sites, then evaluate the results and generate outlines of appropriate environmental responses. One of the three sites will then be selected as the final pilot site. Schoor DePalma will undertake end-use concept planning for that site, using publicly held meetings within the host community to factor public opinion into the final end-use plan for the site.

2.1 What is a "brownfield"?

What is a "brownfield" and what does it have to do with redevelopment? The term originated from the USEPA's Brownfield Pilot Program a number of years ago and generally refers to contaminated commercial or industrial sites. The official definition, taken from the NJDEP Brownfield and Contaminated Site Remediation Act is: "any former or current commercial or industrial site that is currently vacant or underutilized and on which there has been, or there is suspected to have been, a discharge of contamination."

   Contamination that is hazardous to the public's safety and requires remediation in order to create useful property is an impediment to the economic viability of any area. Such contamination can occur in a number of forms, at a variety of locations: from asbestos in old buildings to oil spills; from buried rubble containing hazardous materials to historic industrial hazardous waste discharge. While old, heavy industrial sites such as chemical plants and oil refineries are prototypical brownfields, brownfields can occur much closer to home than one might expect. For example, many municipalities have landfills that have ceased operation, sometimes by order of state environmental regulatory officials, but have not been officially closed because the necessary remediation action plan has not yet been prepared, approved or implemented. There may be a gasoline station on a highway near one's home that has a neglected underground storage tank that has leaked petroleum-based contaminants into the surrounding soil and groundwater. Indeed, brownfields are a common problem faced by many municipalities.

Levels of remediation for brownfield sites vary depending on the planned ultimate use. The New Jersey Industrial Site Recovery Act governs the standards for remediation, and, basically, there are two levels of remediation: residential and nonresidential. Both standards are measured in terms of parts-per-million (ppm) for each of the known hazardous substances. Residential standards are far more stringent than nonresidential standards and, as a result, the cost of cleaning up a site to residential standards is substantial and may affect the economic feasibility of redeveloping the site for residential use.

3 Criteria for Selection

The county's brownfields inventory contains 844 sites, an overwhelming number to say the least. Indeed, the municipalities targeted comprise only 4% of the land in Somerset County while containing 25% of the county's brownfield sites.

3.1

The pilot program contains a number of provisions that help to quickly rule out many of the properties:

3.1.1
The majority of the 844 brownfield sites identified in the updated inventory are petroleum-related contamination sites, presenting with underground storage tanks, which are not eligible for consideration in this initiative because funding for abating underground storage tanks is available under other programs;

3.1.2
Municipal landfill sites are eligible for selection only if they are not already covered under the Resource Conservation and Reclamation Act and they meet other EPA restrictions;

3.1.3
Privately owned brownfield sites are eligible for selection under the program

only if there would be a significant public benefit for using federal funds for environmental response assistance. The county must be prepared to answer these questions if they choose to select a privately owned site for the pilot: Is there a legitimate reason why the property owner cannot redevelop the site without the assistance of public funding? Would the redevelopment of the site create a positive impact beyond the property lines of the site itself? And, what benefits would the community gain by the redevelopment of the site?

3.2

Given the USEPA qualifications for selection of the three pilot sites, and in the interest of adding as much value to Somerset County's brownfields program as possible, Schoor DePalma recommended that the criteria used to select the pilots be designed to yield sites that can be used as models for other sites in the county after the current project is over. The search is on for sites that match one of these three descriptions:

3.2.1
A site that is likely to have very little actual contamination. A Preliminary Assessment might be performed and would indicate that the contamination was more perceived than real. Limited (or no) sampling would be necessary to carry forward with end-use planning. This would demonstrate a means by which communities could quickly eliminate the stigma often associated with abandoned properties;

3.2.2
A site that is known to have contamination and has undergone some level of remedial activity. A preliminary assessment would summarize the existing environmental studies and only a limited site investigation would be necessary to confirm existing data and/or support potential remedial actions (deed notices, classification exception areas);

3.2.3
A site that is known to have contamination and has undergone little or no remedial activity. A preliminary assessment for this site would identify areas of concern that would then be fully investigated in a site investigation.

4 End use planning

At this point, three sites will have been selected and studied to determine the appropriate environmental response. Now, several questions pop up: How will the final pilot site be selected? How will the end-use planning be conducted? What will the final product be? How will it benefit brownfields redevelopment in Somerset County?

The county, based on recommendations by the Brownfields Pilot Steering Committee, will select the final pilot site, which will, in turn, receive a recommendation from the Pilot Site Selection Subcommittee. The end-use planning process (Figure 1) begins when the results of the environmental response options are analyzed to generate a list of potential uses. The next step will be to evaluate the economic feasibility of the proposed uses, given the cost of cleaning the site to appropriate remediation standards (residential or nonresidential).

Once the feasibility of potential land uses is determined, several different concept plans for the site will be developed. These alternative concept plans will be presented to the host community through a series of public forums and meetings with stakeholders (the property owner, neighbors, town officials, etc.). A consensus on the best alternative concept plan for redevelopment of the site will be developed and a final concept plan, including illustrations, will be presented and submitted to the USEPA.

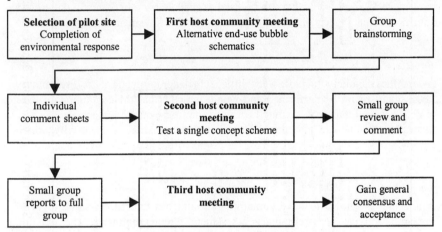

Figure 1: End-use planning process flow chart

5 How will this project benefit Somerset County?

At first glance, the math suggests that, if only three sites of six towns will receive the benefit of a preliminary assessment and site investigation, and only one of those three will receive end-use planning, there is limited benefit to the county as a whole. The benefit of this project, though, lies not in the immediate result, but in the legacy it will leave.

The continued commitment of both the county and the Business Partnership of Somerset County to the economic redevelopment of all underused brownfield sites has led Schoor DePalma to work with them to develop a Brownfields Pilot Website (Figure 2) that will inform the public of the status of site selection, provide opportunities for public involvement in the end-use planning process, offer profiles of the pilot sites, and hold the county's brownfields inventory. The website will be linked to the county's main site and, ultimately, be maintained by the Business Partnership of Somerset County, who will incorporate the website into ongoing efforts to encourage brownfields redevelopment in Somerset County

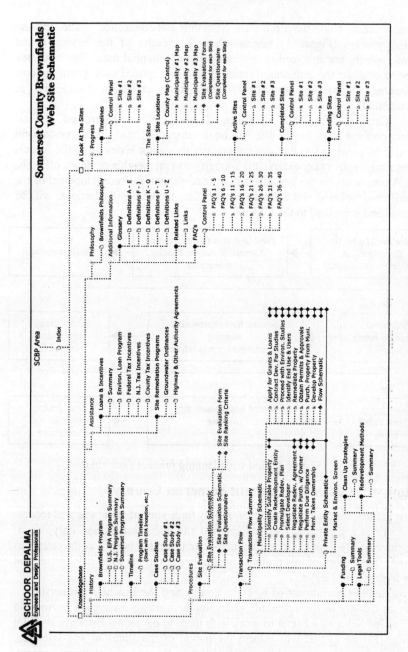

Figure 2: Somerset County brownfields website schematic

Improving the brownfield's timeline for redevelopment

L. Curran[1], D. Kalet[1], D. Marsh[1] & J. Sherrard-Smith[2]
[1]Atlantic Richfield, USA.
[2]BP, United Kingdom.

Abstract

The Brownfield label is applied to properties whenever liabilities are perceived to exceed the value of the asset. The apparent negative asset value impedes the natural process of development. Reduced value may be inferred for a contaminated property because regulatory requirements to control or remove contamination have not been prescribed yet as part of the regulatory process and, therefore, the most expensive remedy with the longest timeline is factored into the property valuation. (If a property can undergo redevelopment without special attention or investment even if contaminated, the Brownfield label typically is not applied.) If an efficient remedy is incorporated into the redevelopment plan, a new focus may be created, engaging many stakeholders to support the change, and providing a means of leveraging resources to accomplish both the community and the property owner's goals.

This paper reviews the factors that have contributed to both the advancement and delay of two redevelopment sites (former refineries) owned by BP in the US. A review of the economics and other incentives from the perspectives of the property owner, the local government, the private investor and other interested parties provides a basis for constructive dialogue among the stakeholders to return the asset to productive use. A well-defined plan with economic support, including resources in addition to the party responsible for cleanup, has shown to reduce the timeline for redevelopment, a critical factor for investors. In these two cases neither the specific remedies nor the reuse plan is as influential in the redevelopment process as is the cooperative approach of the stakeholders. The plans, economics, benefits and the lessons learned from both sites are discussed.

1 Introduction

In the final stage of the life cycle of an asset, facilities are decommissioned and remediated. The question of who determines future land use and hence sets acceptable clean up requirements are critical in terms of planning and executing a successful site exit strategy. In the United States of America, the Resource Conservation and Recovery Act (RCRA) is commonly used by governments to set clean up protocols for oil refineries. While RCRA was intended to focus on the clean up process by establishing very rigid technical protocols, it does not require contemplation of market forces that shape and determine use of an old industrial site. In our analysis of the local markets, we have concluded that these old industrial sites are ideally positioned to support future industrial/commercial activities, meeting broad market needs for commerce vital to the economic sustainability of the community.

In the context of life cycle management, brownfield redevelopment can be defined as returning previously used real estate back to into uses valued by affected stakeholders. In many ways, brownfield redevelopment is the same as development of a typical greenfield site. Where greenfield development is motivated simply by adding value to an undeveloped, or neutrally valued property, brownfield development also contains the element of reducing negative values due to unsightly image and/or environmental impairments.

Converting these negatively valued properties into assets for property owners and communities is not an easy task. Brownfield redevelopment projects generally face challenges of negative public perception, clean up costs, and land use restrictions. These challenges impact the timing of when brownfield properties are available for development, creating financial risks for investors that can be difficult to quantify. In many cases, nearby greenfield properties are immediately available for development, increasing the difficulty of marketing a nearby brownfield property.

However, by not redeveloping brownfield sites, as a society we are actually encouraging urban sprawl and loss of open space. The United States Department of Housing and Urban Development estimates that there are as many as 425,000 Brownfields throughout the United States totaling approximately 5 million acres of abandoned industrial sites in our nation's cities. In recent years, the issue of brownfield redevelopment has increasingly become the focus of both local and national policy debates in the U.S. The environmental and economic benefits of brownfield redevelopment are widely acknowledged. In a recent study done by George Washington University [1], it was found that for every 1 acre of brownfield reused, 4.5 acres of greenspace is saved. According to an independent study conducted by the Council for Urban Economic Development [2], the revitalization of brownfields has created over 22,000 permanent jobs and leveraged $2.48 in private investment for every $1 spent by federal, state, or local governments. Enactment of new US Federal legislation in January 2002, the Brownfield Revitalization Act [3], is expected to stimulate more redevelopment. The Act made changes in federal liability and authorized grants for brownfield redevelopment ($200 m annually for the next five years).

Many mayors, economic development officials and private developers or property owners can point to successful redevelopments where environmental and economic issues have been resolved to the satisfaction of the stakeholders involved. Former manufacturing sites have been converted to attractive riverfront parks, and former military bases have been redeveloped as housing and neighborhood commercial development. Although the size of the property may vary from small lots to tens or hundreds of acres, the issues remain the same. Still, in as many cases, similar sites remain undeveloped and of virtually no benefit to the current owner or the surrounding community and cities of varying sizes across the US continue to struggle to redevelop their properties. What are the factors that contribute to successful redevelopment?

2 Stakeholders

Stakeholders with an interest in the redevelopment of a brownfield property can include the property owner, regulatory agencies, local elected officials, community groups, residents and business owners in the surrounding area, and private financiers and developers. Each of these stakeholder groups may have a different motivation for engaging in the redevelopment of a brownfield property, and their concerns must be balanced to create a successful development project. It is important that all pertinent groups are actively engaged throughout the redevelopment process and that each group's needs and expectations can be assessed, evaluated and balanced for an overall project approach.

The approach to stakeholder involvement often separates a successful brownfield redevelopment project from similar, but less successful projects. In some cases, the issues surrounding a redevelopment cannot be resolved to the satisfaction of all the stakeholders involved and the process stagnates. A successful brownfield redevelopment results in an end use that is beneficial to all stakeholders: property owner, regulatory agencies, the community, potential investors, and potential users.

To begin with, a brownfield project must satisfy the regulatory stakeholders that in the United States are represented by the U.S. Environmental Protection Agency on a national level, and by state or local environmental agencies that are charged with implementing U.S. EPA programs and regulations for their jurisdiction. The U.S. EPA and a number of state governments have made great progress in recognizing that a successful brownfield project does not stop with the completion of a remedial action and the elimination of threat to human health or the environment.

A successful redevelopment often starts with regulatory issues but quickly incorporates the needs and expectations of the local community. Local leaders and residents often bring a range of concerns to the table when a redevelopment project is initiated. They are often driven by the desire to regain the economic benefits, including jobs and tax revenue, that were lost when an industrial facility closed operations in their community. The local community may also be driven by concern over environmental conditions at the former refinery or other industrial property. When a facility is in operation, providing employment and delivering economic development benefits to the community, residents are often tolerant of the environmental impact associated with such a facility. Such facilities were often built before or during the early years of a town's development and upon closure, leave a large vacant space in the center of a town's landscape.

When a site reduces or ceases operations, the community's tolerance can rapidly decrease. In many cases, community leaders often see environmental regulatory action as a tool to prompt action at a former industrial facility and force the property owner to consider redevelopment. Local communities are also motivated to improve the quality of life in their community, and this includes eliminating the unattractive nuisance that an underutilized industrial property represents.

An equally important stakeholder in brownfield redevelopment is the owner of the targeted property. A brownfield redevelopment project must provide a compelling benefit to the owner of the property. The perception remains in the United States that it is cheaper and easier for an industrial property owner to retain ownership of a former industrial property, comply with environmental regulations as necessary and reduce liability by restricting access to the property. There are, in fact, benefits to an industrial property owner being an active participant in brownfield redevelopment. Benefits include reduced costs through lease or transfer of underutilized land, enhanced partnerships with state and federal regulatory agencies, and improved reputation in the local community as the owner becomes a partner in sustainable development for the community. A concern from an owner's perspective is that encouragement in redevelopment at one site may set similar expectations for stakeholders at other sites that do not have the same potential.

Involvement of potential users prior to development can provide assurance to both the property owner and regulators that remediation activity is directed to specific end uses and is both protective and cost effective. Early involvement by potential investors provides guidance in the design and of future institutional controls to both the owners and regulators.

Factoring the concerns of all parties into the design of the remediation approach, rather than discovering new issues after the completion of remediation, can enhance the property value and generate a better solution for all stakeholders. Our research has identified the concerns of stakeholders and are summarized in the following table:

Table 1: Summary of Stakeholder Concerns and Interests

Stakeholders	Their Concerns and interests
Environmental agencies	Assurance of future foreseeable land use Assurance of viable engineered barriers and institutional controls Assurance that use does not pose threat to human health or environment

Table 1: Summary of Stakeholder Concerns and Interests - *continued*

Stakeholders	Their Concerns and interests
Community	Assurance that redevelopment is safe Creation of jobs and commerce opportunities New sources of tax revenues Creation of new image for community Preservation of greenspace and natural areas
Current owner	Reduction of environmental liabilities Reduction of operating expenses Development of a viable exit strategy for excess property Brand image with customers and community Setting realistic expectations with stakeholders
Potential users	Large land parcels near customers and suppliers Access to transportation Access to utilities Availability of labor Price sensitivity Environmental protection
Potential investors	Assurance that investment will not be impacted by remaining contamination Assurance that investment is not subject to future environmental liabilities from prior use

3 Economic Factors

The economics of most brownfield properties are generally similar, that is, the potential cost of remediation outpaces the value of the real estate. This scenario is typical when the next use of the property is undefined and, therefore, the most stringent environmental conditions must be met to satisfy any future use of the land. This condition can be overcome when one or more of the following are incorporated into the development:

A less costly remediation remedy is tied to the future use plan;

A remediation solution is leveraged to supplement development expenditures;

Tax-based incentives are used to supplement the site attractiveness;

The local employment base represents a competitive advantage

The site has one or more features unique to the area and is advantaged over potential greenfield developments

4 Case Examples

In both cases cited in this paper, BP, as a property owner, acts to facilitate redevelopment projects. The Casper, Wyoming and Wood River, Illinois properties are similar in nature: former refineries, operated by BP Amoco (now BP). In Casper, Wyoming, BP is working with the Wyoming Department of

Environmental Quality (WDEQ) to achieve regulatory closure for approximately 3,000 acres of property. Upon regulatory closure, the property will be leased to the City of Casper and Natrona County for redevelopment as a mixture of light industrial, commercial and recreational development, and preserved natural habitat. In Wood River, Illinois, BP has collaborated with the City of Wood River and a private developer to accelerate investigation, remediation and regulatory closure at the former refinery and redevelop the property as a combination of light industrial, commercial, riverfront recreational development and open space.

4.1 Casper, Wyoming refinery

The Casper, Wyoming refinery was founded by Standard Oil (a predecessor of Amoco- now BP) in 1914 to tap the Salt Creek Oil Field north of Casper. Casper, Wyoming is a City of approximately 50,000 residents, located in the largely rural western United States. When the population of surrounding Natrona County is included, the area's population is approximately 66,000. Although Casper is a small urban area, it is the second largest city in Wyoming.

The City first developed as a stop for pioneers traveling to the American west. It marks the point where five major westward trails (the Bridger, Bozeman, Mormon, Pony Express and Oregon Trails) met. Later, Casper thrived as an oil town, and several major refining operations were opened to pump and process the oil found outside the town. The town's identity is closely tied to the petroleum industry and the BP refinery was a major employer for the Casper area before closing its refining operations.

After 77 years of operation, the Casper refinery closed in 1991. BP retained ownership of the entire property and continues to use a portion of the property for distribution operations, but the majority of over 3,000 acres has been underutilized since closure of refining operations. These properties include approximately 350 acres located along the banks of the North Platte River, immediately west of Casper's central business district. They also include approximately 400 acres of former tank farm property, located on the northern edge of the city and several thousand acres of property surrounding and including Soda Lake, a man-made lake originally created by BP for wastewater disposal.

Upon closure of BP's refining operations in Casper, the city and county realized the loss of jobs and tax revenue. The residents of the area were also confronted with the attractive nuisance of an industrial property, located in the heart of their community, that was no longer providing any economic benefit to its residents. Many residents also feared lingering environmental impacts from the refining operations that might spread beyond the physical borders of BP's property.

The closure of the Casper refinery was part of a broader corporate effort to consolidate operations across the company. The oil boom years in Wyoming had passed and many refineries were scaling back operations and closing their doors. However, BP still retained a large acreage of property in Natrona County and continued to bear the burden associated with general maintenance, liability and property taxes (although reduced). There was no precedent for transfer of such a property for use as anything other than a refinery and it seemed inevitable that BP would continue to retain these liabilities.

BP worked with the Wyoming Department of Environmental Quality (WDEQ) to comply with regulatory requirements for the closed Casper facility. The investigation and remediation schedule was driven by the WDEQ and BP focused on compliance with the schedule.

In 1997, the citizens of Casper and Natrona County formed a group to address what they perceived as the problem associated with the former BP refinery property. They voiced the opinion that the refinery property should be returned to productive use quickly and provide some benefit to the community. Without action, the former BP refinery left a large gap in both the economic and physical landscape of Casper and Natrona County and the residents filed a lawsuit to initiate movement towards redevelopment.

A 1998 ruling by the United States District Court resulted in a remedial agreement between BP and the WDEQ. It outlined an approach to investigation and remediation of the former refinery in Casper. The ruling also led to a Reuse Agreement between BP, the City of Casper and Natrona County, which provided a mechanism for redevelopment of the areas of the property that are no longer utilized by BP operations. This redevelopment was to be guided by a collaborative process, including input from the City of Casper and Natrona County, which were collectively represented by a newly formed Joint Powers Board. BP would contribute land for lease, infrastructure improvements and funding for building construction, operations and maintenance and economic development, to seed the redevelopment of the former refinery properties.

At this date, the remedial agreement has been finalized and remediation and redevelopment planning for the former refinery properties are underway. Three separate properties, amounting to over 3,000 acres, are to be leased to the Joint Powers Board for a 99-year term. The Soda Lake property is to be preserved as natural habitat, the former tank farm properties are to be redeveloped as 400 acres of light industrial and warehouse operations, and the 350 acres along the North Platte River, known as the Platte River Commons, is being developed as an 18-hole golf course, office space and a potential hotel/convention center to complement the recreational development at the property.

From BP's perspective, the Casper refinery project has evolved into a successful redevelopment project. What began as a legal fight to resolve conflicting goals of BP, WDEQ and the Casper community developed into a partnership where each party contributed to and realized benefit from the redevelopment of the former refinery property.

BP has built a stronger working relationship with the WDEQ and is on track to achieve regulatory closure at the Casper refinery property. The lease agreement includes deed restrictions that limit BP's liability for future uses by new owners and operators of the properties.

In addition, BP has built an active working partnership with the Casper community, led by the Joint Powers Board. The lines of communication amongst stakeholders have been opened, and BP has worked closely with the Joint Powers Board and the broader community to develop a viable redevelopment plan for the former refinery property. BP has rebuilt its corporate reputation and goodwill in the Casper community, which was diminished when the refinery was closed and jobs for the community lost, through participation in the redevelopment process. This includes aggressive attention to environmental issues at the property, BP participation in community visioning exercises and funding of the redevelopment effort through creation of an economic development fund and construction of infrastructure.

4.2 Wood River, Illinois refinery

The town of Wood River is approximately 11,000 residents, and its history is closely linked to that of the refinery. The town grew up around the refinery, and it was a major employer and contributor to the community. Wood River is located in the River Bend Region of Illinois, a series of 13 small communities, stretching along the Mississippi River. The area developed around large industry, including several major refineries, steel mills and additional manufacturing operations. It was ideal for this type of industry, given its access to transportation routes of the Mississippi and railroad. However, as the U S economy shifted away from heavy industry, and trucking and air replaced rail and barge as leading modes of shipping, the region suffered an economic decline, and many of the area's largest employers and corporate property tax payers reduced operations or completely closed during the last two decades.

The former Wood River refinery includes approximately 800 acres of property along the Mississippi River in Illinois, just 20 miles northeast of St. Louis, Missouri. The refinery was constructed in 1907 and consists of two major parcels: the Main Plant and the Riverfront property. Activities at the Main Plant have included refining, storage and marketing of petroleum and related products. Uses of the refinery's Riverfront property have included various containment ponds, disposal facilities, and product transfer activities at the Mississippi River heavy oils and light oils docks.

Most operations at the Wood River property were closed by 1997; however, petroleum marketing operations, including a terminal in the northwest portion of the Main Plant and operation or lease of over 30 bulk storage tanks, continue. Outstanding environmental issues, including soil and groundwater impacts, are currently being addressed through investigations, remediation and ongoing monitoring.

BP took the lessons learned in Casper, Wyoming and approached the Wood River property with the goal of creating a positive redevelopment project that would reduce environmental liability, create value for the Wood River community, and enhance relationships among BP, the City and state and federal regulatory agencies.

BP approached the Illinois EPA and the City of Wood River with a novel idea: they asked to aggressively approach the investigation and remediation of the

Wood River property, which had been designated a RCRA Corrective Action site, and work together to identify ways to stimulate the reuse of the property. The City of Wood River and Illinois EPA were receptive to the idea.

BP took the initiative in the redevelopment, creating a conceptual plan for redevelopment and conducting market studies to identify potential end-users for the property. The reuse concept plan, developed by BP with input from the City of Wood River and private developers, included several hundred acres of mixed uses, including retail, commercial and industrial development, recreational opportunities along the Mississippi riverfront and preserved natural habitat.

A private developer from the Wood River community stepped forward to express an interest in the redevelopment project. While a developer is a logical participant in most redevelopment projects, their involvement at the Wood River property is somewhat unique in that the developer has committed to the project in the early stages and has been actively involved in planning and identifying viable options for reuse.

BP worked with the Illinois EPA to develop and gain approval for an innovative approach to investigation and remediation of the Wood River property. The property was divided into parcels, based on the redevelopment concept plan, and an investigation plan was built around these redevelopment parcels. Parcels that were known to have little environmental impact, or those with the highest redevelopment potential, were targeted first.

In 2001, BP received a letter of No Further Action from the Illinois EPA for a 7-acre parcel at the northwest corner of the Main Plant. This parcel is located along Madison Street, the south edge of Wood River's Central Business District. It is being targeted for retail development to complement existing development in the Central Business District and bring visitors and sales tax revenue to the City of Wood River. A second, larger parcel located at the northeast corner of the Main Plant is expected to receive a letter of No Further Action within the next year.

The City of Wood River continues to partner with BP toward the common goal of redevelopment. The City acts as an advocate for the project to government agencies and elected officials and has begun efforts to attract end-users to the property through marketing and economic development tools. The Main Plant portion of the property lies within the River Bend Enterprise Zone, which provides tax incentives for job creation and abatements for purchases of materials or equipment purchased within the Enterprise Zone. The City has also committed to adapting its local zoning codes for the property to encourage redevelopment and provide a mechanism for enforcement of covenants and restrictions for future users that will protect BP's liability.

Through this experience, BP has built a strong relationship with the Illinois EPA, as well as the U.S. EPA. The U.S. EPA named the Wood River property one of five RCRA Redevelopment Pilot Projects in 2001. This program is designed to enhance partnerships between property owners and regulatory agencies and develop innovative approaches to remediation and redevelopment at RCRA properties. The program has provided a forum for stakeholders in the redevelopment, including the U.S. EPA, Illinois EPA, BP, the City of Wood

River, and the lead developer, to discuss constraints facing the redevelopment project and develop ways to overcome these.

The Wood River redevelopment is still in progress, but can be considered successful to the extent that it has set reasonable expectations for executing redevelopment.   The project's major stakeholders, including BP, regulatory agencies, the City leaders and residents, and a private developer, have worked together to create a successful project.   BP has built a strong working relationship with the Illinois EPA and been identified by the U.S. EPA as a model for future redevelopment projects through the RCRA Brownfields Prevention Pilot Project.   In addition, BP has partnered with the City and a private developer to share the financial burden of a redevelopment and created a vision for reuse that creates value for the developer, the City and community residents.

5 Summary of lessons learned

At both the Casper, Wyoming and Wood River, Illinois sites, redevelopment continues to progress. In the end, both projects are expected to successfully achieve remediation and reuse of these properties. The difference was in our initial approach. In the Casper project, we first attempted to define and satisfy RCRA clean up requirements and then address other stakeholder needs.   This approach created confusion and inefficiencies in achieving final clean up agreements. Relevant community stakeholders felt disenfranchised and at times frustrated with their perception that the technical discussions between BP and EPA were going too slowly. To correct this, community stakeholders were then brought into the process and participated in setting land use goals.   The community, EPA and BP are now working together to implement those goals and the project is moving forward. In the Wood River project, we applied those lessons learned and began with involving community stakeholders earlier in the clean up process. As a result, remediation issues took less time to resolve as community interests were visibly incorporated into the process, creating comfort for EPA in approving clean up plans.   The lesson that should be taken from a review of the Casper, Wyoming and Wood River, Illinois redevelopment projects is that when key project stakeholders work together, respecting not only the interests of but the concerns of other parties, a project can provide benefit to all parties and result in a brownfield returned to active use. A successful redevelopment project must address a range of issues, not limited only to environmental regulations.

Through redevelopment, the corporate property owner can reduce liability, improve relationships and enhance brand value. The regulatory agencies can achieve regulatory closure, assure protection of human health and the environment, and reduce urban sprawl. Investors can confidently create new wealth. New users are protected from the impacts of past operations and have access to all of the original benefits of the location. The local community can create new jobs and generate new tax revenue while creating a new vision and image by incorporating the latest trends in architecture, recreation, greenspace and infrastructure into the redevelopment.

References

[1] Deason, J.P., G.W. Sherk, G.A. Carroll, The George Washington University, *Public Policies and Private Decisions Affecting the Redevelopment of Brownfields: An Analysis of Critical Factors, Relative Weights and Areal Differentials*, submitted to the Office of Solid Waste, US EPA, September, 2001.
[2] Council for Urban Economic Development, independent study. *The ECS Land Reuse Report 2001*, ECS, an XL Capital company, Exton, PA.
[3] Public Law 107-118 (H.R.2869)- " Small Business Liability Relief and Brownfields Revitalization Act" January 11, 2002.

Section 9:
Cleanup methodologies

Surfactant-enhanced desorption for recalcitrant hydrocarbons contaminated soils

H. S. Yoon & J. B. Park
School of Civil, Urban and Geosystem Engineering,
Seoul National University, Korea.

Abstract

Hazardous substances produced from industrial sectors have caused serious environmental problems and threatened ecological systems dating back to early 1960's in Korea. More than 70% of the domestic wastes are disposed to landfills in Korea; however, a number of abandoned landfills and dump sites are poorly managed and are leaking leachate, which are contaminating surrounding soils and groundwater. This study focuses on the feasibility of recovery of the organic-contaminated Nanji-Do soil by applying surfactant-enhanced desorption technique. Nanji-Do landfill is the MSW(municipal solid waste) dump site under service in Seoul between 1978 and 1992. Surfactant-enhanced desorption technique was studied with nonionic surfactant(Triton X-100) and anionic surfactant(SDS) as desorbing solvents for extracting *p*-Cresol sorbed on soil particles. Sorption characteristics of soil and organic compound were analyzed and the applications of surfactant solution were studied through batch tests and the flexible-wall permeameter tests. The test results show that the surfactant-enhanced subsurface remediation technique can be adequately applicable when the contaminants are hydrophobic and recalcitrant in nature.

1 Introduction

Soils and groundwater exposed to the variety of chemicals and wastes are frequently and heavily contaminated and difficult to be treated and recovered back to normal conditions. This paper focuses on the remediation of the organic-contaminated subsurface by applying the surfactant-enhanced desorption technique. Research over the last fifteen years has shown that surfactant has potential to increase the rate of remediation of ground water contaminated with hydrocarbon compound pollutants. The main goal of surfactant-based soil remediation is enhanced contaminant extraction. Surfactant molecules consist of a hydrophobic and a hydrophilic part and can interact with polar as well as nonpolar surfaces [1]. Surfactant molecules may exist during the soil flushing process in the following forms: monomers, micelles, precipitates and adsorbed on the soil, etc. When the surfactant concentration is lower than the critical micelle concentration (CMC), the surfactant molecules exist in monomer form [2]. These monomers change some physical properties of the liquid, including the surface tension, interfacial tension and osmotic pressure. Above the CMC surfactant molecules aggregate and form micelles with a hydrophobic center in which partitioning of chemicals is possible. The interior region of micelles is composed of the hydrophobic tails of the surfactant molecules while the outer region is formed by the hydrophilic heads. In general, two major mechanisms are responsible for surfactant-enhanced contaminant removal from porous media. Solubilization refers to partitioning of hydrophobic contaminants into the oil-like interior of surfactant micelles [2], [3], [4] and [5]. Mobilization or displacement of the residual and /or separate oil phases can be achieved by significant reduction of the oil and water interfacial tension (IFT) [6], [7].

The objectives of this study were to (1) investigate the feasibility of surfactant-enhanced desorption by evaluating the removal amount of the representative contaminant, *p*-Cresol adsorbed onto soils; (2) accomplish laboratory modeling of the effective stresses and flow velocities found in-situ soil-liquid systems using flexible-wall permeameter. This study was to give further information and essential data for applying the surfactant-enhanced desorption technology for the remediation of *p*-Cresol contaminates sites.

2 Test methods

2.1 Test materials

2.1.1 Test Soil
The test soil was a mixing soil that was homogeneously blended with weathered soil recovered from the vicinity of Nanji-do landfills located near Seoul World-Cup soccer stadium and Jumunjin sand (classified SP by Unified Classification System) and humic soil. For the weathered soil, roots, debris, and stones were removed in the wet condition. The wet soil was then oven dried at 105 °C until the soil was totally dried. The dry soil was sieved through a #40 standard testing seive (opening size of 0.425mm). This soil was mixing with Jumunjin sand to increase the hydraulic conductivity of the test soil, so that permeation tests could be performed within a reasonable period of time with an induced hydraulic gradient of 40-60. And for the adsorption purpose of the contaminants, humic soil was added to the mixing soils. Mixing soil was blended with Nanji-do soil, Jumunjin sand and humic soil by weight ratio of 2:6:1. Mixing soil has a cation exchange capacity of 25.1 meq/100 g,

and organic content of 6.7%, and a specific gravity (G_s) of 2.53. The test soil was classified as SP by Unified Classification System.

2.1.2 Contaminant

The *p*-Crseol was chosen for the research because it is generally considered as one of the recalcitrant halogenated aromatic compounds. It can be generally found in landfill and industrial wastes containing cresols are somewhat difficult to treat by biological methods [8]. It melts at 34.7°C and has a vapor pressure 1.10×10^{-1} mmHg. Its solubility in water is about 2.15×10^4 mg/L and log K_{ow} is 1.94. It has purity greater than 99% and was purchased from Sigma Chemical Co. USA.

2.1.3 Desorbing Solutions (Decontaminants, Surfactants)

Commercially available two surfactants (anionic and nonionic) were selected as washing fluids for the contaminated mixing soil: (1) SDS (sodium dodecyl sulfate); (2) Triton X-100 (octylphenol polyoxyethylene). Table 1 shows the structures, molecular weights, and CMCs of the selected anionic SDS and nonionic TX-100 surfactants tested with various concentrations at their natural pH and room temperature (approximately 20 °C).

Table 1. Physical and chemical characteristic of selected surfactants

Commercial Name	Surfactants	Structure	CMC(mM)	MW	HLB[1]
SDS	Sodium dodecyl sulfate	$C_{12}H_{25}O_2SNa$	8 [9]	288	N/A
Triton X-100	POE(10)octylphenol	C_8PE_{10}	1.7×10^{-1} [3]	646	13.5

C : alkyl chain length (-CH$_2$-)   P : phenol ring (-C$_6$H$_6$-)   E : ethoxylate group
MW : molecular weight   HLB : Hydrophile/Lipophile Balance   N/A : Not Available

2.2 Analysis method

The concentration of *p*-Cresol was analyzed by Gas Chromatography(6890 Series, Hewlett Packard Co. USA) with HP-5(capillary column coated with crosslinked 5% phenyl methyl siloxane, 320nm thickness, 30m length, 0.25mm diameter) and FID(flame ionization detector).

2.3 Test methods

To achieve the objectives of this study, the following tests were performed in the laboratory.

2.3.1 Batch sorption tests

Laboratory batch sorption tests were conducted to evaluate the sorption characteristics of the selected organic contaminant onto mixing soil. Batch adsorption tests included adsorption rate and adsorption isotherm tests. These tests were conducted on duplicate or triplicate samples to maintain the quality assurance of the tests.

Batch adsorption rate test was carried out to establish the equilibrium time required for adsorption of contaminant onto the mixing soil. The adsorption rate test for *p*-Cresol showed that adsorption equilibrium was achieved within 48 hours. Thus, 48-hour contact time was used for later successive reverse isotherm tests and in establishing adsorption equilibrium for preparation of desorption tests.

The adsorption isotherm test method for *p*-Cresol was as follows: 5g samples of mixing-soil were placed in the centrifuge tube(16×100mm Culture c-tube with teflon-faced cap, Wheaton Co., USA) and mixed with aqueous *p*-Cresol solution at various concentrations (5-100mg/L) with head space free. These tubes were stored in the thermohygrostat at 20±1°C on a rotary shaking position. After more than 48 hours contact time, mixture samples were centrifuged (Maraton® 8K, Fisher Scientific, USA), and the supernatant solution were taken in vial and analyzed by GC-FID.

2.3.2 Desorption kinetics studies by successive reverse isotherm (SRI) method

Desorption test was conducted to assess the effectiveness of the extraction solution for the contaminant with the potential desorbing solutions by simple contact shaking of the contaminated soil with an aqueous solution of the extracting agent. This test method is termed the *SRI procedure* [10]. In the SRI test, soil was mixed with contaminant solution (20mg/L). After 48 hours of contacting time, the contaminated soil was washed successively by agitation in the decontaminated solution for more than 30 minutes, centrifuging, decanting the supernatant solution, replacing the decanted solution with fresh decontaminant solution, and repeating the process several times until the contaminant could not be measured in the decanted supernatant solution. The cumulative weights of the desorbed contaminants, determined through gas chromatography, were added to obtain the total weight of contaminant removed. SRI tests were performed to allow a comparison between the washing of the contaminated soil with simple deionized water (DI water) and treatment by the various decontaminant solutions.

2.3.3 Flexible wall permeameter test

The modeling of in-situ flow of a decontaminating solution through a contaminated soil mass requires similitude of hydrological and geotechnical aspects of the sites such as fluid velocity, soil stress conditions etc. Such modeling was performed in this study by employing macrohomogeneous cylindrical compacted soil samples that were confined in flexible-wall permeameter. The soil samples were compacted by Harvard miniature compacter and tamping using pore water with a 100mg/L concentration of the organic contaminant. The test specimens had a coefficient of Darcian permeability, k of 10^{-6} cm/s and were tested under a hydraulic gradient of 50-60. Oven-dried mixing-soil was blended with the *p*-Cresol solutions (the concentration of 100mg/L) at a moisture content of 16%, which is slightly higher than the optimum moisture content. The system consists of a flexible-wall permeameter, an influent reservoir called an accumulator, a buret to collect the effluent, and an air supply to provide pressure. Pressure is applied to the system with compressed nitrogen gas that passes through two manifolds and three pressure regulators before being applied to the permeant. Permeant liquid from the influent accumulator flows upward from the bottom of the soil speciment to the top to help in the removal of air bubbles from the soil pores. The effluent line carries the permeant liquid to the buret, which is also used to measure to flow volume. The confining cell pressure of 0.7 kg/cm^2 (10psi) is representative of an element of submerged soil at a depth of 2-4m.

3 Results and discussion

3.1 Adsorption isotherm tests

For *p*-Cresol, adsorption equilibrium was achieved within 48 hours. The amount of contaminant sorbed onto the mixing-soil after adsorption equilibrium was calculated. From the adsorption isotherm test the linear equilibrium partition coefficients K_p values of *p*-Cresol was found 1.21.

3.2 Desorption tests

To evaluate the desorption of *p*-Cresol which sorbed onto the soil, desorption tests were conducted by solution with several surfactant concentrations. Figure 1 shows that the amount of desorption increased from 25% to 43% as the concentration of TX-100 increased from 0.05% to 0.1% (approximately from 5CMC to 10 CMC). But the desorption amount was from 56% to 60% in case of the concentration of TX-100 increasing from 1% to 2%, and this increasing rate was almost negligible. In case of SDS, 52% of *p*-Cresol was removed by 1% SDS(4CMC) solution, while DI water desorbed only 18% of *p*-Cresol.

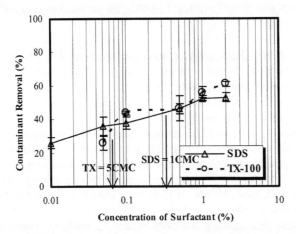

Figure 1: Desorption of *p*-Cresol with TX-100 and SDS

In case of SDS, which is anionic surfactant, the CMC can be lowered by the addition of NaCl. Electrolyte ions stick to the surface of micelle so that the electric double layer is compressed [1]. Then the force of repulsion between surfactant ions is weakened and as a result micelle is stabilized and CMC is lowered. If 0.4M NaCl is added to SDS solution, the concentration of CMC formation changes from 8mM to 1mM. In this study 0.05M, 0.1M, 0.2M, 0.4M NaCl solution was added to 1 CMC SDS solution and a desorption experiment was conducted with each solution. The results are shown in Figure 2. Desorption was promoted by the addition of 0.4M NaCl solution from 41% to 50%. Up to 0.2M NaCl the improvement of desorption was proportional to the NaCl concentration. But the addition beyond 0.4M NaCl to 1CMC SDS solution, precipitation occurred after 2-3 weeks. It is assumed that the precipitation was due to the low solubility of the complexes formed by divalent cations and the anionic surfactant ions, that is $Ca(DS)_2$ [9].

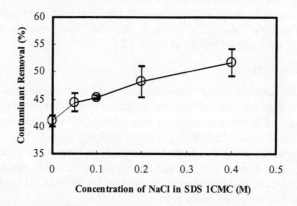

Figure 2: Effect of NaCl addition for *p*-Cresol desorption

The results of the SRI(Successive Reverse Isotherm) test on the mixing soil contaminated with *p*-Cresol are shown in Figure 3 and 4. Figure 3 shows the desorption removal amount with TX-100 at various concentrations. On the succesive desorption experiment with TX-100 solution, desorption amount was 56% after the first washing and 95% after the 7th washing.

Figure 4 shows the results of SRI test with SDS 2% and DI water. In case of SDS 2% solution, desorption amount was 52% after the first washing and 91% after the 7th washing. When treated with water, desorption amount was only 18% by the first washing and 58% after the 7th washing. To investigate the mixing effect of anionic and nonionic surfactants, 2 CMC of each surfactant (TX-100: 2CMC=0.02%, SDS: 2CMC=0.5%) were mixed together. Figure 4 shows that 35% of *p*-Cresol was removed by the first washing and 86% by the 7th washing. This means that the mixed system with anionic surfactant and nonionic surfactant improves the effect of desorption.

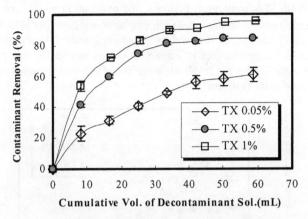

Figure 3: SRI test for *p*-Cresol with TX-100

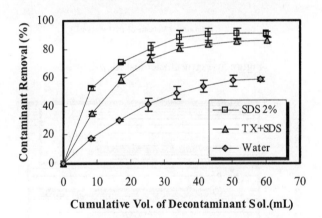

Figure 4: SRI test for *p*-Cresol with SDS and TX-100

3.3 Results of desorption experiments with permeameter test

Flexible wall permeameter tests were conducted to simulate the desorption under the hydrogeological and stress conditions in situ. Figure 5 shows that 87% of *p*-Cresol was removed by 2% TX-100 in the permeation tests, which means TX-100 recovered 1.7 times better than DI water.

Permeameter experiment with DI water shows that the recovery of *p*-Cresol was 18% when the hydraulic conductivity was 8×10^{-6} cm/sec, but it was 51% when the hydraulic conductivity was 8×10^{-7} cm/sec. This means that the contact time between soil and desorbing solution is very important factor when the desorbing solution flows through matrix of soil.

As seen on Figure 7, the total recovery was 75% in flexible-wall permeameter desorption experiment with 0.5% SDS solution. With 2% SDS solution, total recovery was 87% in the permeameter experiment. But in the permeameter experiment with SDS solution in Figure 6, the permeation coefficient declined to 1.5×10^{-7}cm/sec and less as time passed. It can be considered that SDS has a strong dispersion effect that blocks the void by distributing soils [11]. Permeameter and SRI test results are summarized in Figure 7. As you can see in this figure, water desorbed only 50% of the *p*-Cresol, while surfactant solution removed up to 90% of the adsorbed *p*-Cresol onto soil.

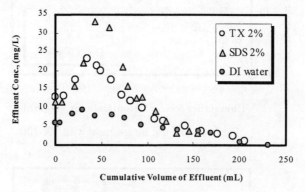

Figure 5: Permeameter test for *p*-Cresol

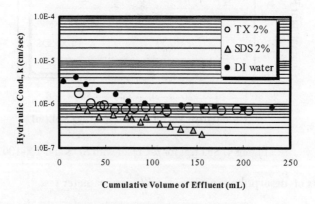

Figure 6: Hydraulic conductivity in permeameter tests

4 Conclusions and further study

Following conclusions are drawn from the test results:
(1) For TX-100, suitable concentration to recover the *p*-Cresol contaminated soil was found to be between 0.1% and 1% in the *p*-Cresol desorption experiment.

(2) For the effect of NaCl to enhance the decontamination of SDS, desorbing capacity changed from 41% to 50% when 0.4M NaCl was added to 1 CMC SDS solution in the batch SRI experiment.

(3) Washing with TX-100 2% solution was 1.7 times higher than washing with DI water in the desorption experiment of contaminated soil using flexible wall permeameter. Recovery in washing with 0.5% SDS solution was 75%, and 87% with SDS 2% solution. But as time passed by, the decline of the hydraulic conductivity coefficient was observed with SDS.

The task of extracting contaminants from contaminated soil is complex; however, the results of this study provide encouraging visions of how to remediate organic-contaminated soils in situ by aqueous solution extraction. Aqueous solution extraction produced physically by applying induced hydraulic gradient can be a promising technique for soil remediation. This method has some limitations and problems. The most serious problem is the secondary contamination by the aqueous solution itself. But this problem can be reduced if biodegradable solutions are used. Further study is necessary for this part of the application.

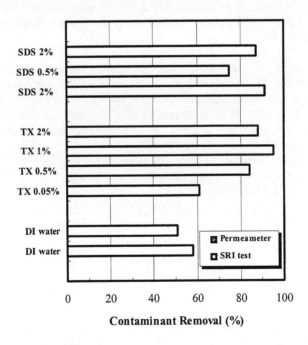

Figure 7: Comparison of SRI and Permeameter tests for *p*-Cresol

References

[1] Rosen, M.J., *Surfactants and Interfacial Phenomena*, 2nd Edition, John Wiley & Sons, Inc., New York., 1989.

[2] Valsaraj, K.T. & Thibodeaux, L.J., Relationships between micelle-water and coctanol-water partition constants for hydrophobic organic of environmental interest. *Water Research*, **23(2)** , pp. 183-189, 1989.

[3] Edwards, D.A., Luthy, R.G. & Liu, Z., Solubilization of polycyclic aromatic hydrocarbons in micellar nonionic surfactant solutions. *ES&T*, **25(1)**, pp. 127-133, 1991.

[4] Abriola L.M., Dekker T.J. & Pennell, K.D., Surfactant-enhanced solubilization of residual decane in soil columns. 2. Mathematical modeling. ES&T, **27(12)**, pp. 2341-2351, 1993.

[5] Sabatini, D.A., Knox, R.C. & Harwell, J.H., *Surfactant enhanced subsurface remediation. Emerging technologies*. In ACS Symposium Series 594. American Chemicl Society, Washington, DC. 1995.

[6] Hunt, J.R., Sitar, N. & Udell, K.S., Nonaqueous phase liquid transport and cleanup 1. Analysis of mechanisms. *Water Resour, Res.*, **24(8)** , pp. 1247-1258, 1988.

[7] Shiau B.J., Sabatini, D.A. Harwell, J.H. & Vu, D., Middle phase microemulsions of mixed chlorinated solvents using food grade (edible) surfactants. *ES&T*, **30(1)**, pp. 97-103, 1996.

[8] Sawyer, C.N., McCarty, P.L. & Parkin, G.F., *Chemistry for Environmental Engineering 4th Edition*, McGraw-Hill International Editions, pp. 227-228, 1994.

[9] Liu, N. & Roy, D., Surfactant-Induced Interactions and Hydraulic Conductivity Changes in Soil, *Waste Management*, **15(7)**, pp. 463-470, 1995.

[10] Park, J.B., O'Neill, M.W. & Symons, J.M., Laboratory Study of Soil Flushing by Aqueous Solutions. *Journal of Geotechnical and Geoenvironmental Engineering*, ASCE, **124(10)**, pp. 989-996, 1998.

[11] Gabr, M.A., Bowders, J.J. & Shoblom, K.J., Flushing of Polyaromatic Hydrocarbons from Soil Using SDS Surfactant. ASCE, *GEOENVIRONMENT 2000*, pp. 1321-1334, 1995.

Recovery of coal tar in 1-D column test using surfactant enhanced aquifer remediation (SEAR)

S. Yoon, K. Kostarelos, & S. Urbanczyk
Department of Civil Engineering, Polytechnic University, USA.

Abstract

This paper presents the results of flow–through column testing using a new surfactant, Alfo–terra 123-8 PO sulfate, for the recovery of coal–tar from contaminated soil. Coal tar is a dense, non–aqueous phase liquid (DNAPL) that is difficult to recover due to its hydrophobic behavior and its high viscosity. Preliminary testing (phase behavior–batch tests) revealed a high degree of coal tar solubilization using a formulation of 12 wt.% Alfo–terra 123-8 PO sulfate surfactant, 4 wt.% secondary butanol, and 200 mg/L calcium chloride (as calcium). The formulation also exhibited rapid equilibration, without forming anisotropic gels. The column test was performed to verify compatibility of the formulation with the field–obtained DNAPL and soil. The column test was successful, resulting in low pressure drop, over 97 % coal tar recovery, with no visible gel formation nor any coal tar mobilized. This experiment showed good compatibility of the selected surfactant using field soil and field DNAPL, and should be studied further and considered for use in the field.

1 Introduction

Generally, coal tar is a by–product from operations at manufactured gas, gas purification, gas storage, gas distribution, and coke production facilities [1, 2]. Coal tar—a mixture of compounds—is a black, gooey, sludge–like material that smells like mothballs [1, 2]. Coal tar has proven difficult to remove from soil due to its hydrophobic behavior and its high viscosity. It dose not easily dissolve in water, yet there is concern for potential ground water contamination because the presence of poly aromatic hydrocarbons (PAHs) that make–up coal tar have been linked to cancer. In the United States, coal–tar contaminated sites are found in such places as South Chattanooga, TN, Columbia, PA, Brooklyn, NY, Newport, NJ, and so on [3,4,5]. Currently, the most common remediation methods for coal tar contamination is pump and treat, solvent extraction, bioremediation, and excavation.

In this column experiment, the selected surfactant solution showed promise for the recovery of coal tar by virtue of solubilizing large quantities when mixed in test tubes and by the absence of viscous gels and other anisotropic behavior. However, before use in the field, the surfactant solution must be tested to ensure compatibility with the field soil as well. Therefore, it is important to note that this objective is accomplished by recording the pressure–drop across the column during the surfactant flood. The second indication of compatibility between surfactant solution, field coal tar, and field soil is found in the recovery profile of the surfactant flood. A high recovery percentage (around 100 %) is an indication of good compatibility.

2 Experimental Equipment and Material

2.1 Experimental Equipment

A Kimble–Kontes® Chromaflex chromatography column (model # 426870-2560), 2.5–cm diameter and 60–cm length, was used in the column experiment. An end piece (model # 426876-0025) with an adjustable bed length was used to keep soil confined after placement. During the flood, a fluid reservoir was used to hold injected fluids (de–aired water and surfactant solution). The reservoir was a 10–liter capacity and was custom–made of Plexiglass. The column set–up was made using Swagelok® valves fittings and tubing with Perforaxy (PFA) tubing of 1/8–inch diameter, fittings and ferrules made of either nylon or stainless steel in the same size. A dual–action piston pump (model # AA 100-S made by Eldex) was used to inject fluids during the experiment. The pump is capable of delivering constant flow rates from 0.2 to 10.0 mL/min and 5000–psi maximum pressure. A fraction collector was assembled with the column set–up for the collecting samples at timed intervals during the surfactant flow experiment. The fraction collector (model # 68-3870-002) has a capacity of 144 tubes (13mm diameter), made by Instrument Specialties Company (ISCO). Corning® glass tubes with a 5–ml capacity and 13–mm diameter were used in the fraction collector.

The pressure drop between the injection and production points of the soil column was measured using a Cole–Parmer® pressure transducer (model # EW07354-22). This transducer uses a solid–state sensor to convert a 24–Volt DC excitation to a 20–mA current. Inside the transducer, a 150–Ohm and a 100–Ohm resistor are connected in series (total of 250–Ohm) between the output terminals and a voltage meter (Cat. No. 22-801 Digital Multimeter made by Radioshack®), measures the transducer response. A DC power supply of 24 Volts by an Omega (model # U24Y101) is used to excite the transducer. After the collecting the samples, a gas chromatograph, an Agilent (formerly Hewlett

Packard) 6890 series equipped with a flame ionization detector (FID), was used for analyzing sample concentration. An HP5 capillary column 30 m–long (diameter 0.32 mm) with a film thickness of 0.25μm was installed in the chromatograph.

2.2 Experimental Material

The surfactant formulation consisted of surfactant, calcium chloride, secondary butanol, and de–ionized water. Alfo–terra® 123-8 PO sulfate, made by Condea Vista Corporation, Austin, TX, was the surfactant used. It is a branched alcohol pro–poxylate sulfate sodium salt. Calcium chloride, made by Fisher Scientific Company, was used as the electrolyte. Secondary butanol, from Fisher Scientific Company, of 99 % purity was used in the surfactant solution as a co–solvent. The water de–ionizer was a Nanopure Infinity UV/Reverse Osmosis System (model # D8971/9011), made by Barnstead Thermolyne Corporation. In the osmosis system, the membrane rejected 93 % of monovalent ions, polyvalent ions, particles, microorganisms and dissolved organics; the UV system's production rate was 18.0–MΩ-cm.

3 Experimental Method

3.1 Mixing Surfactant

Surfactant solutions were made on a percentage by weight basis as–received. Distinction must be made between surfactant concentrations that are reported on an "active" basis or "as received." The column experiment was performed using a solution of 12 wt.% as received (4 wt.% active), 4 wt.% secondary butanol, 200 mg/L calcium chloride (as Ca). The surfactant solution was prepared using two solutions, one consisting of 40 mL secondary butanol with 200 mL water (de–ionized), and a second of 0.735 g calcium chloride in 200mL water. These two solutions were added to a 1L volumetric flask containing 133.3 g Alfo–terra surfactant, and water added to the 1–liter mark.

3.2 Column preparation

The glass column was jacketed and surfactant flood was performed at the aquifer temperature (17 C) using a cold water bath. After weighing the dry, empty column set–up (including end pieces, tubing, and valves), it was packed with the field soil that was contaminated with field NAPL at a ratio of 0.05 g NAPL per 1 g soil. Before contaminating the soil, particles larger than No. 10 mesh and smaller than No. 200 mesh, including organic matter were removed. The end pieces were added and the column re–weighed. Each end piece had two screens—a 60–mesh toward the soil to prevent the movement of fines and then a 120–mesh screen to distribute flow uniformly across the column diameter. The dead volume of the end piece assembly was 2.7 mL.

Once the column was assembled and pressure–tested, it was water–saturated and re–weighed. The data was used to estimate the reduced pore volume of the pack before surfactant injection.

3.3 Surfactant flood

The initial surfactant injection rate was set to 2 mL/min, about 50 ft/day. After examination for any produced NAPL, the samples were covered with aluminum foil and refrigerated to prevent evaporation losses while awaiting GC analysis. GC analysis of the effluent samples was made to determine the concentration of coal tar, and calculate the total mass recovered. During the initial stage of the surfactant flood, pressure measurements were recorded every 2 minutes.

During the initial flood, a shut–in period was made after 5 pore volumes of surfactant were injected into the column. The initial flow rate was determined to be too high to reflect field conditions and there was concern that the contact time between the surfactant solution and the coal tar may be too short to allow the solubilization to reach equilibrium. The shut–in period allowed for greater contact time between surfactant solution and coal tar in order to determine whether equilibrium conditions were reached. After several weeks, the surfactant flood was re–started until complete recovery of the coal tar, 13.59 g, was attained. The new flow rate was 0.13 mL/min, about 3.5 ft/day, which reflects a flow rate that can be reasonably attained in the field. Using this slower flow rate, both sampling and pressure measurement intervals were increased to about 20 minutes. A total of 18.65 PV of surfactant were injected into the column.

After the column test, a water flood was performed to remove any remaining surfactant solution from the pack. Then, the column was re–weighed and the gravimetric data used to compute the mass of coal tar recovered. The column was then opened and the contents extruded into a beaker of methylene chloride. Since coal tar is completely soluble in methylene chloride, we hoped to extract any remaining coal tar from the soil. After stirring for 24 hours to allow any possible coal tar remaining in the soil to dissolve into the methylene chloride, an aliquot was analyzed by gas chromatography.

3.4 Gas Chromatograph analysis

An Agilent 6890 gas chromatograph with flame ionization detector (FID) was used to analyze effluent collected from the column. The GC was equipped with an HP5 capillary column that is 30–m long, 0.32–mm diameter, with a film thickness of 0.25 m. Since the coal tar is a mixture of hydrocarbons that results in several peaks when analyzing with a GC, one peak with a retention time of 4.996 minutes was used as an indicator compound. All coal tar measurements were reported using this indicator compound.

4 Results

4.1 Pressure drop as a function of volume produced

Before contaminating soil, a pressure drop of 0.9 psig was measured for the clean soil. During most of the surfactant flood, the pressure drop across the column remained low, below 2 psig; note that in Figure 1, the pressure drop rose sharply at 450 ml produced after the surfactant solution was shut–in. When the pumps were re–started, the pressure increased temporarily and quickly returned to normal. The final pressure drop across the column was 0.9 psig. This value is the same as the initial pressure drop measured for clean soil.

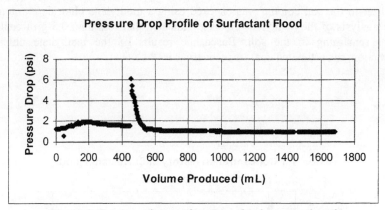

Figure 1: Pressure drop as function of volume produced
(1PV = 89mL)

4.2 Visual observation

During the surfactant flood, no viscous nor anisotropic gels were observed. Also, the samples produced had some color (such as light orange, dark orange, brown, and dark brown). This color is probably related to the degree of solubilization of coal tar in the surfactant solution. The initial samples were light orange, similar to those from the water flood. Then, the samples darkened to brown, and the final sample color was again light orange. It appears that the darker colored samples high concentration of coal tar and the lighter color sample had low concentration of coal tar.

4.3 Coal tar recovery

The produced effluent was analyzed for coal tar and the concentration profile is plotted in Figure 2. The surfactant solution was shut–in after 5 pore volumes (450mL) produced. The surfactant flood was re–started after one month at a flow rate of 0.13 mL/min (3.5 ft/day), which was slower than the initial flow rate of 2 mL/min (50 ft/day). The slower flow rate allowed for longer contact time between surfactant solution and coal tar, and is more realistic of flow rates attainable in the field. During the column experiment, the solubilization of coal tar was quite high coal tar but no coal tar was mobilized. The initial solubilization of coal tar was 30,000 mg/L, about 20 times greater than the aqueous solubility of coal tar. In Figure 2, note that after 5 pore volumes of surfactant flooding, the concentration of coal tar dropped from 30,000 mg/L; however, the concentration remained above 5,000 mg/L until 12 pore volumes. At this point, as can be seen in Figure 3, the recovery percentage was over 80 %, or 13.59 g of coal tar recovered from the soil column. Finally, after 18.65 pore volumes of surfactant flooding, the cumulative recovery of coal tar was over 97 %. After the surfactant flood was stopped, water was used to remove any surfactant solution and the column was re–weighed. Gravimetric data indicated that the remaining amount of coal tar was 0.5 g. Then, the remaining amount of coal tar in the soil column was checked using a methylene chloride extraction. The analysis of the methylene chloride indicated that less than 0.5 g of coal tar were remaining in the soil. Based the results of the methylene chloride extraction, the estimated coal tar recovery is 97 %.

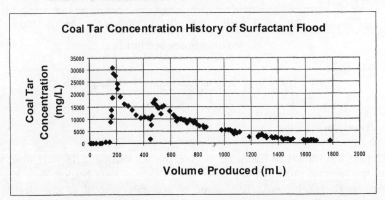

Figure 2: Coal tar concentration as function of volume produced
(1PV = 89 mL)

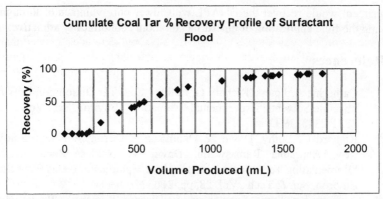

Figure 3: Coal tar recovery (%) during the surfactant flood
(1PV = 89 mL)

5   Conclusions

Based on the results obtained from this experimental study, the following conclusions can be drawn:

A. Pressure drop across the column were low, indicating good compatibility between the selected surfactant solution, the coal tar, and the field soil.

B. No viscous gels or other anisotropic behavior were noted in the produced effluent. These results indicate a suitable formulation for use in a pilot–scale test.

C. The formulation of 12 wt. % surfactant, 4 wt. % secondary butanol, 200 mg/L calcium chloride (as calcium) performed well in the flow–through column test: 1) coal tar was not mobilized, 2) the solubilization of coal tar in the effluent was quite high, initially about 30,000 mg/L (about 20 times the aqueous solubility of the coal tar), 3) pressure drop was low, usually less than 1 psig.

D. The total recovery percentage of coal tar was high, about 97 %, using 18.65 pore volume of surfactant solution. This result leads to the conclusion that the use of Alfo–terra 123-8 PO sulfate surfactant® solution could be an aid in recovering coal tar in the field when compared to a pump–and–treat approach using only water.

The formulation of 12 wt. % surfactant (as received), 4 wt. % secondary butanol, 200 mg/L calcium chloride (as calcium) solution had high solubilization capacity for coal tar in the field soil and good compatibility with coal tar. In the future, an experiment with a surfactant formulation designed to solubilize higher amounts of coal tar will be performed. If successful, this future column experiment would recover the DNAPL with fewer pore volumes of throughput, making the field application of this surfactant more economically attractive.

6 References

1. Catherine A. Peters, Richard G. Luthy, "Coal Tar Dissolution in Water–miscible Solvents: Experimental Evaluation", Environ. Sci., Vol. 27., pp. 2831-2843, 1993
2. Richard G. Luthy, David A. Dzombak, Catherine A Peters, Sujoy B. Roy, Anuradha Ramaswami, David V. Nakles, Babu R. Nott, "Remediating Tar–contaminated Soils at Manufactured Gas Plant Sites", Environ. Sci. & Tech., Vol. 28, pp. 266–276, 1994
3. Agency for Toxic Substances & Disease Registry, "Public Health Advisory for the Tennessee Products Site", 1993
4. Banks, John, "In Situ Removal of Coal Tar", US EPA, [http://www.nato.int/ccms/s13/report/intrm28.html], January 1998.
5. Newhoff, Nancy R., "Meeting slated to discuss cleanup of Waterloo coal site", Waterloo–Cedar Falls Courier, [http://www.wcfcourier.com/metne/917cleanup.html] September 1997.

Selection of alternative sorbent media for contaminated groundwater clean-up

M. R. Boni & L. D'Aprile
Department of Hydraulics, Transportation and Roads, University of Rome, "La Sapienza", Rome, Italy

Abstract

Methods for the remediation of contaminated groundwater have focused, during the past years, on pump-and-treat and containment systems. Many experiences at contaminated groundwater sites and recent studies have shown the inadequacies of this approach and the need for new treatment technologies. The recent development of permeable barriers has presented a potentially viable alternative to conventional systems. Although zero-valent granular iron is the most commonly used reactive medium, a variety of media could be used to treat groundwater contaminants. This paper assesses possible alternatives to conventional sorbent media for metal removal from groundwater. The physical-chemical properties of two alternate materials (compost and expanded clay) were analysed and included particle size distribution, void ratio and dry density. Leaching test were conducted with two different methods, the first suggested by the Italian regulation on solid waste and the second, TCLP, standardized by the U.S EPA, to assess the possible drawbacks of the investigated materials in terms of release of harmful substances. Batch testing was undertaken on the investigated materials to assess their sorption potential for cadmium and nickel. The experimental results show an high removal efficiency of the investigated materials in the tested concentration range, suggesting the possibility of a successful full-scale application in a permeable barrier.

Introduction

At many industrial sites unappropriate plant management and uncontrolled waste disposal have led to widespread contamination of soil and groundwater.

Groundwater remediation technologies usually involve extraction and subsequent treatment of contaminated water. Conventional pump-and-treat systems could be used to capture and treat the plume. However, past experience at contaminated groundwater sites and recent studies have shown the limits of this approach. Also pump-and-treat systems would have to be operated for many years or decades, or as long as the source zone and the plume persist: the associated operation and manteinance costs over several decades can be enormous. Moreover this technology is often inadequate to reach clean-up goals, since subsurface residuals frequently remain at undesirable levels.

The recent research studies on Permeable Reactive Barriers (PRBs) have presented a potentially viable alternative to conventional pump-and-treat systems. These subsurface permeable treatment walls are being considered as a cost-effective in situ remediation technology.

In its simplest form, a permeable reactive barrier consists of a zone of reactive material (such as granular iron, activated carbon, etc.) installed in the path of a contaminated plume. The main advantage of this system is that, generally, no pumping or aboveground treatment is required; the barrier acts passively after the installation, without requiring an external energy source.

Activated carbon is a material widely recognised for its sorption properties and is frequently used to remove aqueous phase organic contaminants. However, the use of activated carbon in permeable reactive wall applications can drive up the costs of this form of remediation. Because of this, alternatives to activated carbon have been investigated in order to estabilish materials with comparative sorptive properties (1).

This paper presents the results of laboratory test that have been undertaken using two possible alternatives to activated carbon, compost and expanded clay. The characteristics of the investigated were analysed and included particle size distribution, void ratio and dry density. Leaching test were conducted with two different methods, the first suggested by the Italian regulation on solid waste and the second, TCLP, standardized by the U.S EPA, to assess the possible drawbacks of the investigated materials in terms of release of harmful substances. Batch testing was undertaken on the investigated materials to assess their sorption potential for cadmium and nickel. The experimental results show an high removal efficiency of the investigated materials in the tested concentration range. However the results of the leaching test show potential drawbacks for the compost in terms of release of harmful substances, thus advising against a possible in situ application.

Materials and methods

Tha materials selected for analysis were: compost derived from the organic fraction of MSW (supplied by a MSW treatment plant in Italy) and expanded clay (Leca S.p.A, Italy). The biological stability of the compost was evaluated through dynamic respirometric test. Three granulometric fractions of expanded clay were chosen for the test: 8-16 mm, 2-4 mm, 0,125-0,25 mm. The main characteristics of the three fractions are reported in Table 1.

Fraction	Dry density (g/cm^3)	Bulk density (g/cm^3)	Void index	Porosity (%)
8-16 mm	0,292	2,554	7,746	88,566
2-4 mm	0,298	2,554	7,570	88,320
0,125-0,250 mm	0,812	2,554	2,145	68,196

Table 1: Main characteristics of the expanded clay fractions

Table 2 shows the main characteristics of the organic fraction compost.

Dry density (g/cm3)	Bulk density (g/cm3)	Void index	Porosity (%)
0,428	1,783	3,167	76

Table 2: Main characteristics of the organic fraction compost.

Aqueous solution with the contaminants of interest (cadmium and nickel) and all the reagents for the leaching tests were prepared using all chemicals as received. Batch tests were conducted in 40 ml amber glass vials. Each vial, containing 3,8 g of solid (compost or expanded clay) was filled with 38 (solid/liquid ratio = 1:10) ml of the contaminated solution, and kept at constant agitation (IKA KS 260, 300 rpm) until sampling time. The investigated initial concentrations were 5, 10, 20 mg/l of cadmium and 20, 40, 80 mg/l of nickel. Samples of the liquid for the analysis of the investigated metals were taken after 10, 30 min, 1 h, 5 h, 24 h, 28 h, 27 h, 7 days, 10 days. The concentrations of metals in the samples from batch and leaching test were determined by Atomic Absorption Perkin Elmer 3030B.

Leaching Tests

Two different leaching tests were performed on the selected media. The first test is suggested by the Italian regulation on solid waste (D.Lgs. 22/97, "Decreto Ronchi", Appendix 3), performed in deionised water with sequential extractions; the second test, TCLP, Toxicity Characteristics Leaching Procedure is an EPA standard test performed in acid environment and usually applied in the United States for the chemical characterization of the reactive material used in PRBs. The results of both the leaching test methods performed on the organic fraction compost (Table 3) show concentration over the regulatory limits for iron, copper, manganese, nickel, lead and total chromium.

Leaching test suggested by the Italian regulation (D.Lgs 22/97, "Decreto Ronchi, Appendix 3)						
Sampling time	Average concentration (mg/l)					
	Fe	Cu	Mn	Ni	Pb	Cr
2 h	7,75	4,17	0,50	0,58	0,27	1,08
8 h	3,21	1,73	0,17	0,22	0,18	0,40
24 h	7,37	2,18	0,42	0,26	0,38	0,59
48 h	2,36	1,26	0,17	0,17	0,19	0,35
72 h	4,10	1,93	0,23	0,28	0,23	0,50
102 h	2,20	1,04	0,15	0,09	0,06	0,30
7 d	1,62	0,95	0,12	0,07	0,19	0,19
16 d	3,30	0,37	0,26	0,07	0.04	0,20
Regulatory Limit (mg/l)	0,2	1	0,05	0,02	0,01	0,05
TCLP						
Sample #	Average concentration (mg/l)					
	Fe	Cu	Mn	Ni	Pb	Cr
1	2,11	1,27	4,22	0,27	0,36	0,38
2	2,96	1,40	4,24	0,26	0,36	0,45
3	1,87	1,53	4,15	0,25	0,34	0,39
Regulatory Limit (mg/l)	0,2	0,005	1	3	0,05	0,02

Table 3: Results of the leaching test on the organic fraction compost

The same test performed on the three expanded clay fraction show concentration below the regulatory limits for the investigated metals (iron, cadmium, copper, zinc, manganese, nickel, lead, chromium).

Sorption potential

The sorption potential of the investigated materials for cadmium and nickel was evaluated using a linear approximation for the sorption isotherm (2):

$$q_c = \frac{X}{m} = C_e K_d$$

Where q_c is the amount of solute sorbed onto the solid (mg/g) and C is the equilibrium concentration of the solute (mg/l) (1).

The distribution coefficient, K_d, represents the partitioning of the compound between liquid and solid.
Retardation factors (R_f) were calculated using laboratory values for K_d, dry density and porosity and the following equation (1):

$$R_f = 1 + \frac{\rho_d K_d}{\theta}$$

where ρ_d is the dry density and θ is the porosity.
Figures 1, 2 and 3 show the trend of the concentration of cadmium in solution versus time for the investigated fractions of expanded clay (8-16 mm, 2-4 mm, 0,125-0,25 mm). Figure 4 shows, as an example, the linear approximation for the sorption isotherm of the 8-16 mm fraction. The values of K_d, and R_f are reported in Table 4.
Figures 5 and 6 show the results related to the concentration of cadmium in solution versus time and the linear approximation for sorption isotherm for the organic fraction compost.

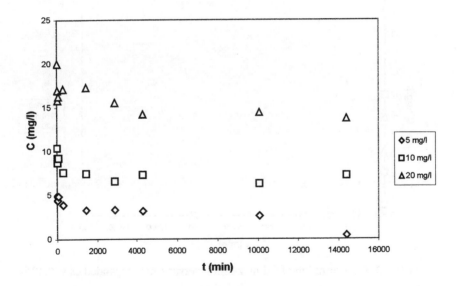

Figure 1: Concentration of Cd in solution versus time (expanded clay, 8-16 mm)

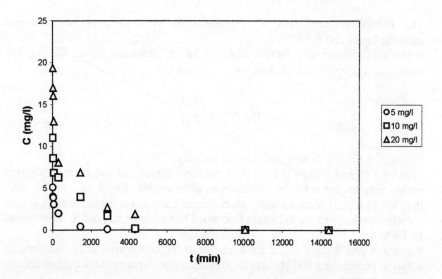

Figure 2: Concentration of Cd in solution versus time (expanded clay, 2-4 mm)

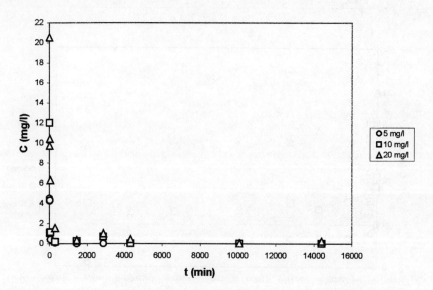

Figure 3: Concentration of Cd in solution versus time (expanded clay, 0,125-
0,25 mm)

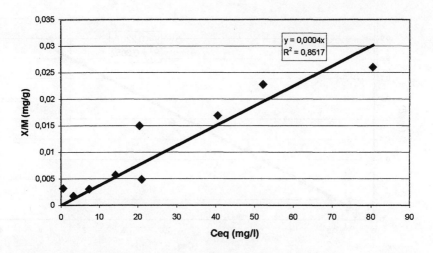

Figure 4: Linear approximation for the sorption isotherm of the 8-16 mm fraction of expanded clay for Cd

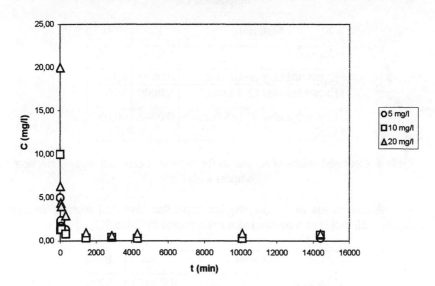

Figure 5: Concentration of Cd in solution versus time (organic fraction compost)

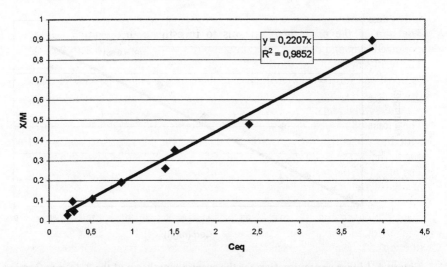

Figure 6: Linear approximation for the sorption isotherm of the organic fraction compost for Cd

Material	K_d	R_f
Compost	0,2193	124,50
Expanded clay (8-16 mm).	0,0004	1,13
Expanded clay (2-4 mm).	0,0009	1,30
Expanded clay (0,125-0,25 mm).	0,0748	90,05

Table 4: Obtained values of K_d and R_f for expanded clay and organic fraction compost with Cd

The obtained values of K_d and R_f for expanded clay and organic fraction compost with nickel as a contaminant are reported in Table 5.

Material	K_d	R_f
Compost	0,0188	11,59
Expanded clay (8-16 mm).	0,0003	1,01
Expanded clay (2-4 mm).	0,0031	2,04
Expanded clay (0,125-0,25 mm).	0,0048	6,71

Table 5: Obtained values of K_d and R_f for expanded clay and organic fraction compost with Ni

Conclusions

The aim of the performed test was to investigate the sorption potential of expanded clay and organic fraction compost as alternatives to activated carbon. The results of batch test show an high removal efficiency for the organic fraction compost and for the smaller fraction of the expanded clay (0,125-0,25 mm).

However the values of K_d and R_f (Table 4 and 5) show a lower adsorption potential when nickel was chosen as a contaminant of interest.

The effect of particle size is particularly evident in the case of expanded clay with an increasing of the adsorption rate when the particle size decrease. This is due to the fact that the surface area increases with decreasing particle size (3).

The effect of the initial concentration in solution has a clear effect only in the results of the test performed on the larger fraction of the expanded clay (see Figure 1) in which the adsorption rate is slower.

Even if the organic fraction compost seems to have better sorption properties (higher retardation factor) if compared to the expanded clay, even in the smaller fraction, the results of the leaching test show some environmental drawbacks for this material due to the potential release of harmful metals (copper, manganese, nickel, lead and total chromium) in solution, thus indicating more useful applications, in permeable reactive barriers, for the expanded clay.

References

[1] Parker, F., Smith D.W., Hitchcock P.W., The use of alternatives to activated carbon for the clean up of contaminated groundwater, Proceedings of Geoeng 2000, Melbourne (Australia), 2000

[2] Hinz, C. Description of sorption data with isotherm equations, Geoderma 99, pp. 225-243, 2001

[3] Mellah, A., Chegrouche, S., The removal of zinc from aqueous solutions by natural bentonite, Wat.Res. Vol. 31, n°3, pp.621-629, 1997

In-situ chemical oxidation of a chlorinated groundwater plume at a brownfields site

P. Kakarla, R. Greenberg & T. Andrews
In-Situ Oxidative Technologies, Inc., USA

Abstract

Groundwater at a former manufacturing facility to be converted to retail use located in Bergen County, New Jersey, USA was contaminated with chlorinated solvents principally consisting of Trichloroethene (TCE), Vinyl chloride (VC) and Cis-1,2-Dichloroethene (Cis-DCE). A patented Fenton-based oxidation process known as the ISOTEC[SM] process was utilized to perform remedial activities at the site. A proprietary blend of chelated iron catalyst and mobility control agents that function in the pH range 5-7 was applied along with stabilized hydrogen peroxide to promote in-situ generation of the potent hydroxyl radical and other free radical oxidants, which occur as intermediaries in Fenton's reaction. Laboratory comparison experiments conducted in soil columns revealed significantly higher mobility and superior sorption characteristics for ISOTEC[SM] catalysts when compared to a conventional Fenton's catalyst such as acidified iron (II) solution. For the highest column length used in the experiments, greater than 95% of iron in acidified iron (II) catalyst was adsorbed to the soil as opposed to less than 14% adsorption using the ISOTEC[SM] catalyst. Preliminary laboratory-scale research was conducted to test the process efficacy towards in-situ chlorinated solvent remediation. Based on successful experimental results that indicated greater than 99% reduction of TCE, VC and cis-DCE; a field-scale treatment program was initiated utilizing a plurality of injection wells installed in a grid fashion throughout the site. Results of treatment indicated over a 98% reduction in total chlorinated contamination detected in the site groundwater within five months, which increased to greater than 99% decrease after one year.

Introduction

In situ application of Fenton's process was not considered feasible for a long time due to the instability of oxidizing and catalytic reagents when introduced into the subsurface. Several biotic and abiotic catalysts that are native to the soil were found to quickly decompose hydrogen peroxide to oxygen and water [Schumb et al. (1955)]. More over, conventional Fenton's catalysts were found to be effective only under acidic conditions [Watts et al. (1990)], which are impractical to replicate during "real world" application. Traditionally, a Fenton catalyst introduced into the subsurface was always found to be consumed within inches from the point of injection and its pH buffered by native soil from the desired acidic conditions to more natural circum-neutral conditions. These natural high pH conditions in turn lead to precipitation of iron that is the key component needed to promote Fenton's reaction, rendering the process ineffective for in-situ application. Hence, a need was realized for effective Fenton's catalysts that maintain iron in soluble form under natural conditions while simultaneously promoting hydroxyl radical generation.

In-Situ Oxidative Technologies, Inc. (ISOTEC[SM]) developed catalysts that are effectively chelated to prevent precipitation under natural subsurface conditions. The catalyst is prepared at natural circum-neutral pH conditions that prevent any sudden pH changes when introduced into the subsurface due to native soil buffering. In addition, the ligand molecules within the chelated catalyst have greater affinity towards the iron compared to other metals, thereby preventing iron loss through adsorption to the soil. The resultant catalytic reagent was experimented with and found to have significantly higher mobility compared to conventional Fenton's catalysts. In addition, hydrogen peroxide mixed with stabilizing compounds and introduced at lower concentrations (<10%) was found to be more effective towards in-situ remediation compared to using at high concentrations presumably due to free radical scavenging by excess peroxide. Not only did stabilized low concentrations of peroxide have a longer lifetime relative to high concentrations, but were also easier to handle during in-situ application.

The purpose of this article is to demonstrate the laboratory and field application of the ISOTEC[SM] process towards remediation of mobility and pH chlorinated contaminated groundwater at a former manufacturing facility. In addition, laboratory experiments, which compare the sorption, characteristics of ISOTEC[SM] catalysts with conventional acidified iron (II) catalyst, are presented.

Part I: Laboratory experiments

Mobility/ sorption experiments

The soil columns used to test catalyst mobility and sorption characteristics were identical to those described in another article (Kakarla and Watts, 1997). The columns (3.2-cm I.D.) were prepared from Schedule 40 polyvinyl chloride

(PVC) used in lengths of 30 cm each for experimental purposes. The top of the PVC column was left open while the underside was fit with perforated PVC caps to allow for eluent drainage. The columns were internally lined with chemically inert Tedlar (PTFE) sheeting (Du Pont, NY). A typical soil column was packed using a high organic soil obtained from northern New Jersey in increments of 50 grams with exactly 10 taps given to the column between each increment. This ensured uniformity in the columns packed during comparison experiments.

Experiments to compare the sorption and mobility characteristics of ISOTEC^{SM} catalyst with acidified iron (II) catalyst were conducted using three pairs of columns packed at three different depths. The column pair #1 was packed using 100 g of soil, column pair #2 using 200 g of soil, and column pair #3 using 300 g of soil. The corresponding post-packing depths of each column pair were determined to be approximately 6.2 cm, 13 cm and 20 cm, respectively. The catalyst solutions were prepared to obtain an initial concentration of 12.5 mM iron (or 687.5 mg/L as iron) in each catalyst. Exactly 120 ml of the catalyst solution was introduced at the top and allowed to permeate through the depth of the column. One column in each pair received acidified iron (II) catalyst as influent while the second column in each pair received ISOTEC^{SM} catalyst as influent. The eluent collected at the base was analyzed for residual iron concentration (using orthophenanthroline spectrophotometric method) and final pH value. Results for each column pair were compared to evaluate the sorption and mobility characteristics of each catalyst.

Contaminant treatability experiment

Contaminant treatability experiments were conducted on groundwater samples collected from the most contaminated area of the subject site. The experiments were conducted in 140-ml reaction vessels, which were sealed with aluminum caps containing rubber septa. The ISOTEC^{SM} catalytic reagent and stabilized hydrogen peroxide were sequentially introduced in cycles into the reaction vessels using syringes. Each treatment cycle consisted of a concentration and volume of reagents in the optimal combination determined in previous experiments (unpublished; procedures discussed in Greenberg et al 1998). One of the reaction vials was initially isolated to serve as control sample and received equivalent doses of distilled water to compensate for reagent volumes injected. The reaction vessels were left undisturbed for at least 24 hours or until such time that the peroxide is completely consumed. The reaction was quenched with a drop of catalase to decompose any residual peroxide. All samples were analyzed for volatile organic compounds (VO+10) by USEPA method 624.

Results and Discussion

Mobility and Sorption Experiments

The results of these experiments have been plotted in Figures 1 and 2. Figure 1

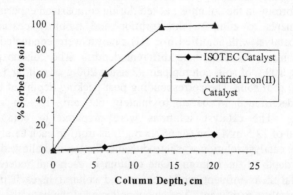

Fig. 1 Catalyst Sorption in Soil Columns

shows the iron adsorption as a function of column depth for both ISOTEC[SM] catalyst as well as acidified iron (II) catalyst. Nearly 100% of iron in acidified iron (II) catalyst is adsorbed to the soil when passed through a 300-g column (or 20-cm column). In comparison, only 13.6% of iron present in ISOTEC[SM] catalyst was adsorbed. As expected, the percent sorption for both catalysts increased with the depth of the columns. While this increase is only marginal for ISOTEC[SM] catalysts, it is steep for acidified iron (II) catalyst, indicating that the mobility of this catalyst is very limited when injected into a "real world" environment. Using the concentration of residual iron in the eluent as the criteria, it is evident that ISOTEC[SM] catalysts are at least 5-6 times more mobile than acidified Fe (II) catalyst.

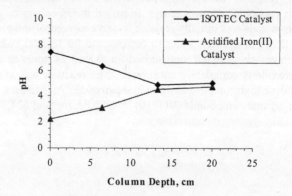

Fig. 2 pH Variation in Soil Columns

The pH variation of the catalytic reagents as a function of the column depths has been plotted in Figure 2. The plot demonstrates a clear buffering effect of the soil on the injected catalysts as they permeate through the soil column. The pH of the acidified iron (II) catalyst gradually increased from its low values towards the pH range 4.75-5.5 where as the circum-neutral pH of the ISOTEC^SM catalyst gradually converged towards the pH range of 5-5.5 as the column depth increased. This increase in the pH of the acidified iron (II) catalyst may have significantly contributed to the increased adsorption of iron to the soil through precipitation at high pH values, as was evident in Figure 1. The chelates present within the ISOTEC^SM catalyst have a greater affinity towards iron compared to the native soil and hence, minimized iron loss through adsorption during the permeation process.

Contaminant treatability

The results of bench scale study performed on contaminated groundwater procured from the former manufacturing facility site have been tabulated as below (Table 1). Results indicate over a 99% decrease in concentrations of TCE, Cis-1, 2-DCE and VC within the treated samples when compared to the control sample. Even though these results provided only minimal information about the site-specific effectiveness of the ISOTEC^SM process, they demonstrated the effectiveness of the ISOTEC^SM catalysts towards oxidation of chlorinated contamination of concern under controlled conditions. Also, when combined with the data on mobility and sorption characteristics presented in Figures 1 and 2, the contaminant treatability results demonstrated the enormous potential of the ISOTEC^SM process towards in-situ field application.

Table 1 Results of Contaminant Treatability Experiments.

	VC	TCE	Cis-DCE	Total VOC's	% Reduction
Initial/Control	696 µg/l	904 µg/l	187,000 µg/l	188,600 µg/l	-
Treated	ND (<10) µg/l	12.9 µg/l	110 µg/l	123 µg/l	99.9%

ND = Analyzed for but not detected at the method detection limit (MDL), µg/l = micrograms per liter, VC = Vinyl Chloride; TCE = Trichloroethene; Cis-DCE = Cis-1,2-Dichloroethene; VOC = Volatile Organic Compound.

Part 2: Field implementation

The subject site located in Bergen County, New Jersey, USA consisted of a former manufacturing and distribution facility for metal fasteners that used and generated hazardous chlorinated solvents and wastes that was to be converted for retail use. Based on the results of laboratory experiments, a field-scale treatment program was implemented to reduce the overall chlorinated VOC contamination in the subsurface aquifer. General site geology consisted of layers of gravel, fine sand, sandy silts, clayey silts, and clays in the top 1.5-3 m below grade surface and very fine-grained sandy silts at depths ranging from approximately 3 m to 12 m below grade. Groundwater was encountered in unconfined conditions at depths ranging from approximately 1 m to 1.4 m below site grade. Bedrock was not encountered during site investigations. A site map is included as Figure 3. Site contamination principally consisting of TCE, Cis-DCE and VC originated from a former wash pit area within the warehouse building (see Figure 3). Based on delineation activities, the groundwater plume traveled in an easterly direction towards the NJ Transit railroad tracks. Majority of site contamination existed in the depth range of 3-12 m below grade surface.

A total of thirteen deep injection wells (IP) and four shallow injection wells (SIP) were installed to encompass the delineated area of the plume (Figure 3). Each deep injection well was constructed to a depth of 12 m separated by a spacing of approximately 1.5-3 m between the points. The injection wells were screened from a depth of 6-12 m below grade surface to ensure uniform vertical distribution of injected reagents. The location of the shallow injection wells corresponded to the former source area where the chlorinated solvent releases occurred. The four shallow injection wells were constructed to a depth of 3 m and were screened from a depth of 1.5-3 m bgs.

Each injection well was retrofitted with a 10 cm diameter bolt down seal connected to 1.9 cm PVC drop tubes that extended to a depth between 11-12 m (for deep wells) and 2-3 m (for shallow wells) in order to enable reagent introduction at the bottom of the aquifer where majority of contamination has been found to exist. A valve-controlled vent opening was provided at the top of the well seal to release reaction pressures and off gases, when. Because of the low permeability associated with the subsurface, when the reagent chemicals were introduced at the bottom of the injection point, they traveled upward along the length of the screen interval and dispersed into the aquifer under an applied pressure of 130-200 kPa.

Field activities were performed over two separate phases for a total of thirty-two (32) days during the period September 2000 to November 2000. A site-engineered injection apparatus consisting of injector heads, pumps, chemically inert hoses, gas release and liquid flow valves was used to control flow of oxidizer and ISOTEC catalyst into the contaminated subsurface. Approximately 56,000 liters of ISOTEC reagents were injected into the designated points with associated average catalyst and oxidizer injection flow rates of 3.08 liters per minute (lpm) and 2.04 lpm, respectively. Groundwater samples were collected from the monitoring wells in August 2000 to provide baseline VOC data. In addition, sampling was performed in between two phases in October 2000, following second phase in February 2001 and approximately one year later in July 2001 at the same locations.

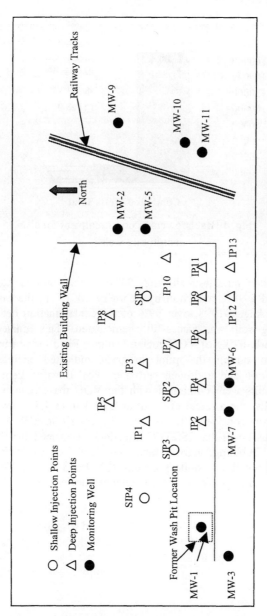

Fig. 3 Site map with location of injection and monitoring wells

Results and discussion

Results of baseline and post-treatment contaminant concentrations detected in all the monitor wells have been averaged, their cumulative mass calculated and plotted in Figure 4.

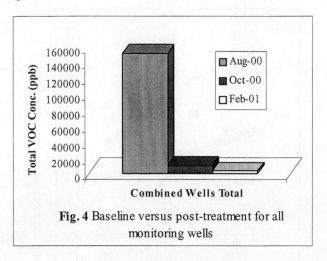

Fig. 4 Baseline versus post-treatment for all monitoring wells

The data plotted indicates a site-wide 93% reduction in total VOC mass by October 2000 and a 98% reduction by February 2001. In the one-year time period following August-1998, over 99% of the contamination detected in the on-site monitoring wells was successfully remediated. This equated to a mass destruction of nearly 62 kg of VOC contamination. Field monitoring for radial effects conducted during the pilot program indicated increases in the concentration of iron and hydrogen peroxide in all downgradient monitoring wells. These results were combined with the VOC mass reduction noted in downgradient wells, and a radial effect of at least 3 m to 4.5 m away from the injection points was considered reasonable. Especially in MW-6 and MW-7, visual observations noted that ISOTEC injections performed in IP-2 and IP-4 caused vigorous bubbling activity and increased water levels. Similar observations were noted for monitoring wells MW-2 and MW-5. Figures 5a and 5b delineates the baseline versus post-treatment concentrations on a per well basis for total VOC's.

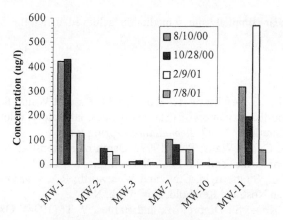

Fig. 5a Baseline vs post-treatment VOC
concentrations on a per well basis

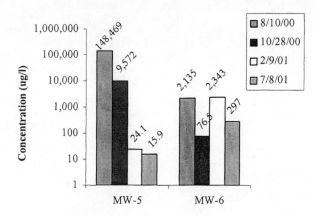

Fig. 5b Baseline vs post-treatment VOC
concentration for MWs 5 & 6 (Log Scale)

It may be noted that VOC concentrations in MW-6 and MW-11 indicated baseline to post-treatment increases in February 2001 but decreased substantially by July 2001. Such a trend may have occurred due to the presence of small pockets of untreated contamination in the vicinity of these wells that was gradually depleted through desorption followed by subsequent transfer to the water phase. As contamination transferred to the aqueous phase, it was oxidized by ensuing injections. Therefore, even though February-2001 sampling indicated an increase, the VOC concentrations in wells MW-6 and MW-11 decreased considerably by July 2001 (Figures 5a and 5b). Based on the dramatic VOC reductions noted in all the monitoring wells as well as site wide contaminant mass destruction (greater than 99% decrease), it is clear that majority of contamination present at the former manufacturing facility has been remediated. Subsequent sampling event conducted approximately one year later in July-2001 confirmed the near-complete site remediation achieved using the ISOTEC^{SM} process.

References

1. Greenberg, R.S.G.; Andrews, T.; Kakarla, P.K.C.; and Watts, R.J. (1998). In-situ Fenton-like oxidation of volatile organics: laboratory, pilot, and full-scale demonstrations. 29-42. *Remediation*, Spring Issue.
2. Kakarla, P.K.C. and Watts, R.J. (1997). Depth of Fenton-like oxidation in remediation of surface soil. 11-17. *J. Environ. Engng.*, **123**(1).
3. Schumb, W.E., Stratterfield, C.N., and Wentworth, R.L. (1955). *Hydrogen Peroxide*, Van Nostrand Reinhold, New York, NY.
4. Watts, R.J., Kong, S., Dippre, M., and Barnes, W.T. (1994). Oxidation of sorbed hexachlorobenzene in soils using catalyzed hydrogen peroxide. 33-37. *J. Haz. Mat.* **39**(1).

Enhanced in situ bioremediation of chlorinated solvents

N. M. Rabah[1] & D. E. Lekmine[1]
[1]Remedial Technology and Engineering, (RTE), Inc. of PMK Group, U.S.A.

Abstract

Anaerobic enhancement of in *Situ* bioremediation of a shallow, confined, silty sand aquifer has resulted in a substantial biodegradation of a dissolved tetrachloroethene (PCE) plume at an abandoned metal finishing facility located within residential area in northern New Jersey. The injection of a proprietary time-release hydrogen compound rapidly converted ambient aquifer aerobic conditions to anaerobic. This enhancement effectively induced and facilitated the successive reductive dechlorination of PCE and daughter products. Field results confirmed this anaerobic biodegradation enhancement within the overburden aquifer resulted in substantial degradation of PCE and its daughter products in the underlying sandstone bedrock aquifer.

The pre-injection dissolved PCE levels of 5 to 8 mg/l at the source declined by 90 to 96% within 2 months of injection. Degradation rates of 96% to 100% were obtained for TCE and t-DCE, whereas the degradation rates of c-DCE were slower and smaller at 30 to 70%. The potential presence of residual product and larger adsorbed solvent mass in the source area resulted in seemingly smaller but faster PCE and TCE overall degradation rates in comparison to downgradient areas. The c-DCE mass degradation during the same period was apparently greater and faster near the source than in the downgradient area.

Of interesting significance was the rapid degradation of VC, which was scarcely detected during the monitoring period. Carbon dioxide and methane levels increased by more than 10 fold indicative of VC degradation under both aerobic and anaerobic conditions. Local changes in the REDOX state and in oxygen levels coincided with preferential reduction of competing species (*e.g.,* sulfate), and apparently facilitated the degradation of VC.

Introduction

The 1½-acre project site is located in a northern New Jersey and is surrounded by residential dwellings. The site operated since 1947 for nearly 30 years as a metal plating and finishing facility [1]. The underlying overburden aquifer consists of 10 feet (3.3 m) of silty sand below a clay layer and is under confined conditions as illustrated in figure 1.

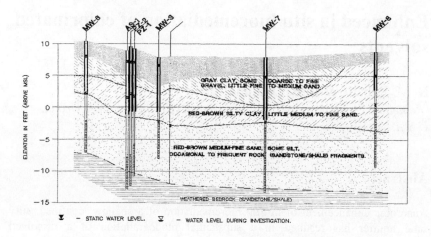

Figure 1: Geologic cross section

     The overburden is underlain by bedrock that consists of a Triassic age sandstone, siltstone and shale [1]. A layer of residual soil made up of silt, clay and weathered bedrock fragments separates the overburden from the bedrock. Groundwater within the bedrock is also confined.

     The depth to groundwater within both the overburden and bedrock wells varied between 0.2 feet and 4 feet (0.1 to 1.3 m) below existing grade levels. The piezometric levels of groundwater within the shallow bedrock zone appear to be generally higher and more seasonally stable than those in the overburden aquifer indicative of a downward groundwater flow potential. Laterally, the groundwater flows in a northerly to easterly direction. The average hydraulic conductivity within the overburden aquifer was estimated [1] to be on the order of 1-3 ft/day (4×10^{-4} to 1×10^{-3} cm/sec). Accordingly, for an average lateral hydraulic gradient of 0.01 at the site, the lateral groundwater velocity varies between approximately 0.02 to 0.2 feet/day (0.006 to 0.06 m/d).

Nature and extent of the contamination

Available data [1] indicated past on-site operations had impacted the soil and groundwater under the site. These impacts primarily consisted of chlorinated volatile organic compounds (VOCs) including tetrachloroethene (PCE) and its breakdown daughter products. Subsequent groundwater investigations [1] revealed the presence of groundwater dissolved PCE plume at levels above applicable groundwater quality criteria. The plume had originated in the southeastern portion of the site, in the vicinity of a former dry well, and migrated along the general groundwater flow direction in a northerly-easterly direction as shown in figure 2. The chlorinated solvent plume had migrated more than 600 feet (200 m) under the nearby downgradient properties.

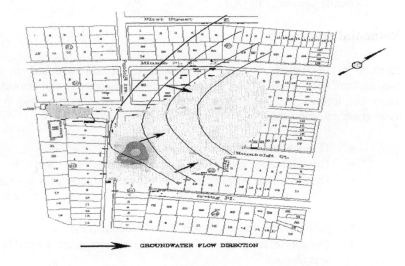

Figure 2: Site groundwater flow/plume map

The total mass of PCE and its breakdown daughter products was estimated [1] to be on the order of 17 to 20 lbs. It was further estimated that nearly 85% of the contaminant mass might be present on-site, and 15% distributed off-site.

Closure Criteria

The groundwater remedial targets are the applicable New Jersey Department of Environmental Protection (NJDEP) Groundwater Quality Criteria (GQC). Table 1 presents the closure criteria.

Table 1. NJDEP Closure Criteria

Contaminant	Remedial Target (μg/l)
Tetrachloroethene (PCE)	1
Trichloroethene (TCE)	1
Cis-1,2-dichloroethylene (cis-DCE)	70
Trans-1,2-dichloroethylene (trans-DCE)	100
Vinyl Chloride (VC)	5
Total Volatile Organics (TVOC)	500

Remedial Approach

Pursuant to the initial screening, in *Situ* chemical oxidation and in *Situ* bioremediation technologies were selected for further assessment and evaluation. These two technologies were primarily selected for further analysis due to site hydrogeologic conditions:

- The overburden aquifer is confined;
- The overburden aquifer is of limited thickness and extent;
- The plume has laterally migrated off-site and most of it is situated underneath residential homes and right of ways (i.e., streets, sidewalks)
- Relatively moderate to low hydraulic conductivity;
- The subsurface physico-chemical characteristics are amenable for degradation.

The assessment included:

- An oxidation bench-scale treatability study using hydrogen peroxide; and
- An anaerobic biodegradation bench-scale treatability study using a proprietary hydrogen release compound (HRC®).

A qualitative comparison was performed based on the above screening and compatibility of each alternative with the following assessment criteria:

C-I.   Remedial effectiveness and protection of human health & environment
C-II. Efficiency (remedial duration);
C-III. Technical feasibility and implementability;
C-IV. Public response;
C-V.   Regulatory response; and
C-VI. Estimated costs.

The comparative evaluation revealed that enhanced bioremediation using HRC, was the preferred remedial alternative.

Pilot study program

A field pilot test injection program was performed [1]. HRC solution was injected into 2-inch (5 cm) diameter boreholes throughout the saturated zone to the top of bedrock at depths of 15 to 20 feet (5 to 6 m) below surface grade. The boreholes were installed using a truck-mounted Geoprobe drill rig. The HRC was pumped through the unit and controlled by a flow restriction valve, intended to maintain a relatively constant pumping rate of approximately 6 gallons per linear foot.

Two injection areas were selected based on concentrations of VOCs and other chemical parameters including sulfate and pH levels. The first area represented the source area, whereas the second area represented downgradient conditions. Injection was completed in a grid pattern with a higher grid intensity and injection rate in the source test area. Different grid patterns from 7 to 10 feet (2.3 to 3.3 m) and injection rates of 4 to 6 lbs/foot were tested.

HRC is a time-release polylactate ester that provides a lasting hydrogen source for anaerobic biodegradation of chlorinated solvents [2,3]. It facilitates and accelerates the anaerobic microbial activities and reductive dechlorination of these chlorinated solvents [2]. Native anaerobic microbes ferment the lactic acid provided by HRC, and in the process release hydrogen. Microbes utilize hydrogen to degrade chlorinated solvents (by successively removing and replacing chlorine by hydrogen) into non-toxic end products (*i.e.,* ethene, carbon dioxide, and water) as follows:

$$PCE \longrightarrow TCE \longrightarrow DCE \longrightarrow VC \longrightarrow Ethene \longrightarrow CO_2 + H_2O$$

Table 2 summarizes the analytical results obtained during the pilot study for one of the test wells. A comparative assessment of pre-treatment and post-treatment analytical results indicated [1] that injection of HRC facilitated and augmented anaerobic biodegradation of PCE and its breakdown products.

Table 2. Groundwater laboratory results - HRC

Sample ID	MW-2								RW-2		
	Pre HRC Pilot Study	Post HRC Pilot Study				Pre Full-scale HRC	Post full-scale HRC		Pre Full-scale HRC	Post full-scale HRC	
Sampling Date	9/1/1999	10/4/99	11/2/99	12/2/99	1/4/00	3/2/00	10/11/01	1/10/02	1/31/01	10/12/01	1/9/02
Laboratory ID. No.	1922331	1925755	1928354	1931533	1000188	1005198	11J0551-01	12A0439-05	254897	306866	326760
Volatile Organic Compounds (mg/l)											
Tetrachloroethene	5,800	296	410	1,050	828	330	4.7	6.4	1,900	44	15
Trichloroethylene	35.3	5,320	3,670	883	1,600	201	7.5	3.5	340	3.3	0.8
cis-1,2-Dichloroethylene	7.5	6,180	7,960	1,520	2,430	2,390	62	21	150	7.2	9.6
trans-1,2-Dichloroethylene	ND	106	85	ND	52.1	ND	19	17	ND	ND	ND
Vinyl Chloride	ND	ND	ND	ND	ND	45.6	27	ND	ND	ND	ND
TVOC	5,842.8	11,979.2	12,125.0	3,453.0	4,910.1	2,966.6	120.2	48.8	2,390	56.3	26.4
Permanent Gases (mg/l)											
Ethane	ND	ND	ND	ND	ND	ND	0.3	0.021	NS	NS	NS
Ethylene	ND	ND	ND	ND	ND	ND	0.021	ND	NS	NS	NS
Carbon Dioxide	0.896	4.04	1.48	7.18	9.88	10.7	9.35	21	NS	NS	NS
Methane	ND	ND	ND	0.0166	0.4210	1.7200	4.82	3.23	NS	NS	NS
General Chemistry Parameters (mg/l)											
Dissolved Oxygen	0.3	0.7	0	1.14	0.7	8.36	2.3	NM	1.41	0.4	6.65
pH	6.53	6.48	5.72	8.05	8.05	5.67	6.75	6.51	8.45	7.43	7.61
Alkalinity as CACO₃	NA	172	207	570	NA	864	260	507	NS	NS	NS
Sulfate	48	47	3	12	9.5	1	7.9	2.1	NS	NS	NS
Total Organic Carbon	NA	8.7	133	1620	1470	1300	54.3	99.3	NS	NS	NS
Oxygen Reduction Potential	204	- 160	NM	NM	3	268	18	NM	- 1306	141	- 141

Legend:
TVOC - Total Confident Concentrations of Volatile Organic Compounds
mg/l - milligram per liter
mg/l - microgram per liter
330 - Concentration in excess of NJDEP Groundwater Quality Criteria for Class IIA Aquifers (GQC)

Concentrations of all daughter by-products (*i.e.,* TCE, DCE, and VC) also significantly decreased following the initial increase due to production. The total mass loss of dissolved PCE, TCE, and DCE due to the pilot injection was estimated to be approximately 0.3 lbs. The total mass loss from both adsorbed and dissolved phases of PCE, TCE and DCE was estimated to be more than 2 lbs. The mass removed during the pilot test was estimated to correspond to about 10% of the initial contaminant mass within the entire plume.

Concentrations of dissolved carbon dioxide and methane, pH, alkalinity, metal (iron and manganese), TKN, and TOC increased and sulfate and nitrate concentrations decreased due to anaerobic/anoxic conditions. These observations along with the relatively stable to increasing levels of metabolic acids confirmed the sustainability of ongoing anaerobic biodegradation.

Full-scale remedial design and construction

Based on the pilot study, a full-scale bioremediation program of HRC injection was completed during August 2001 in order to achieve the site remediation goals. The bioremediation scheme optimized access to the majority of the overburden plume both on-site and off site (*i.e.,* roads, residential dwellings and streets). HRC was injected both in a grid pattern at the site and in a barrier pattern at off-site locations. Injection rates varied from approximately 3 to 5 gallons per linear foot with grid spacing of 7 to 10 feet. A total of 449 injection points were installed within the on and off-site grids and barriers. A number of overburden and adjacent bedrock wells were monitored closely following the full-scale injection.

Results and discussions

Table 2 summarizes field and laboratory analytical results for one overburden well and an adjacent bedrock well. The concentrations trends of PCE and its daughter products in each of these wells are shown in figure 3 and 4. PCE concentrations declined significantly due to treatment in the all the source wells following the full injection. The decrease of PCE concentrations varied between approximately 99.9% at the source area and 45% at the downgradient area. Also, TCE, DCE and VC degraded significantly following injection. VC was only detected in wells near the source at low concentrations within about 5-6 months after injection and then rapidly depleted. Localized and temporary fluctuations of dissolved oxygen levels were observed likely due to changes in REDOX potential and concurrent biochemical reactions involving competing species including (*e.g.,* sulfate reduction). VC continued to degrade and fully deplete under both anaerobic/anoxic and aerobic conditions as both methane and carbon dioxide have been detected. Available literature [4] supports these VC degradation patterns.

The HRC enhancement of anaerobic conditions in the overburden aquifer was observed to induce substantial degradation of contaminants within the underlying shallow bedrock aquifer. The significant decline of PCE levels from 1,900 mg/l to 40 in the shallow bedrock well RW-2 shown in figure 4 supports that such in-situ enhancement of anaerobic biodegradation conditions would be effective in remediating the dissolved impacts within the bedrock aquifer.

Figure 3: VOCs results at overburden monitoring well MW-2

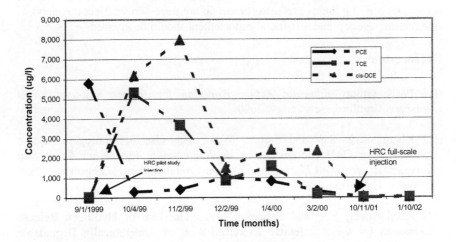

Figure 4: VOCs results at shallow bedrock well RW-2

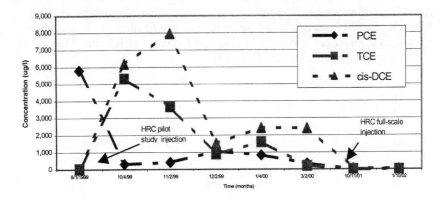

Conclusion

This project demonstrates that enhanced bioremediation of chlorinated solvents aquifers is possible with minimal Site disturbance to the residential community.

Levels of chlorinated solvents in the groundwater have been declining since the implementation of the enhanced bioremediation process. It is anticipated that a second full scale HRC application will be implemented to further the reductive dechlorination process and ultimately achieve effective site remediation within the shallow aquifer and regulatory closure of the case.

As part of an aggressive Brownfields Redevelopment, this abandoned Site, which stood as an eyesore to the residential community for years was mitigated to address imminent health risks associated with previous site operations and was successfully developed with a park.

The NJEDA HDSRF program provided funding for the site remedial investigations. Opinions presented herein do not necessarily reflect the views of the funding agency, and no official endorsement should be inferred.

References

[1] PMK Group, *Remedial Action Workplan for former Keystone Metal Finishers*, Cranford, New Jersey, 2000.

[2] Koenigsberg, S.S. and Farone, W.A., The Use of Hydrogen Release Compound for CAH Bioremediation, *In Engineering Approaches for In Situ Bioremediation of Chlorinated Solvent-Contamination*, Regenesis, San Clemente, California, 1999.

[3] Koenigsberg, S.S. and Sandefur, C.A., The Use of Hydrogen Release Compound for the Accelerated Bioremediation of Anaerobically Degradable Contaminants, Remediation, John Wiley and Sons, New York, 1999.

[4] Cornuet, T.S., Sandefur, C.A., Eliason, W.M., Johnson, S.E. and Serna, C., Aerobic and Anaerobic Bioremediation of cis-1,2-Dichloroethene and Vinyl Chloride, Proceedings *of the 2nd International Conference on Remediation of Chlorinated and Recalcitrant Compounds*, Monterey, California, May 22-25, 2000. Battelle Press, Columbus, OH., 2000.

Integrated remediation of brownfield sites using batch and continuous thermal desorption combined with physiochemical processes

S. Nusimovich
Tox Free Solutions Ltd, Australia

Abstract

The use of an integrated approach to manage environmental cleanups, based on the use of thermal desorption and other processes, is ideally suited for the management of brownfield sites. Processing of sludge, sediment, and contaminated soil using continuous indirect thermal desorption based on a vertical retort achieves the desired cleanup levels for contaminants such as organic compounds, pesticides, and PCB. This process when connected to a high-temperature ceramic filtration unit and a full-condensing gas treatment train, in combination with destruction processes such as plasma destruction or recycling processes for organic compounds, offer a range of resources that, when used in an integrated way, can deliver the desired results whilst minimising environmental and health risks and achieving a one stop solution. The effectiveness of this system is supported by data from industrial-scale applications of the processes in the treatment of PCB-contaminated soil and tetrachloroethylene-contaminated material.

1 Introduction

Progress and industrialisation have brought about immense improvements in our quality of life. This development has produced a number of impacts, both negative and positive. Amongst the negative impact of this development, the adverse effects on the environment caused by the release of chemicals are considered to be the most significant. The discharge of contaminated liquids has impacted waterways and aquifers with a variety of chemicals, including both halogenated and non-halogenated organic compounds.

It is the contamination of soil and groundwater that is considered by experts to be "amongst the most complex and challenging environmental problems faced by many countries" [1]. Soil and groundwater are complex matrices making both the delineation of the extent of the contamination and its removal difficult tasks. The definition of clean up targets is also made extremely difficult by the technical and economic limitations of the available remediation technologies, as well as the complexity of evaluating potential long term health risks, modelling the evolution of contaminants and their breakdown products in the different media, and the use, both present and future, of the affected land.

The search for a portfolio of technologies to address these problems commenced as soon as the first major cases of soil and groundwater pollution were detected. The application of in-situ technologies offered a number of advantages over ex-situ options. However, complete removal or destruction of the pollutants present in the contaminated soil has proven an elusive target. It is the excavation of the contaminated soil and its processing that has proved to be a more efficient operation, that can offer guarantees of performance. Removal of the contaminated soil allows for the specific dimensioning of the problem, the thorough sampling and characterisation of the soil, and, more important, for the processing under controlled conditions and the analysis of all produced streams.

Amongst the technologies used in the remediation of contaminated soils, thermal technologies have found a wide range of application. The use of heat to evaporate, desorb, oxidize, or incinerate compounds were soon seen as processes that could be applied to the treatment of contaminated soil. The separation of compounds from the soil or their destruction using thermal processes has been applied extensively to the remediation of contaminated soil. Amongst these technologies, thermal desorption has been used as a separation process and has evolved since its early uses, in the 1970s.

2 Review of the technology

Thermal desorption is a physiochemical separation process consisting in the direct or indirect heating of a mass of sludge or solids to a certain temperature, either in a batch or a continuous process. Thermal desorption separates the volatile and semi-volatile compounds. Thermal decomposition of the contaminants does not generally take place, although some degree of oxidation or pyrolysis may occur in localised areas [1].

Thermal desorption units may be directly or indirectly fired. In direct systems, heat is transferred by convection, radiation, or from an open flame, to the contaminated material, whilst in indirect systems, heat is transferred by conduction. A number of systems have been developed over time and applied to the decontamination of soil, sediment, and sludge. These systems include rotary dryers, heated screws, thermal blankets, batch desorbers, and hollow disc dryers [2]. Desorption temperatures vary with the system used and the source of heat, with systems classified into low temperature and high temperature processes. Low temperature processes operate at 120-200°C, whilst high temperature desorption processes are designed to operate above 350°C. Given the nature of the process, it is suited for the separation of contaminants that are volatile or semi-volatile, with boiling points within the range of operation of the units. Organic compounds, both halogenated and non-halogenated, pesticides, hydrocarbons, polychlorinated biphenyls (PCBs), and polycyclic aromatic compounds have all been separated using these processes. The technology has been successfully used to process large volumes of contaminated materials in the last twenty years, with projects completed in Europe, the U.S., Asia, and Australia.

Treatment of contaminated material from contaminated sites has been the preferred remediation option for Superfund sites in the U.S. since 1982, where the environmental authorities have selected this option over passive measures [3]. Although there has been an increase in the selection of source control measures such as institutional controls, monitoring, and relocation of affected populations, these methods are still a minority. Ex situ technologies, where the contaminated materials are removed and treated, represent the majority of the projects [3]. Outside solidification and off-site incineration, thermal desorption has been the preferred technology in the period 1982-1999, treating a total volume of over one million tonnes. The application of incineration technologies has followed a decreasing trend since 1985, from a high of 38 % of all projects in 1985 to 9 % in 1999 [3].

The performance of the thermal desorption process is dependent on three main parameters: temperature, residence time, and the operating pressure. These three parameters affect the evaporation or desorption of the contaminants from the solid or semi-solid matrix. Root [4] describes the drying process as consisting in three steps. The first step involves the addition of heat to the material until the boiling point of the liquid is reached. The second stage, once the boiling point is reached, involves the evaporation of the liquid at a certain rate. The last stage commences once a critical moisture point is reached, and the drying rate starts to fall. This process, when applied to desorption of a mix of compounds with different boiling points, trapped in a complex heterogeneous matrix, becomes complicated. The rate of heat transfer becomes critical in such a scenario, with agitation providing the necessary force to increase thermal conductivity, and enhancing contact of the particles to heat.

Residence time varies depending both on the characteristics of the product, such as particle size, and the contaminant. This variable, along with temperature, offer the operator the capability to control the outcome of the process. It is important to identify the particular parameters that affect processing of contaminated soils, materials, and sludge. Treatment of these materials presents significant engineering and project management challenges.

Given the heterogeneity of the materials to be processed in remediation projects, engineers are presented with two options: the design of a multi-purpose system that is able to process a wide range of products, or the design of a specific process for a particular product. Temperature control, residence time, and the selection of the gas treatment technology are all affected by the contaminants to be desorbed. The selection of the gas treatment technology and desorption temperature will vary with the compounds targeted for desorption. Careful selection of the gas treatment system is important as either total destruction of the pollutants or complete recovery are targets to be attained. Authorisation by the relevant authorities and achieving performance targets depend on the efficiency of this system. Examples of the variability in the application of thermal desorption are presented in Table 1.

Table 1. Application of thermal desorption

Contaminant	Desorption temperature range	Gas treatment technology
Hydrocarbons	Low-medium	Condensation Thermal oxidation
Chlorinated compounds	Medium	Condensation Carbon adsorption Chemical absorption
PCB	High	Condensation Carbon adsorption Chemical absorption

In most cases, separation of particles from the vapour phase is necessary before other gas treatment technologies are used.

The use of thermal desorption technology in remediation projects requires the consideration of two other important issues: the management of all streams separated by the process, and the siting of the plant and its permitting. The first issue relates to the destination of all the products obtained as a result of the process. Usually, in the case of remediation projects, desorbed soil and a liquid concentrate will be produced. The composition of these streams will obviously vary depending on the feed material and process conditions. These materials will require final disposal or recycling, with the environmental and technical specifications for these operations requiring careful analysis.

The siting of the thermal desorption plant will depend on the magnitude of the project, with the design of a transportable plant to be set up on site as an option to be used in projects requiring the treatment of over 10,000 tonnes. The construction of a central facility to process contaminated materials can be considered when, either the complexity of the plant, or the location of the contaminated sites, makes it an economically viable option.

3 TDPP Technology

Tox Free Solutions Ltd (Australia) developed thermal desorption technology using a novel system based on the use of a vertical retort and a high temperature filter. This was the result of many years of development, and the experience gained from several thermal desorption projects in Australia. The proprietary technology has been used to build a plant, TDPP-III, which was commissioned in January 2002 as a fixed facility to process contaminated soil. Previous plants had been successfully used in desorption of organochlorine pesticides, coal tars, chlorinated compounds, and hydrocarbons. Analysis of the problems posed by contaminated soil identified the preferred criteria used in the selection of a suitable technology for the remediation of contaminated soil. These criteria are presented in Table 2.

Table 2 – Criteria for the selection of remediation technology

Criteria	Preferred features of the remediation technology	Expected results
Applicability	Wide range of applications for the remediation of contaminants	Potential to process wide range of contaminants
Environmental performance	Minimisation of emissions and discharges Maximisation of recycling Minimisation of environmental risks during operation and in the future	Minimum environmental impact Recovery of products for reuse Complete destruction of contaminants
Economic performance	Cost effectiveness	Good return on investment
Public acceptance	Clean technology Flexibility for siting the plant or process	Public acceptance

Design of TDPP-III followed the criteria of Table 2. The configuration of a high temperature thermal desorption unit was combined with a modular gas treatment system that offered great versatility in the treatment of different contaminants. The design of an indirectly heated closed system, operating under a slight negative pressure, under continuous Nitrogen purging minimised the decomposition of contaminants whilst preventing the formation of hazardous atmospheres and emissions. Recovery of products was maximised through the design of a full condensing system that allowed for the collection of different fractions. The system is presented in Figure 1.

Figure 1 – Process flow diagram of TDPP-III

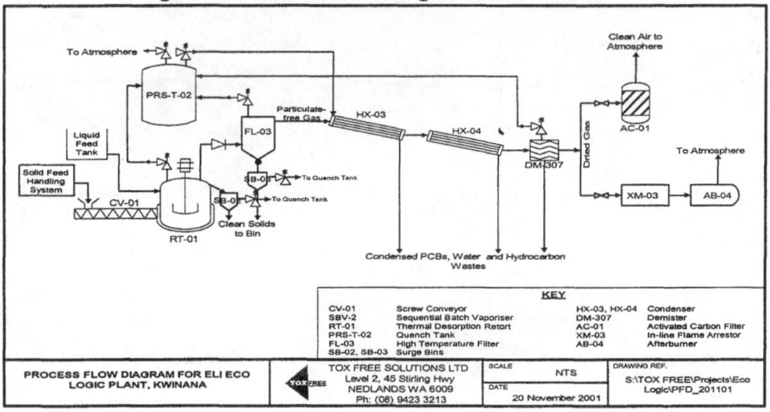

The system consists of a thermal desorption unit which can reach temperatures of 600°C. Thermal desorption takes place in a vertical retort, filled with an inert medium agitated by a helicoidal stirrer. Solid contaminated material is fed using an auger. Vaporised pollutants are filtered thorugh a high temperature heated filter. The particle-free gas then passes through a series of condensers operating at ever-decreasing temperatures. The two condensers separate the majority of VOC and SVOC, that are collected in tanks for recycling or destruction. As the system operates under Nitrogen, the gas exiting the last condenser is mostly this inert gas with traces of compounds that have not been condensed. This gas stream can then be circulated through an activated carbon filter for final polishing or, alternatively, it can be thermally oxidised. The use of the thermal oxidation option is applied when materials are contaminated with hydrocarbons, whilst the full condensing mode with activated carbon is used when materials are contaminated with halogenated compounds. The system includes a secondary containment system to prevent release of materials even in the case of overpressurisation.

The plant has been designed with a high level of instrumentation and control, to be operated through a computer-based logic, whilst registering all process parameters in real time. A typical screen during processing of contaminated soil is shown in Figure 2.

Figure 2. Process control screen of TDPP plant

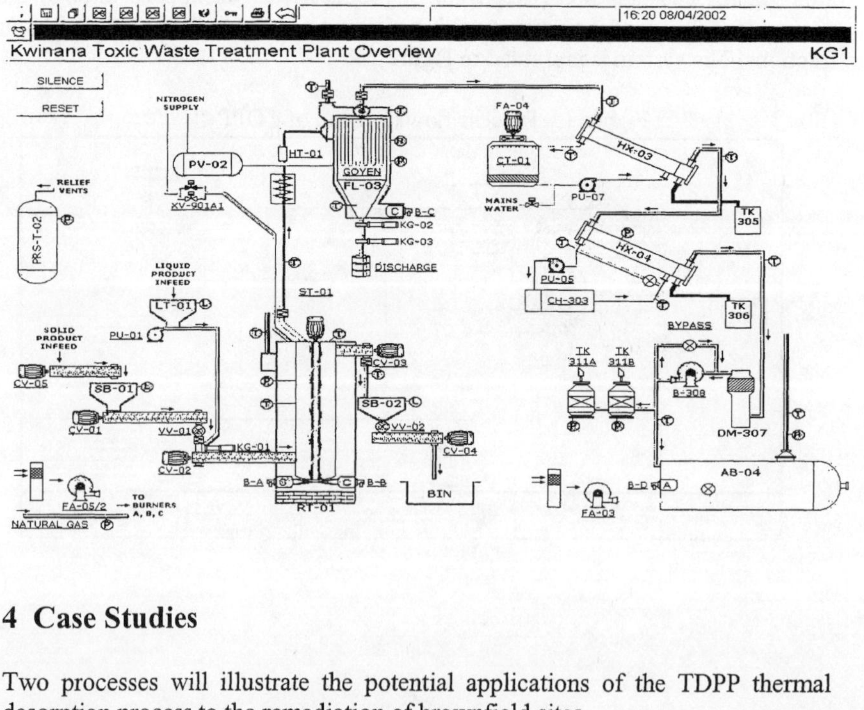

4 Case Studies

Two processes will illustrate the potential applications of the TDPP thermal desorption process to the remediation of brownfield sites.

4.1 Thermal desorption of PCB-contaminated soil and debris

Soil and debris from several contaminated sites was treated by thermal desorption using plant TDPP-III. The material was produced during the demolition of concrete slabs and removal of contaminated soil at several power plants, in Western Australia. The areas had been affected by spills and leaks from electrical transformers filled with PCB. The material was also contaminated with petroleum hydrocarbons and was heterogeneous in particle size, characteristics, and moisture content. PCB concentrations were above 50 mg/kg, with various levels of contamination.

Table 3. Concentration of hydrocarbons and PCB in soil

Sample reference	Concentration of PCB (mg/kg)	Concentration of total petroleum hydrocarbons (mg/kg)
SF-104	3,400	580
SF96	200	580
SF177	160	190
SF181	360	190

The soil was processed by thermal desorption at a temperature of 430-450^{0}C, with the gas treatment train in full condensing mode, and the activated carbon filter AC-1 as the final stage before release of non-condensable gases to atmosphere. A total of 216,512 kg of soil were processed during the first stage of the project, between January 25 2002 and March 14 2002. The plant ran discontinuously, with feed rates of 700 kg/hour, producing 12,800 kg of condensate, composed of PCB, water, and hydrocarbons. The desorbed soil, equivalent to 94 % of the starting material was disposed of at a licensed landfill. In all cases, concentrations in the desorbed soil were below 0.2 mg/kg of PCB and below detection limits for hydrocarbons. The detection limits for hydrocarbons were 0.2 mg/kg for C_{6-9} and C_{10-14}, and 0.4 for C_{15-28} and C_{29-36}, respectively. All collected condensate is to be destroyed by plasma destruction.

An independent party [5] completed sampling and analysis of emissions from the treatment process. These results are summarised in Table 4.

Table 4. Results of the emission monitoring program

Compounds	Concentration (mg/m^3, 0^0 and 760 mmHg, on a dry basis)
PAHs	All compounds <0.015
PCBs	All compounds <0.001
Total SVOCs	**< 0.15**
Chloromethane	0.2
Dichloromethane	0.85
Benzene	0.001
Bromomethane	0.01
All other VOCs	<0.003
Total VOCs	**1.1**

All emissions were well below the limits established by the environmental authorities for the process.

The use of the TDPP plant allowed the operator to offer a system that could manage desorption of both hydrocarbons and PCB in one step, with a reduction of over 94 % in volume that required destruction. Desorption of debris can be achieved by batch desorption or crushing of the material for feeding through the TDPP system.

4.2 Thermal desorption of tetrachloroethylene sludge

Tetrachloroethylene is a common solvent used in dry cleaning and other applications. Numerous sites contaminated with chlorinated solvents require considerable budgets to attempt complex remediation projects. As an indication of the dimensions of the problem, recent pilot work carried out at a site in the U.S., contaminated with tetrachloroethylene, with an affected area of less than 6,500 m^2 put the total remediation figure at $3 million to $25 million [6].

Sludge produced by the dry cleaning industry is currently processed at the TDPP plant by thermal desorption. A study of the process, completed in 2001, tested the use of the thermal desorption unit to attempt the recovery of tetrachloroethylene from the sludge for reuse as solvent. Analysis of samples of the organic phase of the condensate, obtained at retort temperatures of 180^0C and 221^0C, showed consistent levels of tetrachloroethylene in the range of 98 % to 98.7 %. The composition of the organic phase of the condensate is presented in Table 5.

Table 5. Composition of organic phase of condensate samples

Tetrachoroethylene	98.7 % v/v
Other chlorinated solvents	1.2 % v/v
Hydrocarbons	<0.1 % v/v
Hydrochloric acid	<0.001 % v/v

The presence of other chlorinated species, absent in the starting material indicate that a certain degree of decomposition of the tetrachloroethylene takes place, as it would be expected under the process conditions.

The final solid product still shows detectable levels of hydrocarbons and polycyclic aromatic hydrocarbons. This is a significant finding, as it shows that contaminants present in the starting material, with boiling points above that of tetrachloroethylene would require a higher processing temperature.

The gas produced during the process was also sampled and analysed. The condenser was shown to have an efficiency of 98.5 % for removal of tetrachloroethylene from the gas stream. Other compounds detected in the gas stream included other chlorinated compounds, ketones, and hydrogen chloride.

This application demonstrates the feasibility of using the process to recover organic compounds with a high efficiency and the potential for reuse, not only of the desorbed soil as clean fill, but also of the liquid fraction.

5 Potential applications of the technology as an integrated modular system

It is clear from the two cases described in previous sections and the more than 60 Superfund projects completed in the U.S., that thermal desorption continues to offer tremendous potential for remediation of contaminated sites. This is especially so, in the case of brownfield sites, where the level of cleanup achieved by thermal desorption can offer the possibility of reuse of the desorbed soil on site.

The combination of the TDPP technology with a destruction technology and the design of a gas treatment system specified for the control of the target pollutants for a particular project is a cost-effective strategy that provides a fully integrated solution that can be implemented on-site. It is possible to treat the off gas or the condensate using a destruction process directly. An example of such an application is the design of an integrated system, where packaging material is separated to be desorbed in a batch desorber and all condensed liquids are directed to a PLASCON® unit to be destroyed by a plasma arc. This application would be especially suitable for halogenated materials such as PCB, dioxins, and chlorinated pesticides.

The reduction of costs, risks, and environmental impacts by implementing this strategy are significant. The elimination of the need for transporting the separated contaminants to the disposal facility by destroying them on-line reduces risks and costs, whilst the installation of gas treatment equipment specifically designed for control of the target pollutants minimises environmental impacts. The on-site recycling of the processed soil not only eliminates the cost of transport of that material, but it offers a safe alternative that reduces development costs of the brownfield site.

6 Conclusions

The TDPP thermal desorption technology is a high temperature separation process that can treat a wide range of contaminated materials. Its greatest potential derives from the fact that it offers a cost-effective high-performing means to separate contaminants from a complex matrix. The use of the technology for desorption of PCB-contaminated materials and tetrachlorethylene-contaminated sludge are examples of successful industrial-scale applications of the process.

The use of TDPP technology combined with a destruction process as an integrated solution for on-site treatment of contaminated materials results in a cost-effective option to reclaim contaminated soil in brownfield development projects. The engineering of the gas treatment system as a modular system that can be assembled to target specific pollutants minimises environmental impacts and risks. These combined strategies result in the design of an integrated system that can be implemented on-site to maximise recycling of materials and minimise costs and environmental risks.

References

[1] U.S. Environmental Protection Agency, *Phase II Final Report NATO/CCMS Pilot Study: Evaluation of Demonstrated and Emerging Technologies for the Treatment and Clean Up of Contaminated Land and Groundwater*, USEPA Office of solid Waste and Emergency Response, Technology Innovation Office, Washington D.C., 1998.

[2] Sullivan T. P., Thermal Desorption: The Basics. *Chemical Engineering Progress*, **95(10)**, pp 49-56, 1999.

[3] U.S. Environmental Protection Agency, *Innovative Treatment Technologies: Annual Status Report*, Tenth Edition, USEPA Office of Solid Waste and Emergency Response, Technology Innovation Office, Washington D.C., November 2001.

[4] Root W. L., Indirect Drying of Solids. *Chemical Engineering*, May 2, pp 52-64, 1983.

[5] U.S. Environmental Protection Agency. Surfactant Flooding used to Remove PCE DNAPL", *Ground Water Trends*, EPA 542-N-01-007, Cincinnati, October 2001.

[6] Environmental Consultancy Services, *Tox Free Solutions Limited Eli Eco Logic Site Monitoring Program January 2002*, Report R01194, Perth, January 2002.

The coupled effect of electrokinetic and ultrasonic remediation of contaminated soil

H. I. Chung[1], I. K. Oh[1] & M. Kamon[2]
[1]*Department of Civil Engineering, Korea Institute of Construction Technology, Korea*
[2]*DPRI, Kyoto University, Japan*

Abstract

The laboratory tests were conducted using specially designed and fabricated devices to determine the effect of sonication and electrokinetics on seepage as well as on contaminant removal in contaminated soil. The electrokinetic technique was applied to remove mainly the heavy metal and the ultrasonic technique was applied to remove mainly organic substance in contaminated soil. Test soil was Jumunjin(East beach of Korea) fine sand; the test specimen was prepared at loose density. Diesel fuel and Cd were used as a surrogate contaminant for soil flushing test.

A series of laboratory experiments involving the simple, ultrasonic, electrokinetic, electrokinetic+ultrasonic flushing test were carried out. An increase in permeability and contaminant removal rate was observed in electrokinetic and ultrasonic flushing tests. Some practical implications of these results are discussed in terms of technical feasibility of in situ implementation of electrokinetic ultrasonic remediation technique.

1 Introduction

The contaminated ground can be removed and replaced with non-polluted soils. This method requires installations of sheet-pile walls to protect the surrounding ground. The execution of this method is costly and may also interrupt the ongoing business. Therefore, it is always more desirable to clean up the pollutant in situ than to remove the contaminated soil. In response to the demand for developing effective and economical cleanup techniques, numerous studies have been conducted over the years. Several clean-up techniques have been developed; examples include the pump-and-treat method, the soil vapor extraction method, and the bioremediation method, to name a few. The pump-and-treat method is ineffective due to the requirement of a large equipment with high energy and slow removal of contaminants. The oil vapor extraction method is not applicable to remove contaminants from saturated soil deposits and ground water. The effectiveness of bioremediation depends greatly on suitable microorganism and nutrients in the subsurface. Therefore, much still remain to be done in order that a generally accepted methodology can be developed for a broad range of applications.

Natural concentrations of heavy metals on soil deposits are not high; however, studies have indicated that many areas near urban complexes, metalliferous mines or major roads display abnormally high concentrations of these elements. Cadmium is a silver-white metal. In nature, it is usually found combined with other elements such as oxygen (cadmium oxide), chlorine (cadmium chloride), or sulfur (cadmium sulfate, cadmium sulfide). Cadmium does not corrode easily and has many uses. In industry and consumer products, it is used for batteries, pigments, metal coatings, and plastics. Cadmium or its compounds do not have a definitive taste or odor. Cadmium gets into the environment from the weathering of rocks and minerals that contain cadmium. Exposure to cadmium can occur in industries that commonly use or produce cadmium such as mining or electroplating. Cadmium exposure can also occur from exposure to cigarette smoke.

A variety of options may exist to select a cleanup remedy at a site, however the efficiency and costs of these options may vary widely. Most of the existing remediation technologies are limited to soils with high hydraulic conductivities and are not effective in removing heavy metals adsorbed on soil particles, particularly fine-grained deposits. There exists a need to introduce cost-effective, innovative, and preferably in-situ remediation technologies.

Electrokinetic(ek) soil processing is a new, innovative, and cost-effective remediation technology that employs conduction phenomena under electric currents for transport, extraction, and separation. A low level direct electric current (or electric potential difference) is applied across contaminated soil deposits through inert electrode placed in holes or trenches in the soil filled with processing fluid. The driving mechanisms for species transport are ion migration by electrical gradients, pore fluid advection by prevailing electroosmotic flow, pore fluid flow due to any externally applied or internally generated hydraulic potential difference, and diffusion due to generated chemical gradients. As a result, cations are accumulated at the cathode and anions at the anode, while there is a continuous transfer of hydrogen and hydroxyl ions across the medium. Various lab-scale studies on the feasibility of the process have shown that heavy metals and other cationic species can be removed from the soil specifically when process enhancement techniques are, such as electrolyte conditioning, are employed. The feasibility and cost effectiveness of electrokinetics for the extraction of heavy metals such as copper, zinc, and cadmium from soils have been demonstrated through bench-scale laboratory studies (Runnels and Larson 1986; Hamed 1990; Pamukcu et al. 1990; Acar et al. 1992 and 1993). Acar et al.

(1994) demonstrated 90% to 95% removal of Cd^{2+} from bench-scale kaolinite specimens with initial concentration of 99-114   /g. Acar and Alshawabkeh (1993) and Acar et al. (1994) showed higher removal rates of charged species can be achieved by electric migration rather than electroosmotic flow. Other uses of electrokinetics in environmental geotechnics include injection of grouts, cleanup chemicals or nutrients for growth of microorganisms essential to biodegradation of specific wastes, contaminant detection, monitoring the physicochemical soil profiling (Acar and Gale 1986; Mitchell 1986l; Acar et al. 1989; Acar and Hamed 1991; H. Lee 2000; S. Han 2000), and the use of eletrophoresis for sealing impoundment leaks (Yeung et al. 1994).

Numerous researchers have proposed possible mechanisms for the effect of stress waves on fluid flow through porous media. Simikin and Verbitskaya (1989) proposed that cavitation and capillary forces are principally responsible for the movement of fluid in porous media. Capillary forces play an important role in liquid percolation through fine pore channels. They suggested that the liquid films that are adsorbed onto pore walls during the percolation process could be destroyed by mechanical vibrations. This study results show that seismoacoustic fields lower the capillary pressure, and that seismoacoustic wave affects the stratum resulting in increasing oil saturation in the upper part and the water saturation in the lower part of the stratum. According to Suslick (1988) and Frederic (1965), the mechanisms responsible for the observed increase in transport rates and unit-operation processes due to ultrasonic energy can be divided into two categories: (1) first-order effects on fluid particles involving displacement, velocity, and acceleration, and (2) second-order phenomena including radiation pressure, cavitation, acoustic streaming, and interfacial instabilities. Usually, one or more if the second-order effects are responsible for the enhancement in the transport process. Murdoch et al. (1998) conducted falling-head permeability tests on sandy clay samples in ultrasonic cleaning chambers, and found that hydraulic conductivity of the samples increased abruptly about 370% after 30 seconds of ultrasonic excitation. They later conducted tests using a sonication probe, and observed an increase in hydraulic conductivity. Iovenitti et al. (1995) studied the effect of acoustic excitation on the porous media. They summarized the potential effects of acoustic wave on the porous grain framework of the soil and pore fluid on Figure 1 and 2. They showed the enhancement of the contaminant recovery by acoustic waves through a series of soil flushing tests.

Undisturbed Soil   After Application of Acoustic Energy

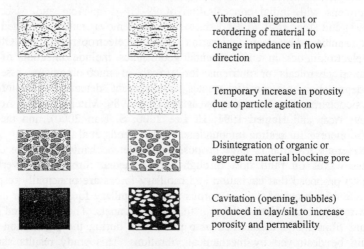

Vibrational alignment or reordering of material to change impedance in flow direction

Temporary increase in porosity due to particle agitation

Disintegration of organic or aggregate material blocking pore

Cavitation (opening, bubbles) produced in clay/silt to increase porosity and permeability

Figure 1: Potential effects of acoustic excitation on porous grain framework (After Young Uk Kim, 2000)

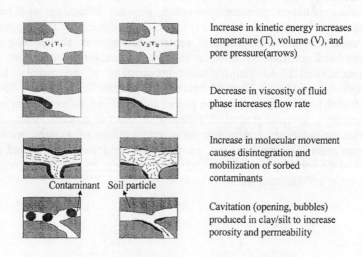

Increase in kinetic energy increases temperature (T), volume (V), and pore pressure(arrows)

Decrease in viscosity of fluid phase increases flow rate

Increase in molecular movement causes disintegration and mobilization of sorbed contaminants

Cavitation (opening, bubbles) produced in clay/silt to increase porosity and permeability

Contaminant   Soil particle

Figure 2: Potential effects of acoustic excitation on pore fluids (After Young Uk Kim, 2000)

2 Experimental methodology

The soil flushing tests were conducted by using the setup shown in Figure 3. The test chamber is made of a plexiglas cylinder having an insider diameter of 10cm with a height of 30cm. The cylinder was filled with contaminated sand. In the side part of the cylinder are installed inlet and outlet tubes. The inlet tube is connected to a reservoir of de-aired water, which is connect to the water tap; and the outlet tube is used to maintain constant heads by allowing overflow of excess water. Outlet tube is connected to a burette for measuring the outflow quantity.

The elecrokinetic flushing processor consists of three parts: anode electrode, cathode electrode, and electric power supplier. The graphite electrode is used and situated on the top (cathode part) of soil specimen and the bottom (anode part) of the cylinder. A constant voltage gradient of 1.0 V/cm was applied to the anode and cathode electrodes. Two sheets of filter paper were placed on the upper and lower graphite electrode. The ultrasonic flushing processor consists of three parts: a generator, a converter, and an acoustic horn. The transmitting acoustic horn, which is mounted on top of the soil sample, is used for generation ultrasound. The ultrasonic processor has a maximum power output of 200W with a frequency of excitation equal to 28kHz. The test setup used in this study was made to combine the electrokinetic flushing processor and the ultrasonic flushing processor. Photography for test setup is shown in Figure 3, and more detailed schematic view of the test setup is shown in Figure 4.

The test soil was sand, a fine aggregate, and a natural soil obtained Jumunjin, East beach of Korea. Diesel fuel and Cd were used as a surrogate contaminant to demonstrate the soil contaminated by organic substance and heavy metal. Some physical properties of test soils are shown in Table 1. The flushing tests were conducted for four conditions: simple flushing, electrokinetic(ek) flushing, ultrasonic flushing, electrokinetic(ek)+ ultrasonic flushing. Test conditions are shown in Table 2. Hydraulic gradient for the flushing process was constant to 1.0. In the tests, the soil specimens were thoroughly mixed with diesel fuel of 770ml and Cd of 500ppm(385mg in test soil). The test specimen was then subjected to ultrasonic waves at 28kHz frequency from ultrasonic test setup and to electric power at 1.0V/cm from electrokinetic test setup. Tests were continued to maximum 100 minute.

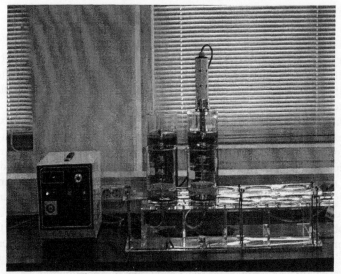

Figure 3: Photography for test setup

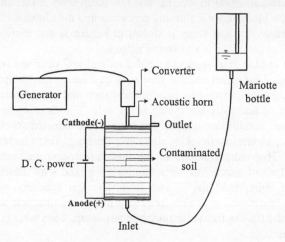

Figure 4: Test setup for soil flushing experiment

Table 1. Physical properties of test materials

Sand			Diesel fuel		
Specific gravity	Maximum dry density	Minimum dry density	Specific gravity	Solubility in water	Centistoke (cSt, mm^2/sec)
2.62	1.60	1.40	0.80	0.075 ~31.3	0.564

Table 2. Test conditions for soil flushing experiment

Test	Contaminants	Hydraulic gradient
Simple flushing		
Electrokinetic(ek) flushing	diesel fuel + heavy metal (cadmium)	i = 1.0
Ultrasonic flushing		
ek+ultrasonic flushing		

3 Results and Analysis

Under the actions of hydraulic gradient, ultrasonic waves, electroosmosis and electromigration, the pore water was allowed to flow upward through the specimen and contaminant was allowed to migrate upward from the specimen. The effluent was collected in a 500ml polypropylene cylinder. The effluent in the cylinder was allowed to stand overnight for gravitational segregation of oil from water. The volumes of the separated water and oil were then measured. Also, Cd concentration of effluent was measured by laboratory chemical analysis facilities.

Firstly, the test results for the accumulated flow volume with time are presented in figure 5. The figure shows the accumulative water flow is varied and increased with time. The accumulated flow volume with time is higher in the case of ultrasonic flushing and ek+ultrasonic flushing tests than in the case of simple and electrokinetic flushing tests. And the accumulated flow volume with time is almost same in the case of simple and electrokinetic flushing tests. Also, the accumulated flow volume with time is almost same in the case of ultrasonic flushing and ek+ultrasonic flushing tests. It means that the ultrasonic process has a role to increase the liquid outflow due to sonication effects, but electrokinetic process has not due to short test duration and sandy soil. Normally, electroosmosis by electric power is not or few developed in a short time and in sandy soil.

The test data shown in figure 5 were used to compute the discharge velocity. The computed discharge velocity was used to compute the coefficient of permeability of the test soils. The figure 6 shows the variation of permeability with time. According to figure 6, the permeability of specimen is considerably increased by applying the ultrasonic process. This can be attributed to the combined effects of hydraulic gradient and sonication. The permeability is high in the tests used the ultrasonic process compare with in the tests not used the ultrasonic process.

The mean permeability throughout the total test time is calculated from the results in Figure 6. The calculated mean permeability is shown in figure 7. In this figure, we can see that the mean permeabilities of test soils are 2.64 E-04cm/sec for simple flushing, 2.68 E-04cm/sec for electrokinetic flushing, 4.70E-04cm/sec for ultrasonic flushing, 5.14 E-04cm/sec for electrokinetic+ultrasonic flushing, respectively.

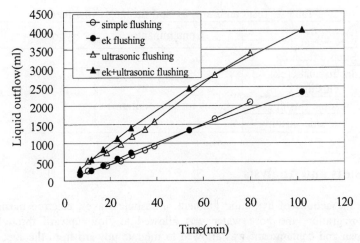

Figure 5: Accumulated flow volume with time

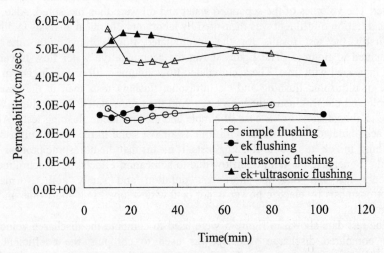

Figure 6: Permeability with time for test sample

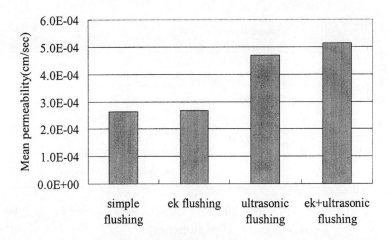

Figure 7: Mean permeability of test soils

The quantity of oil removal is measured from the liquid outflow, and oil removal rate is calculated. The results of calculated oil removal rate is shown in figure 8. The Cd concentration is measured from the liquid outflow and shown in figure 9. Reddi et al (1993) also investigated the effect of ultrasonic energy on enhancement of the permeability of clayey soils. They observed an increase in the permeability of all tests. They attributed the increased permeability to the removal of particles smaller than clay and colloidal size particle in the test specimens due to sonication. Therefore, for the test soils, the increased permeability due to sonication can be attributed primarily to particle agitation and dislodging.

In the figure 8, it is suggested that the oil contaminant is removed with time elapsed, and the percent of oil removal is increased with time and reached a constant after about 24 minute. According to this figure, it is shown that the maximum percentages of oil removal rate for all tests are approximately 65%. The removal velocity of oil is high in the case of ultrasonic process for a very short time.

In the figure 9, it is demonstrated that Cd concentration of outflow is increased with time. The Cd contaminant in soil specimen is steadily removed with time elapsed. The Cd concentration of outflow is high in the cases of electrokinetic and electrokinetic+ ultrasonic flushing process by comparing with in the cases of simple and ultrasonic flushing process. The heavy metal contaminant such as Cd is migrated and removed by electromigration phenomena induced from electrokinetic process. Electrokinetic process is most effective technique to remove heavy metal in contaminated soil.

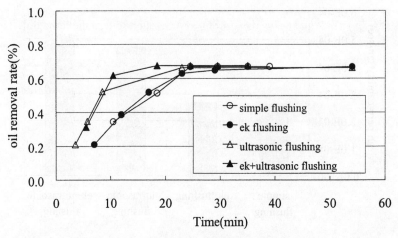

Figure 8: Oil removal rate with time

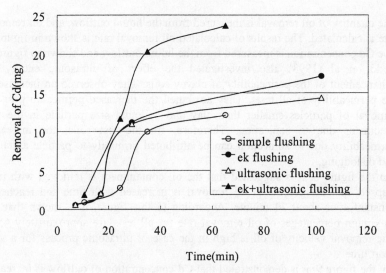

Figure 9: Removal of Cd with time

4 Conclusions

In this study, the coupled effect of electrokinetic and ultrasonic technique is demonstrated by laboratory tests. A series of tests were conducted for simple flushing, electrokinetic flushing, ultrasonic flushing, electrokinetic+ ultrasonic flushing. The test results showed that the ultrasonic technique can enhance the removal of diesel fuel oil from contaminated soils, the electrokinetic technique can enhance the removal of heavy metal from contaminated soils. And the combined new technique of these two techniques can be effective in removal of organic substance and heavy metal from contaminated soils at the same time.

References

[1] Young Uk Kim, Effect of sonication on removal of petroleum hydrocarbon from contaminated soils by soil flushing method, The Pennsylvania State University, The graduate School, Department of Civil and Environmental Engineering, Thesis, Ph. D, 2000

[2] Hyun-Ho Lee, Electrokinetic remediation of soil contaminated with heavy metal and hydrocarbons, Department of Chemical Engineering, Korea Advanced Institute of Science and Technology, Thesis, Ph. D, 2000

[3] Lakshimi N. Reddi, S. Berliner, and K. Y. Lee, Feasibility of ultrasonic engineering of flow in clayey sands, Journal of Environmental Engineering, Vol. 119. No. 4, 1993

[4] Sang-Jae Han, Characteristics of electroosmosis and heavy metal migration in contaminated soil by electrokinetic technique, Department of Civil Engineering, Chung-Ang University, Seoul, Korea, Thesis, Ph. D, 2000

Accelerating pollutant release from mud bed by sand drains.

C. J. Lai[1], J. M. Leu[1] & H. C. Chan[1]
[1]*Department of Hydraulic & Ocean Engineering, Cheng-Kung University, Tainan, Taiwan 70101 R.O.C.*

Abstract

It is proposed to use sand drain systems in a channel mud bed to accelerate the pollutants release rate of the mud bed, based on the fact that the hydraulic conductivity of mud in the horizontal is usually larger than that of the vertical. Experiments are conducted in model channel bed system to assess this concept. This model is filled with various mud depths and has different sand drain arrangements. Ammonium nitrogen (NH_4-N) and dissolved oxygen (DO) concentrations at the water body adjacent to the mud surface are measured. Experimental results show that the ammonium nitrogen concentration increases with the test time. It's release rate increases with intensity of the sand drains. Deployment of a sand drain system is proved to be useful in accelerating the releasing of mud- trapped pollutant.

1 Background

To restore both the water body and the bottom to a better condition that some engineering measures, such as dredging or consolidating the bottom material, are usually needed. Owing to the property of the accumulation and consolidation of the sediment and some side effects of the measures, the required purification objective may not be achieved. The slow release of the embedded pollutants is part of the reason for the failure of the measures. A proposal is made that uses sand drain system to accelerate the release of the embedded pollutant is depicted in Fig. 1. This is based on the fact that the horizontal conductivity in mud is usually larger than that of the vertical. The system collects the horizontal pollutant flow and diffuses it upward through the pore water in the sand drain.

Further actions can be taken to purify the water body and eventually both water and the polluted mud are restored to a better condition. This paper reports some results for a model test of this proposal.

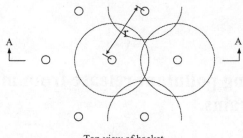

Top view of backet

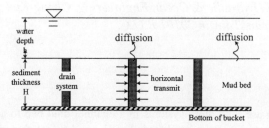

Aː̵DA section

FIG.1: Layout of sand drain system in a water-mud environment.

2 Test criteria

The following criteria have been used in the evolution of the tests:
(1) Mass Flux
The Ammonium nitrogen release rate from the sediment-water interface to the fluid is determined by differentiation of the mass flux. The average Ammonium nitrogen mass flux per unit interfacial area is written as:

$$J = \frac{(C_{i+1} - C_i) \times V}{(t_{i+1} - t_i) \times A} \tag{1}$$

$$R = \frac{dJ}{dt} \tag{2}$$

where C_i is the bulk ammonium nitrogen concentration at time i, V is the water volume of the overlying water, and A is the sediment-water interfacial area.
(2) Water-Sediment Region Oxygen Demand (WSOD)
WSOD is tentatively used in this study and is different from the traditional definition of the Sediment Oxygen Demand (SOD). It is the value of the measured bulk DO concentration drop near the mud-water interface. It is

standardized to the 20   value for comparison of different cases. The relation is:

$$(S_B)_T = (S_B)_{20}(\theta)^{(T-20)} \tag{3}$$

where $(S_B)_T$ is the SOD at T   , $(S_B)_{20}$ is the SOD at 20   , and is the temperature coefficient,   , has a range of 1.040 to 1.130. A value of 1.065 is used.

3 Experimental apparatus and procedure

The tests were performed in the National Cheng Kung University Ecohydraulic Laboratory in Tainan, Taiwan. The environment of the Laboratory was controlled so that the physics of mass transfer at the sediment-water interface could be studied.

A glass tank, 4×4×0.5m (long × width× height), loaded with three 0.2×0.8m and 0.5×0.65m (inside diameter× height) test buckets, was set in the laboratory. The tank was filled with water to eliminate any influences aroused from temperature variation. The model mud in each bucket, which was highly organic, was taken from the bottom of a Canal at Tainan city. The sand drain system was designed to collect the horizontal flow so pollutant can diffuse upward more easily through the pore water. The sand drains used in this study had an average diameter of 1.7 cm. They were made of circular wire meshes filling with sands (d90=2mm). Length of each sand drain was set to fit the sediment thickness.

Instruments used in this study included a WTW multiple-parameter water quality probe, Model OXI330, for bulk DO and temperature measurements, and a WTW Model Photometer S12 for bulk NH4-N concentration measurements.

4 Results

A total of 16 tests were conducted. Each test was conducted in a period of 208 hours and the ammonium nitrogen concentrations were measured at 8-24 hours intervals. Table 1 summaries the experimental conditions and results of the tests, in which h, H and r, as those also depicted shown in Fig.1, are the depths of water, depth of mud bed and distance of drains, respectively. Test number 1-1, 1-4, 2-1 and 3-1, at the drain distance column marked with "—", are the cases without drain deployment. The ammonium nitrogen release rates R_{64} and R_{208} that are calculated from Eq.3 for all the tests are also listed. Subscripts 64 and 208 represent that they are the averaged values for 64 or 208 hours. The WSOD is determined from the drop in bulk DO concentration inside the buckets and the values in Table have been calibrated to the WSOD at 20   , $(S_B)_{20}$, from Eq.3.

From the data and analyzed results, the variation of Ammonium nitrogen concentration, the averaged ammonium nitrogen mass flux per unit interfacial for No.1-1, the relations of the release rate R_{208} with sand drain distance and $(S_B)_{20}$ are plotted in Fig.2~5, respectively. Two features are observed and discussed as follows:

4.1 Ammonium Nitrogen Concentration

It is shown in Fig.2 that the ammonium nitrogen concentrations for all the tests increase rapidly at the first stage, 0-64 hours, and then increase gently at the second stage. As one has expected, the concentrations for the test cases that have no sand drains deployment, tests No.1-1, 1-4, 2-1 and 3-1, are smaller than the others. The dense the arrangement of the sand drains, the higher the concentration tends to be.

TABLE 1. Summary of measured and derived experimental parameter.

No.	Bucket diameter (cm)	h (cm)	H (cm)	r (cm)	t=0~64hr		t=0~208hr		$(S_B)_{20}$ (g/m^2/day)
					R_{64} (mg/m^2/day)	R-square	R_{208} (mg/m^2/day)	R-square	
1-1	20	10	10	—	185.1	0.99	99.3	0.86	0.091
1-2	20	10	10	7.5	397.0	0.94	121.0	0.80	0.082
1-3	20	10	10	10	189.3	0.71	118.1	0.82	0.078
1-4	50	10	10	—	187.5	0.72	67.40	0.84	0.061
1-5	50	10	10	7.5	509.8	0.96	115.0	0.61	0.107
1-6	50	10	10	10	286.0	0.87	114.3	0.86	0.070
2-1	50	10	20	—	302.0	0.96	115.2	0.86	0.078
2-2	50	10	20	7.5	419.6	0.90	113.2	0.64	0.123
2-3	50	10	20	10	386.0	0.93	121.0	0.79	0.119
2-4	50	10	20	15	453.5	0.94	124.3	0.71	0.100
2-5	50	10	20	20	351.6	0.94	111.7	0.87	0.084
3-1	50	10	30	—	214.6	0.97	101.8	0.92	0.063
3-2	50	10	30	7.5	258.6	0.71	140.0	0.86	0.100
3-3	50	10	30	10	417.7	0.86	109.5	0.71	0.073
3-4	50	10	30	22.5	423.5	0.72	95.90	0.70	0.109
3-5	50	10	30	30	477.5	0.74	110.0	0.67	0.088

4.2 Mass Flux and WSOD

The averaged ammonium nitrogen mass fluxes per unit interfacial for No.1-1, shown as the dashed lines in Fig.3, clearly demonstrate the differences in pollutant release rates at the first and second stages. The solid line shows the overall trend. Slopes of these lines are the release rates given in Eq.2. Similar trends have been observed in all other test cases. The relation of the overall ammonium nitrogen release rate R_{208}, for cases that satisfy Eq.2 more closely (R-squared 0.7), with sand drain distance shown in Fig.4 indicate a trend that R_{208} increases as the sand drain distance decreases. Examining R_{208} in Table 1, it indicates that R_{208} increases 19~71%, 5~8% and 8~38% for the tests that has mud depths of 10, 20 and 30 cm respectively, if the tests have sand drain deployment. The WSOD given in Table 1 has values in the range between 0.061 and 0.123 g/m^2/day. These values are comparably in the same order as the field measured values of some previous studies [4]. Owing to the turbulent mixing near the sediment-water interface in the field the field measured value should be higher. The WSOD in Fig.5 shows that it has a trend to increase with R_{208} but the data are rather scattered.

5 Discussion

The differences in the release rates at different stages may be interpreted by: (i) In the initial stage the ammonium nitrogen contains in the sediment diffuses into the overlying water, the rate of denitrification is bigger than the rate of nitrification and the ammonium nitrogen concentration increases at a faster speed. (ii) In the later stage the organic nitrogen content in the sediment becomes lesser and lesser and settling of particulate forms is very small, the rate of denitrification becomes slightly larger than the rate of nitrification; the change of the ammonium nitrogen concentration is slower and the system in the bucket is in a dynamic balance condition.

Insertion of sand drain system in a polluted mud bed is shown to be useful in accelerating the release of pollutant trapped inside the mud bed. However, the space limitation of the mud model in bucket and the gaps between mud and the bucket edge may be the causes of the data scattering of the R_{64} and R_{208} regressions. Ammonium nitrogen could emit through these gaps and the measured value would have included these emissions. Future experiments should consider reducing this the boundary effect.

6 Conclusions

This study proposes to use sand drain systems to accelerate the pollutants that are trapped in a polluted channel mud based on the fact that the hydraulic conductivity of mud in the horizontal is usually larger than that of the vertical. By measuring the ammonium nitrogen release rate at bed water-sediment interface, the following conclusion may be drawn:

(1) Deployment of a sand drain system is useful in accelerating the releasing of mud- trapped pollutant. In the tests, the releasing rate increases 19~71%, 5~8% and 8~38% for the models that have mud depths of 10, 20 and 30 cm respectively.

(2) The ammonium nitrogen release rate can be distinguished into two stages. The first is from 0-64 and the second is from 64-208 hours. The water-sediment oxygen demand (WSOD) also has a trend to increase with ammonium nitrogen release rate in all the experiments.

Acknowledgment

This study was found by the National Science Council of R.O.C. under the contract NO: NSC90-2211-E-006-123.

Reference

[1] Joseph H.W. Lee, Kuang C.p., Yung K.S., Analysis of three-dimensional Flow in a Cylindrical Sediment Oxygen Demand Chamber, Applied Mathematical Modeling. Vol. 24, Issue 4, pp. 263-278, 2000.

[2] Orgunrombi, J.A., and W. E. Dobbins, The effect of bethel deposits on oxygen resources of natural streams. J.WPCF., Vol. 42, No. 4. pp. 538-552, 1970.
[3] Orhon D. and N.Artan, Modeling of Activated Sludge System, Technomic, Pennsylvania, U.S.A. 1994.
[4] R.V. Thomann, J.A. Muller, Principles of Surface Water Quality Modeling and Control, Harper & Row, New York, 1987.
[5] Rutherford J.C., Boyle J. D., Elliott A.H., Hatherell T.V.J. and Chiu T.W, Modeling Benthic Oxygen Uptake by Pumping, Journal of Environmental engineering. Vol. 121, No. 1. pp. 84-95, 1995.
[6] Steinberger N. and Midhat H., Diffusional Mass Transfer at Sediment Interface, Journal of Environmental engineering. Vol. 125, No. 2, pp.192-199, 1999.
[7] Wang W., Kinetic of Sediment Oxygen Demand, Water Research, Vol. 15. pp. 475-482, 1981.

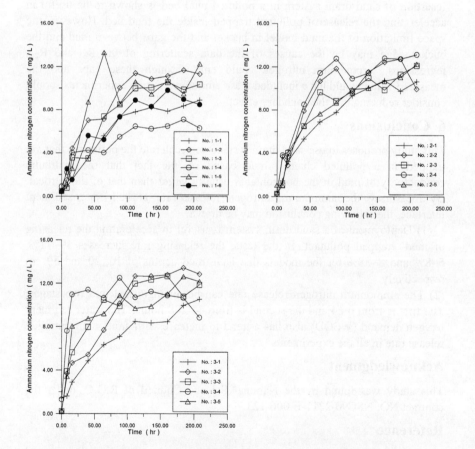

FIG.2: Variation of ammonium nitrogen concentration with time.

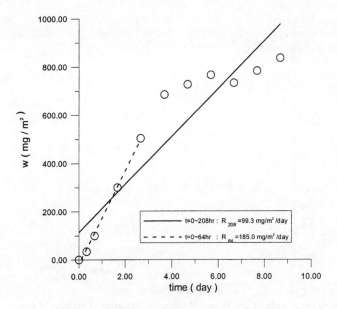

FIG.3: Averaged ammonium nitrogen mass flux in case No.1-1.

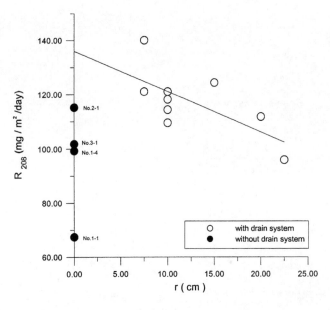

FIG.4: Ammonium release rate decrease with drain system distance.

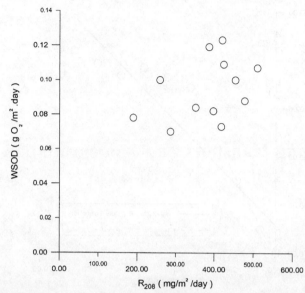

FIG. 5. Experimental data for Water-Sediment Region Oxygen Demand (WSOD) and overall ammonium nitrogen release rate (R_{208})

Modeling transport and biodegradation of PCE in sandy soil

T. Sato[1], Y. Kimura[1] & K. Takamizawa[2]
[1]*Department of Civil Engineering, Gifu University, Japan*
[2]*Department of Bioprocessing, ditto.*

Abstract

Transport of microorganism activating tetrachloroethylene (PCE) dechlorination was well described by mobile-immobile two-region model from laboratory column tests. Biological reaction was identified as first-order kinetics from batch tests where the bacterium strain was mixed with halogenated aliphatic compounds of PCE or TCE (trichloroethylene) as initial pollutant. PCE dechlororenation to dichloroethylene (DCE) via TCE was mathematically expressed as Michaelis-Menten equation while microbial growth rate did not show good performance due to Monod function .

1 Introduction

It has been reported more than one thousands of sites were polluted by halogenated aliphatic compounds, such as tetrachloroethelene (PCE), trichloroethylene (TCE), dichloroethylene (cis-1,2-DCE, trans-1,2-DCE and 1,1-DCE), vinyl chloride(VC), etc., in the environment white paper of Japanese Government. These chemicals were commonly used for industrial and domestic purposes until the use was prohibited by the regulation of water quality protection in 1989. Halogenated aliphatic compounds are characterized by its large amount of unit weight, small viscosity and volatility. It easily infiltrates into ground when its being spilled away on the surface. Several rehabilitations have been reported as success manner for remediation of soil and groundwater contaminated by PCE or TCE.

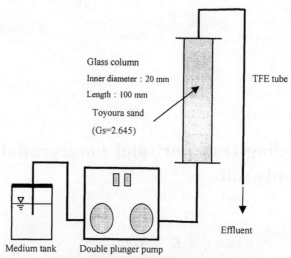

Figue 1: Loboratory column set up.

Table 1: Column test conditions.

Run	Column length (L)	Bacterial concentration (C₀)	Discharge rate (Q)	Darcy velocity (q)	Degree of saturation (Sr)
	cm	mg protein/L	cm³/min	cm/min	%
1	10	4.98	3.5	1.115	100
2	10	9.60	3.5	1.115	100
3	10	11.91	3.5	1.115	100
4	10	28.81	3.5	1.115	100
5	10	3.55	0.1	0.032	100
6	10	9.60	0.1	0.032	100
7	10	35.74	0.1	0.032	100

Bioremediation is one of inexpensive rehabilitation practices to be available for removal of pollution at low concentration close to the Japanese environment standard, however, with spreading large area. There are several barriers to be overcome in realization of biological techniques as in-situ remediation. We have to have know-how about distribute manner of bacteria and/or nutrient into wide area of ground, and to know microorganism and nutrient transport characteristics through complicated soil pores. There is still lack of well findings on behavior of bacteria and nutrient in actual flow region.

The purpose of this study is to stimulate basic understandings of microorganism transport and fundamental description of biodegradation of PCE via TCE to DCE for setting up bioremediation practices. One of the authors has succeeded in isolating PCE-degrading bacterium from mixed culture under anaerobic condition. In the present study, bacterial transport and biological dehalogenation was investigated from laboratory tests using the bacterium identified as *Clostridium bifermentans* DPH-1 [1].

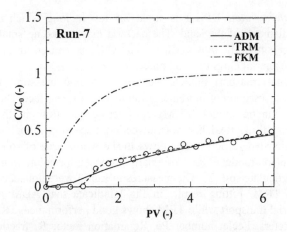

Figure 2: Measured breakthrough curve and models computation.

Table 2: Parameters of TRM determined from the best fittings.

Run	Bacterial concentration (C_0)	Discharge rate (Q)	Darcy velocity (q)	Stanton number (ω)	Fraction rate of mobile water (β)	Peclet number (Pe)	Retardation factor (R)
	mg protein/L	mL/min	cm/min	–	–	–	–
1	4.98	3.5	1.115	0.241	0.449	52.0	2.7
2	9.60	3.5	1.115	0.603	0.106	50.8	12.5
3	11.91	3.5	1.115	–	–	–	–
4	28.81	3.5	1.115	1.412	0.143	51.0	8.3
5	3.55	0.1	0.032	0.698	0.577	57.0	2.0
6	9.60	0.1	0.032	0.759	0.128	50.1	8.5
7	35.74	0.1	0.032	1.590	0.140	61.0	9.5

2 Bacterial Transport through Saturated Sand

Transport of the bacteria was characterized from laboratory column tests with steady state flow condition in fully saturated Toyoura sand of which dry density is 1.55g/cm³. Constant discharge rate was maintained at 0.1 or 3.5 mL/min by using a double plunger pump during the tests. The column size is 10cm in length and 2 cm in inner diameter as shown in Figure 1. Bacterium solution was supplied to the bottom and flow out the top through the sand column. The test conditions were shown in Table 1.

Breakthrough curve in Run-7 was shown in Figure 2 as relationship between relative concentration (C/C_0) describing the ratio of concentration to the initial and pore volume (PV) describing the ratio of total discharge to water contained within soil pore. Measurements are open circles. The knowledge of miscible displacement tells that breakthrough curve passes through $C/C_0=0.5$ at PV=1.0. The bacterial transport consumes more PV to reach at $C/C0=0.5$. Fluctuation at large amount of PV may take place due to pores clogging by microorganism. The size of

Clostridium bifermentans DPH-1 is about 2.5 μm in long which is larger than average pore diameter of the sand. The bacteria may choke up small pore where water does not smoothly move. Change of C/C_0 is caused from buildup and breakdown of bacterial clogs in small pore.

The conventional transport models were fitted to the measures to check the applicability to assessment of bioremediation in actual subsurface. Dashed line was the first-order kinetic model (FKM) considering the flow in the column as complete mixing flow. Real line was advection-dispersion model (ADM) with chemical reaction where pore-water flows in the whole area of soil pore. Dotted line was two-region model (TRM) originally proposed for solute transport through unsaturated particulate soil [2]. The measures in Run-3 were not successfully fitted by ADM and TRM. Fitting results in Fig.2 indicate that ADM is difficult to simulate bacterial transport while TRM shows good performance. TRM is featured by four parameters, Peclet number Pe, Retardation factor R, fraction of mobile water β and Stanton number ω. The parameters were identified from the best fitting to the measures by the least square method as shown in Table 2.

Peclet number expresses spatial distribution of the bacteria in the sand column. It takes almost constant independently on discharge rate and the bacteria concentration in the solution. This is identical to miscible displacement using non-reactive tracer. The amount of dispersion coefficient depends linearly on discharge rate. Peclet number relates to dispersion coefficient as follows;

$$Pe = \frac{vL}{D} \tag{1}$$

in which *v, L* and *D* are flow velocity in mobile region, column length and dispersion coefficient, respectively. Peclet number is constant when dispersion coefficient linearly depends on velocity.

Retardation factor tends to increase as the initial bacteria concentration increases. On the contrary, fraction rate of mobile water decreases with increase of the initial concentration. These findings indicates retardation factor and fraction rate of mobile water strongly depends on mass of microorganism in pores. These are independent of pore water velocity. The fraction rate takes a constant value of 0.14 when the initial concentration of bacteria solution becomes more than 10 mg protein/L.

Stanton number is defined as;

$$\omega = \frac{\alpha L}{q} \tag{2}$$

in which q, α is discharge velocity (Darcy velocity), mass transfer coefficient, respectively. The equation (2) defines ratio of advection time scale to bacteria diffusion into micro-pore via mass transfer. There is no evidence enough to show dependence of Stanton number on discharge velocity. It linearly increases with the initial bacteria concentration. It takes long time in convection when initial bacteria concentration becomes large. Clogs of microorganism prevent smooth flow in mobile water and diffusion into micro-pore.

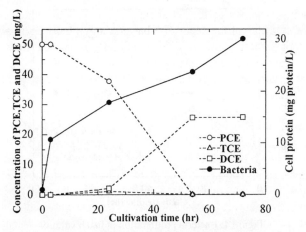

Figure 3: PCE degradation and bacteria growth in batch culture.

3 PCE and TCE Dechlorination

Performance of a PCE and TCE-degrading bacterium from the highly enriched culture was checked using batch procedure. *Clostridium bifermentans* DPH-1 isolated from ditch sludge [1] was used as the source of dechlorinating bacterium. Batch procedure was done by adding PCE or TCE solution to mix culture with growth medium, yeast extract and the bacterium. Cultivation was performed at 30ºC. Measurements were made for identifying PCE or TCE dechlorinating speed and the bacteria growth rate. The tests were carried out at 5, 50 and 150 mg/L of initial concentration of PCE and 2, 20 and 60 mg/L of initial concentration of TCE. PCE degrading was shown in Figure 3 of which case is 50 mg/L of the initial PCE concentration. The figure indicates PCE degradation, via TCE to DCE and increase of cell protein. PCE concentration decreased at 0.022 mg/L in 54 hours. The Japanese environment standard is less than 0.01mg/L for PCE. The test results will pass the standard when operating time is expanded.

For the case of 50 mg/L of PCE concentration, TCE decreased to zero in 54 hours while it increased at 1.1 mg/L in 24 hours later. Other cases showed that TCE concentration smoothly became to zero in 72 hours. TCE degrading speed is rapid rather than PCE and DCE. The mass balance was confirmed from decrease of PCE and increase of DCE. The amount of DCE concentration measured in 72hours was 25.7 mg/L while that estimated from decrease of PCE concentration 29.2 mg/L. The difference comes from the measurement by ECD-GC for PCE in 72 hours and FID-GC for DCE.

The bacteria concentration was shown in Figure 4, which showed that Clostridium bifermentans DPH-1 does not increase with the increase of PCE concentration. The bacterial concentration was estimated from optical density measured by UV spectrophotometer.The concentration, however, gets to be larger than the control. Then, the figure showed the bacterial increase at least relates to PCE concentration.

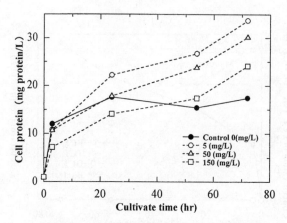

Figure 4: Bacteria proliferation in batch culture.

PCE dechlorination kinetics to DCE via TCE was studied in the batch test series. Activity of PCE dechlorination was assumed to be effective in 60 minutes. The kinetic parameters for PCE degradation were estimated from the Lineweaver-Burk transformation of the Michaelis-Menten equation. Michaelis-Menten equation is given by

$$v = v_{max} \frac{[S]}{K_m + [S]} \tag{3}$$

in which v is degrading rate, v_{max} is maximum degrading rate, K_m is Michaelis constant and $[S]$ is substrate concentration.

For PCE, v_{max}, K_m was 0.591 mg/hr/mg protein, 242.12 mg/L, respectively. For TCE, v_{max}, K_m was 0.222 mg/hr/mg protein, 13.84 mg/L, respectively.

The growth rate of the bacteria could not be described in Monod type function [3]. The empirical description was extracted from the batch tests as follows;

For PCE,

$$\mu_p = -0.0015[PCE] + 0.5064 \tag{4}$$

and for TCE,

$$\mu_t = \frac{0.163}{[TCE]} + 0.435 \tag{5}$$

in which μ_p is bacteria increase rate in PCE-degrading, μ_t is bacteria increase rate in TCE-degrading, $[PCE]$ is PCE concentration and $[TCE]$ is TCE concentration. The bacterial increase rate in PCE-degrading was different from TCE-degrading in the present study.

4 Numerical Bio-assessment Applications

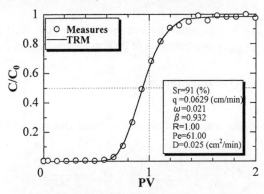

Figure 5: Breakthrough curve for NaCl and fitting by TRM.

4.1 Modeling microorganism transport

Bacteria transport was modeled by TRM in accordance with considerations on laboratory column tests in the preceding section. For mobile zone,

$$\theta_m \frac{\partial x_1}{\partial t} + \frac{\partial}{\partial z}\left(\theta_m v_m x_1\right) = \frac{\partial}{\partial z}\left(\theta_m D_1 \frac{\partial x_1}{\partial z}\right) + \mu_1 x_1 - \mu_{d1} x_1 - \alpha(x_1 - x_{1im}) \qquad (6)$$

and for immobile zone,

$$\theta_{im} \frac{\partial x_{im1}}{\partial t} + \rho_d \frac{\partial S_1}{\partial t} = \mu_2 x_{1im} - \mu_{d2} x_{1im} - \alpha(x_1 - x_{1im}) \qquad (7)$$

in which θ_m, θ_{im} is volumetric water content of mobile zone, immobile zone, v_m is average pore water velocity of mobile zone, μ_1, μ_2 is bacterial increase rate which is given by the preceding section, μ_{d1}, μ_{d2} is decrease rate by death, respectively.

4.2 PCE transport with biodegradation

PCE transport characteristics were determined from NaCl tracer test by the column used in bio-reactor. TRM was fitted to measured breakthrough curve to identify the parameters as shown in Figure 5.
For mobile zone,

$$\theta_m \frac{\partial C_1}{\partial t} + \frac{\partial}{\partial z}\left(\theta_m v_m C_1\right) = \frac{\partial}{\partial z}\left(\theta_m D_2 \frac{\partial C_1}{\partial z}\right) + V_{max1} \frac{C_1}{K_{s1} + C_1} x_1 - \alpha_2 (C_1 - C_{pce}) \qquad (8)$$

For immobile zone,

$$\theta_{im}\frac{\partial C_{pce}}{\partial t}+\rho_d\frac{\partial S_2}{\partial t}=-V_{max12}\frac{C_{pce}}{K_{sp}+C_{pce}}x_{1im}+\alpha_2(C_1-C_{pce}) \qquad (9)$$

in which V_{max11}, V_{max12} is maximum degradation rate for PCE which is given by preceding section, C_1, C_{pce} is concentration of PCE in mobile, immobile zone.

4.3 TCE transport with biodegradation

TCE and DCE transports were modeled as the same description of PCE. For mobile zone,

$$\theta_m\frac{\partial C_2}{\partial t}+\frac{\partial}{\partial z}\left(\theta_m v_m C_2\right)=\frac{\partial}{\partial z}\left(\theta_m D_3\frac{\partial C_2}{\partial z}\right)+V_{max21}\frac{C_2}{K_{s2}+C_2}x_1+V_{max11}\frac{C_1}{K_{s1}+C_1}x_1-\alpha_3(C_2-C_{tce}) \qquad (10)$$

For immobile zone,

$$\theta_{im}\frac{\partial C_{tce}}{\partial t}+\rho_d\frac{\partial S_3}{\partial t}=-V_{max22}\frac{C_{tce}}{K_{st}+C_{tce}}x_{im}+V_{max12}\frac{C_{pce}}{K_{sp}+C_{pce}}x_{1im}+\alpha_3(C_1-C_{tce}) \qquad (11)$$

in which V_{max21}, V_{max22} is maximum degradation rate of TCE, C_2, C_{tce} is concentration of TCE in mobile, immobile zone.

4.4 Transport of DCE with biodegradation

For mobile zone,

$$\theta_m\frac{\partial C_3}{\partial t}+\frac{\partial}{\partial z}\left(\theta_m v_m C_3\right)=\frac{\partial}{\partial z}\left(\theta_m D_4\frac{\partial C_3}{\partial z}\right)+V_{max21}\frac{C_2}{K_{s2}+C_2}x_1-\alpha_4(C_3-C_{tce}) \qquad (12)$$

$$\theta_{im}\frac{\partial C_{dce}}{\partial t}+\rho_d\frac{\partial S_4}{\partial t}=V_{max22}\frac{C_{tce}}{K_{st}+C_{tce}}x_{im}+\alpha_4(C_3-C_{dce}) \qquad (13)$$

Dehalogenation of DCE to EC (ethylene) via VC(vinyl chloride) may take place in this biodegradation process but it was not detected in the present study. Then the modeling biodegradation is limited to PCE dehalogenation to DCE. The model will be improved in the future when DCE-degrading bacteria is isolated and equipment is developed for easily measuring concentration of VC and EC.

4.5 Application to reactor column

A continuously-fed up-flow, anaerobic column reactor was developed using a pure organism. The reactor employed a glass column of 22 cm in length and 4.5 cm in diameter filled with immobilized ceramic beads and MY medium (Fig.6). The reactor was percolated in an up-flow mode (ascending flow) under anaerobic

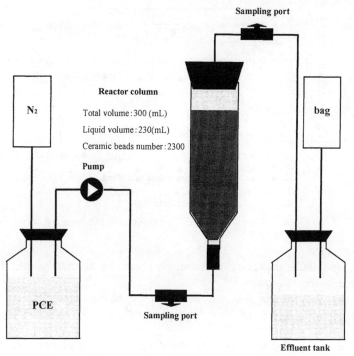

Figure 6: Schematic diagram of fixed-bed reactor system.

conditions. PCE solution with 5mg/L was supplied at the bottom of the column. The operation was conducted at 0.3mL/min of flow rate which is resident time within the column is about 12hours. The measurement of optical density was done at 660nm using a UV spectrophotometer. PCE, TCE and DCE were identified and quantified by static-headspace analysis using a gas chromatograph.

The reactor performance was simulated using the preceding model. There were no measures of microbial decrease factor by death. The model adopted 0.185 (1/hour) as μ_{d1} and μ_{d2}. Simulation was described in Figure 7. PCE degradation to DCE via TCE was well computed. The measured TCE takes a small concentration during the tests series while the simulation gives a little large. The maximum DCE concentration in the simulation occurs at different time from the measures. The model, however, shows a good agreement with the reactor performance.

5 Concluding Remarks

The bacteria transport and tetrachloroethylene (PCE) and trichloroethylene (TCE) dehalogenation have been studied using laboratory column and batch test results. The numerical simulation was done for checking the applicability in bioremediation of brownfields polluted by PCE or TCE. The study concludes that;

(1) Transport of bacteria was well described by two-region model while conventional adevection-dispersion model did not show satisfactory

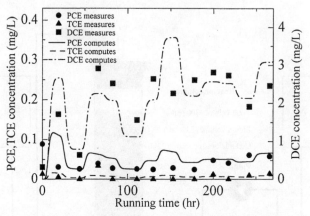

Figure 7: Simulation of PCE dechlorination in bioreactor.

performance.

(2) PCE and TCE degrading characteristics can be accurately quantified by Michaelis-Menten equation.

(3) Bacteria growth rate was not described by Monod type function. It was mathematically expressed by proportion to PCE and by inverse proportion to TCE.

(4) Numerical model showed a good performance of PCE dehalogenation measured in fixed-bed reactor column.

References

[1] Chang, Y.C., Okeke, B.C., Hatsu, M. and Takamizawa, K.: In vitro dehalogenation of tetrachloroethylene (PCE) by cel-free extracts of Clostridium bifermentans DPH-1, Bioresourse Technology, 78, 141-147, 2001.

[2] Van Genuchten, M.Th. and Wierenga, P.J.: Mass transfer studies in sorbing porous media I. analytical solutions, Soil Science Society of American Journal, 40(4), 473-480, 1976.

[3] Wood, B.D., Dawson, C.N. Szecsody, J.E. and Streile, G. P.: Modeling contaminant transport and biodegradation in a layered media system, Water Resources Research, 30(6), 1833-1845, 1994.

The *i*-BO™ process and related treatments for mine waste remediation

K. McEwan[1] & D. Ralph[2]
[1] *Micron Research Pty. Ltd, Australia*
[2] *Division of Science, Murdoch University Australia*

Abstract

Acid mine drainage is an ongoing problem associated primarily with sulphidic mine wastes. Remediation of the mine sites is expensive. Micron Research Pty Ltd has developed the *i*-BO™ Process for treatment of a particularly difficult to remediate tailings dumps located near Tennant Creek in the Northern Territory of Australia. Application of the *i*-BO™ Process to the tailings dams results in an increase of the rate of natural weathering and hence acid production. Years of weathering are condensed into months allowing the acid drainage to be controlled and collected. The contained gold, cobalt and copper are then available to downstream metallurgical recovery processes making the remediation process profitable.

1 Introduction

Centuries of mining and metallurgical treatment of ore bodies have left a legacy of billions of tonnes of tailings and low-grade waste. The sulphur contained in tailings is thermodynamically unstable and exposure to oxygen and water results in the oxidation of the sulphur containing minerals to sulphuric acid, generating acidic solutions that contain a cocktail of base metals and impurities. Hence, the remediation liability is exacerbated by seasonal parameters and natural oxidation of the exposed mineral, creating additional environmental problems in the form of acid and base metal drainage into the surrounding land, eventually polluting the waterways. The high cost of the remediation of mine sites frequently results in the liability being passed onto successive operators, or being ignored.

Micron Research Pty. Ltd. focuses on developing processes that profitably recover precious and base metals from these low-grade resources, making the rehabilitation of mine wastes economically attractive. Our proprietary process for the *in situ* bio-oxidation of low-grade mineral resources, the *i*-BO™ Process, increases the rate of the natural weathering, using naturally occurring lithotrophic bacteria to oxidise the minerals[1]. Contained precious metals are then available to conventional recovery processes.

2   The Peko Tailings

The *i*-BO™ Process was developed to rehabilitate a refractory and impermeable tailings dump, located at Tennant Creek, in the Northern Territory of Australia. The 4 Mt of mine tailings at this site contain mostly magnetite ($\sim$ 80 %), clay minerals and small amounts of mineral sulphides with the most common being pyrite (FeS_2), chalcopyrite ($FeCuS_2$) and cobaltiferous pyrite ($FeCoS_2$)[2,3]. Gold is also present (1.2 ppm average)[4] and the combined value of the base metals and gold is approximately \$A 80 per tonne. Despite the contained metal value, the refractory, impermeable nature of the tailings makes conventional recovery options uneconomic and remediation expensive.

An aerial photograph of the dump is presented in Figure 1. Fine black acidic dust blows from the dump into the town of Tennant Creek and local indigenous communities.

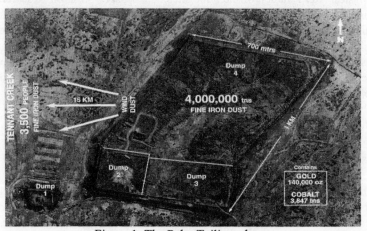

Figure 1: The Peko Tailings dump

The dump contains the tailings from two processing plants, which operated from 1954 to 1976 and treated the Peko ore body and that of four nearby mines. The rehabilitation obligation was passed between subsequent owners of the mining tenements.

The area is subject to inundation from monsoonal rainfall. The rainwater is contaminated by acid and base metal drainage from the dump and subsequently pollutes the local waterways.

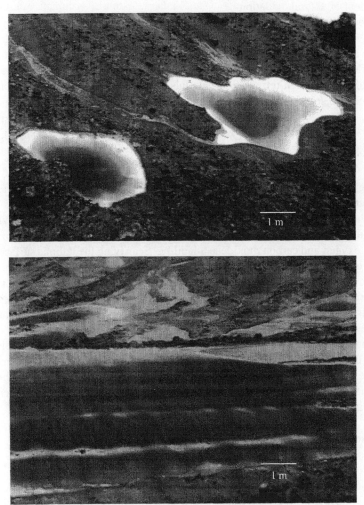

Figures 2 & 3: Drainage from the tailings

3   The remediation strategy

The tailings dump was comprehensively sampled and assays indicated that natural weathering had oxidized the outer surface layers to a depth of ca. two meters (ca. 380 ktonne). The gold recovery from the oxidized surface layers was 65% to 95%, with cyanide consumption of 2 kg/t and lime consumption of 5 kg/t; whereas the gold recovery from the lower layers was 17% to 50% with cyanide and lime consumption of 10 kg/t and 27 kg/t, respectively. The surface layer was acidic, with a pH of 3.5, the pH increases with depth, the bottom being pH 8.2. The method used for depositing the tailings has resulted in very low vertical percolation rates for water and air resulting in high runoff volumes and a limit to the extent of natural oxidation.

These observations provided some insights into the behavior of the tailings dump. It was clear, that the metal sulphides contained in the surface layers, which are exposed to water and air, enabling the oxidative activity of naturally occurring bacteria had become oxidized by natural weathering. The gold bearing minerals become non-refractory during this process, the base metals and acid are released into solution. The chemical reaction is described below:

$$MS_x + yO_2 + H_2O \rightarrow M(SO_4)_z + H_2SO_4 \qquad (1)$$
$$(M = Cu, Co, Fe)$$

The treatment strategy that appeared most promising was to maximize the rate of natural weathering that has already oxidized the outer layer of the tailings heap over the last 25 years. It was postulated that this could be done by providing large numbers of lithotrophic bacteria and optimizing the conditions of pH, water, oxygen and carbon dioxide availability. Scarifying the tailings surface to a depth of 0.5 m (assuming a 'treatment' area of 2.5×10^5 m^2) provides a broken layer containing 0.3 Mt, or ca. 150 days supply for a conventional CIP plant. Oxidizing enough of the sulphur in this layer to give a high recovery of precious and base metals imposes only a modest oxygen and carbon dioxide demand to be supplied by diffusion over 150 to 200 days.

Water could be added by an irrigation system to maintain adequate moisture levels in the broken layer while collecting the drainage would enable base metals to be extracted. The strategy also included a means to increase the number of bacterial cells present in the tailings without relying on the natural increase occurring within the broken layer.

3.1 Inoculum Generator (IG)

To provide large numbers of lithotrophic bacteria, cells were grown in an external vessel and transferred to the tailings oxidation test pads (also called farms). Draining the IG vessel gave a flow of fluid that was acidic (pH 2), contained dissolved iron (mainly as the ferric form) and a large number of cells. This fluid was used to inundate the pad of tailings. Collecting the run-off after a few hours showed that:

1. Cells contained in the IG fluid tended to attach to the tailings and not report to the run-off sump.
2. The ferric ions in the IG fluid were reduced to ferrous form by contact with the tailings in the farm

3. The acid in the IG fluid created a tailings slurry in the Farm of pH favorable to bacterial growth.

3.2 Tailings oxidation farms

The tailings oxidation Farm allowed a thin layer of tailings to be exposed to favorable oxidation conditions so that the extent of the oxidation and the recovery of base and precious metals could be quantified. The laboratory scale farms were inundated (with IG fluid) and 'ploughed' on a regular basis to maximize their exposure to air.   Tailing from all levels within the heap were tested in the oxidation 'farms'.

4 Materials and methods

4.1 The Inoculum Generator (IG)

The waste rock used in the IG was obtained from the Geko mine waste pile near Tennant Creek. This sulphide mineral is stockpiled and exposed to air and water, generating continuous acid drainage.   The Geko mine waste rock is a rehabilitation liability to the current owners of the mining tenements on which it is situated. This mineral (averaging 3% S) was crushed and sieved. The -12 mm fraction was loaded directly into the IG column (1600 mm length by 300 mm diameter) while the large fraction was returned.   Sulphuric acid was added to 50 L of water until the pH was 1.7 and small amounts of ammonium sulphate and potassium hydrogen phosphate added and the fluid used to fill the column.   The air sparge was set to add ca. 100 mL per minute to the base of the column.   This basic design was scaled up by a factor of 10 (in 4 separate units) to provide inoculum for the full scale of farm test-work. Conventional cyanidation test work was conducted on milled and unmilled material both before exposure to the bacterial cultures and after 450 days of inoculum generation.

4.2 Tailing Farms

The tailings were sampled in a comprehensive program of drilling.   Samples collected at the various depths were stored in airtight bags until used.   The tailings collected at different levels in the dump (e.g. 0 – 2 m) were weighed, sampled and spread in a plastic tray (1200 x 500 x 150 mm) to give a layer of ca. 10 cm. The tailings (usually 15 – 25 kg) were inundated with fluid from the IG. After 24 hours, the fluid was drained and the cycle of flooding and scarifying described below was begun.

1. Tailings were irrigated with fluid from the IG.
2. The excess fluid drained naturally to a sump and was collected.
3. The tailings bed was allowed to dry to a particular moisture content.
4. The tailings bed was then ploughed to allow access of air.

This cycle of farm treatment was continued for 150 to 250 days and solid samples were withdrawn at intervals to test for copper and cobalt removal and gold recovery after conventional cyanidation. Lights arranged above the surface of the tailings (between 200 – 300 mm) gave a radiant energy of 1000 W m^{-2} and were operated for 12 of every 24 hour cycle. These lights were set up to simulate the effect of the sun in evaporating moisture from the tailings farm surface

5 Results and discussion

5.1 Inoculum Generator

Fluid from the inoculum generator used to inundate the tailings farms was replaced with tap water. After a period, the acid, ferric ions and cell numbers in the diluted vessel settled to constant values that were a function of the dilution rate. The Geko mine waste rock contained 2ppm Au, 3.3% S^{2-}, 2.5% Cu and 17.5% Fe. An IG column containing 35 kg of Geko rock and 50 L of fluid, yielded ca. 1 L of fluid containing 1 - 2 g of sulphuric acid (pH 1.9 – 2.1), 2 g of ferric ions and ca. 1 x 10^8 cells each day over a period of ca. 450 days. Over this period there was a mass loss of 5.3% and 88% sulphide mineral oxidation, there was no longer an acid drainage problem from the spent mineral from the IG.

The results of conventional cyanidation treatment of milled and unmilled rock, both before and after exposure to the bacterial inoculum are presented in Table 1.

Table 1: Gold dissolution from the Geko mine waste rock before and after residence in the Inoculum Generator

Treatment		Gold Dissolution (%)	Cyanide Consumption (kg/t)	Lime Consumption (kg/t)
Untreated	Unmilled	24	3.5	3.4
	Milled	63	9.0	5.9
450 day IG residence	Unmilled	84	1.7	3.4
	Milled	89	3.3	8.5

Gold dissolution from the unmilled and milled waste rock increased by 250% and 41% respectively, following the 450 day treatment period. There was no substantial change in cyanide consumption. Lime consumption for the cyanidation of the milled rock increased from 5.9 kg/t in the untreated samples to 8.5 kg/t in the oxidized samples. Recovery of gold from this waste rock used in the IG makes its rehabilitation process profitable.

5.2 Tailings Farms

The Tailings Farms were used to test tailings extracted from all levels of the heap and each was treated with the same methodology.   A four day cycle of inundation of the tailings with solution containing bacteria from the inoculum generators and aeration by scarifying was carried out and representative solid samples were extracted regularly.   Conventional cyanidation test work was conducted on the samples.

Since the tailings originate from several different mines, their composition varies with their location in the dumps. The high grade tailings contain 7g/t Au, 0.9% Cu and 18% Fe. The lower grade material contains 1.2g/t Au, 0.4 to 0.8% Cu and 40% Fe. The S^{2-} concentration ranges from 1 to 7%.

The tailings from various locations within the dumps were treated in farms until the drainage of acid and base metals ceased. Between 50 to 60% of the cobalt and copper was solubilized as a result of this process. The recovery by proprietary techniques of valuable base metals from these solutions is currently being developed at Micron Research.

The effect of the treatment on gold dissolution is presented in Table 2 for both the high grade and low grade materials.

Table 2: The effect of the *i*-BO™ Process on gold dissolution

Treatment Time (days)	Gold Dissolution (%)	
	Low Grade	High Grade
0	40	17
28	36	39
63	44	46
91	56	Not Available
119	65	79
207	65	87

Gold dissolution from the low grade material improved from 40% to 60% over a four month period. Gold dissolution from the high grade material improved from 17% to 87% over a 7 month period.

After application of the *i*-BO™ Process to the tailings and subsequent gold recovery by CIP, the tailings residue will no longer leach acid and base metals and will be disposed of in a nearby open cut mine, covered with top soil and re-vegetated. The gold and base metal recovery from the tailings makes their rehabilitation profitable and rehabilitation of the area economically attractive.

6 Conclusions

The *i*-BO™ Process was developed as a low cost, viable treatment option for the Peko tailings dumps. By using the Geko mine waste, this material is also remediated. Furthermore, the tailings can then be used as an inert fill to rehabilitate a nearby open cut mine. The oxidation process solubilizes 50% to 60% of the contained cobalt and copper and renders 65% to 87% of the contained gold recoverable by conventional cyanidation techniques. The low input cost for the recovery of the cobalt, copper and gold makes the application of this process and hence the rehabilitation of the Peko tailings dumps, Geko mine waste rock and a nearby open cut mine, attractive to the owners of the mining tenements on which they are situated.

References

[1] Crundwell, F., Holmes, P.R., & Fowler, T.A., The Mechanism of Bacterial Leaching of Pyrite by *Thiobacillus Ferrooxidans Applied and Environmental Microbiology*, **65** pp. 2987-2993, 1999
[2] Henley, K., Amdel Report Number G6404/86.
[3] Radke, F., Amdel Report Number G259PO98.
[4] Mujdrica, S., & Hatcher, M., *Peko Tailings Resource Report, Peko Tailings Project Tennant Creek* Tennant Creek Library No. 97065, Kent Town Library No. 97065, 1997

Acknowledgements

The authors wish to thank Peko Rehabilitation Project Pty Ltd for permission to publish this article.

Biological degradation of PCBs in soil. A kinetic study

M.A. Manzano, J.A. Perales, D. Sales & J.M. Quiroga
Department of Chemical Engineering, Food Technology and Environmental Technologies
Cadiz University, Spain

Abstract

In this paper a kinetic study is made of the biodegradability of Aroclor 1242 in sandy soil employing a mixed culture of acclimatized bacteria. The assays were done in stirred tank reactors, and the biodegradation process was monitored by High Resolution Gas Chromatography (HRGC) with Electron Capture Detector. These results are supported by other indirect measurements and indicators of the existence of microbial degradation process, as well as the parameters for the control of the process.

The biodegradation occurred as a first order process and it proved most effective in respect of dichlorinated (100% removal), followed by trichlorinated (92%) and tetrachlorinated biphenyls (24%).

1 Introduction

Awareness of the toxicity of PCBs has led to increased research into the development of PCB waste treatment technology. Although incineration is currently the most frequently used method of dealing with waste containing a high concentration of PCBs, waste products containing a large proportion of inert material such as soils and sediments require other alternatives [1].

One of the options for this type of waste are biological treatments, made attractive by their low cost and operative simplicity. The aim of this research has been to study the kinetic of the aerobic biodegradation of PCBs adsorbed to soil particles by employing a mixed culture of acclimatized bacteria.

2 Materials and methods

The mixed culture of PCB degrading bacteria was acquired from New York State Centre for Hazardous Wastes Management (Buffalo, N.Y.). The mixed culture of microorganisms was isolated by biphenyl enrichment from PCB contaminated sediment. This culture consisted mainly of gram negative strains exhibing Type II organism PCB biodegradation patterns. Type II organisms are Pseudomonas that biodegrade similar congener profiles as the Pseudomonas strain LB400. The mixed culture was grown on biphenyl, aqueous PCB solutions, and a phosphate-buffered mineral nutrient solution.

The type of soil employed was sandy quartz (X-Ray Diffraction, Phillips PW 1830) from Guadalete river (South-West of the Iberian Peninsula) containing very low levels of organic material (< 0.05 % weight). Sand was selected because the surface soils -the region most affected by an accidental spill- is often predominantly made up of sand [2]. Futhermore, due its low natural organic composition, this sand was expected to have low interaction with PCB and thus results in better interpretation of the experimental data. Using Aroclor 1242 (SUPELCO), and following the method described by Barriault and Sylvestre [3], the soil was contaminated to a concentration of 100 mg/Kg (dry weight).

The experiments in stirred tank reactors (STR) were performed in 2.5L Pyrex glass vessels, covered with aluminum foil and stirred with steel agitators at 200 rpm, at a temperature of $23\pm3°C$. The experiments were performed in duplicate with controls to evaluate abiotic losses (400 ppm of $HgCl_2$ to ensure cessation of biological activity).

Evaluation of the active and total number of aerobic microorganisms was determined with the Epifluorescence Microscope (Nikon AFX-DX) using the following reagents as fluorochromes [4]: 5-cyano-2,3-ditolyl-tetrazolium chloride (CTC) and 4,6-diamido-2-phenylindole, respectively (DAPI).

The PCB were analyzed by capillary gas chromatography using an electron capture detector (ECD). All samples were run on a Perkin Elmer Autosystem HRGC equipped with a 30 x 0.32 mm I.D. fused silica colum (SPB-5, Supelco Inc., Bellefonte, Pa). The chromatographic protocol employed was that of Ofjord et al. [5] employing hexachlorocyclohexane as an internal standard. PCB extraction was performed using the method developed by Quensen et al. [6].

And finally, to determine pH, temperature, and dissolved oxygen concentration, selective electrodes were used in accordance with standard methods [7].

3 Results and discussion

In STR assays, the m/V ratio was 1/10 g/mL and the cosubstrate (biphenyl) was added every 2 days with 100 mg/L. Fig. 1 shows the evolution of residual Aroclor 1242 in the experiments.

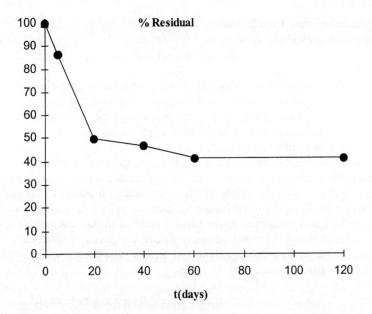

Figure 1: Evolution of residual Aroclor 1242 in soils in the experiments using mixed culture of PCB degrading inoculum.

The results show that Aroclor levels in the slurry decrease rapidly ,14% at five days and 50% over the first 20 days of the experiment. After 120 days, the duration of the experiment, the biodegradation rates achieved was 60%.

The biodegradation rates obtained for Aroclor 1242 with the acclimatized culture are higher than those obtained by Brunner et al. [8] (13% removal with *Ancinetobacter* P6 over 210 days) and by Barriault and Sylvestre [6] (4.5% removal with *Pseudomona Testoteroni* B-356 over 90 days). On the other hand, the results obtained in these experiments are somewhat lower than those recorded by Fotch and Brunner [9] who noted 75% biodegradation after 49 days with *Acinetobacter* P6.

An increase in the level of chlorination resulted in a lower extent of biological degradation (100% DiCBs, 92% TriCBs, 24% TetraCBs and non appreciable degradation for PentaCBs after 60 days.

With regard to the evolution of total and viable microorganisms, the results from the experiment show that initially the bacterial population was small (3.5×10^7 total cells/mL and 10^6 active cells/mL), that it increased with each application of biphenyl, and stabilised after 20 days at values for total and active cells ranging between $4\text{-}5 \times 10^8$ y $1.5\text{-}3 \times 10^7$ cells/mL, respectively (5% viable). The pH, temperature and inorganic anion values all remained within an optimum range for bacterial development.

Middelton et al. [10] postulated an empirical model to evaluate the levels of polycyclic aromatic hydrocarbons (PAH) in soils via bioremediation. The model, also known as the general bioremediation model, is based on soil/sludge bioremediation data from laboratory tests, pilot and full scale field studies. The proposed equation is as follows:

$$C_t = C_r + (C_o - C_r)\, e^{-kt}$$

Where:

C_t = concentration of the organic compound at time t, mg/Kg

C_o = initial concentration of the organic compound, mg/Kg

C_r = concentration of the organic compound which is resistant to biodegradation or has no bioavailability, mg/kg

K = first order rate constant, t^{-1}

The model, which has been used to predict biodegradability in matrices with a high degree of adsorption, is based on two important premiss; (a) that a residual concentration of substrate exists which is resistant to biodegradation due to its non bioavailability and (b) that biodegradation occurs as a first order process.

This model has been applied to the results obtained in this research. In Table 1 the results deriving from the adjusted model are given for the experiments relating to the aerobic biodegradation of PCB-contaminated soils in the STR using acclimatized bacteria.

Table 1: Parameters adjusted to the Middelton kinetic model

	DiCB	TriCB	TetraCB	Total
C_r **(mg/Kg)**	0.12	2.62	25.41	40.7
	(0.09)	(2.7)	(25.9)	(39.8)
k**(days^{-1})**	0.0823	0.0654	0.044	0.074
r^2	0.982	0.967	0.966	0.987

* The data in brackets are the values derived from the experimental results.

The model was applied successfully, as is evident from the correlation coefficients obtained. The results also show how an increase in the number of congener chlorines is accompanied by an increase in relative residual amounts and a decrease in the kinetic constants of the degradation rate. Fig. 2 shows the evolution of total and homolog PCB concentrations, including those obtained by experiment (represented by bullets) and those predicted by the model (represented by a continuous line). Their similarity is evidence of the validity of the model.

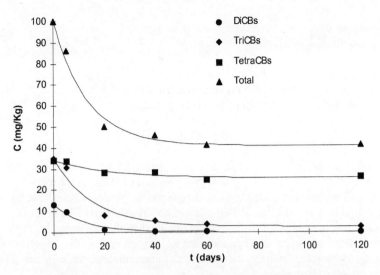

Figure 2: Evolution of overall and homolog PCB concentration.

4 References

[1] Watts RJ, Udell MD, Monsen RM. Use of iron minerals in optimizing the peroxide treatment of contaminated soils. *Water Environ Res* 65: 839-45. 1993.

[2] Ravikumar JX, Gurol MD. Chemical oxidation of chlorinated organics by hydrogen peroxide in the presence of sand. *Environ Sci Technol* 28: 394-400. 1994.

[3] Barriault D, Sylvestre M. Factors affecting PCB degradation by an implanted bacterial strain in soil microcosm. *Can. J Microbiol* 39:594-602. 1993.

[4] Winding A, Binnerup SJ, Sorensen J. Viability of indigenous soil bacteria assayed by respiratory activity and growth. *Appl Environ Microbiol* 60:2869-2875. 1994.

[5] Ofjord GD, Puhakka JA, Ferguson JF. Reductive dechlorination of Aroclor 1254 by marine sediment cultures. *Environ.Sci.Technol* 28:2286-2294. 1994.

[6] Quensen JF, Boyd SA, Tiedje JM. Dechlorination of four commercial polychlorinated biphenyl mixtures (Aroclors) by anaerobic microorganisms from sediments. *Appl Environ Microbiol* 56:2360-2369. 1990.

[7] APHA-AWWA-WPFC. Standard Methods for the Examination of Water and Wastewater. 1989.

[8] Brunner W, Sutherland FH, Focht DD. Enhanced biodegradation of polychlorinated biphenyl in soil by analog enrichment and bacterial innoculation. J Environ Qual 14:324-328. 1985.

[9] Fotch D, Brunner W. Kinetics of byphemil and polychlorinated biphenyl metabolism in soil. Appl Environ Microbiol 50:1058-1063. 1985.

[10] Middelton AC, Nakles CV, Linz DG. The influence of soil composition on bioremediation of PAH-contaminated soil. Remediation Autumn:391-406. 1991.

Understanding the outcomes of brownfield clean-up programs

R.A. Simons[1], K. Winson-Geideman[1] & J. Pendergrass[2]
[1] *Levin College of Urban Affairs, Cleveland State University*
[2] *Environmental Law Institute*

Abstract

This study provides preliminary evidence regarding the clean up and development of previously contaminated properties in the Chicago area that have completed the State of Illinois Pre-Notice Site Cleanup Program (1989 through 1995) and Site Remediation Program (1995 to present). The program offers differing levels of liability relief and allows caps and other on-site remediation strategies. Real estate and environmental databases are merged to allow analysis of development outcomes for a large sample of entering sites, which are followed through the program and into the marketplace.

1 Introduction

The concept of brownfield redevelopment by volunteers is relatively new, having evolved over the past decade. The United States Environmental Protection Agency (USEPA) originally served as the government agency responsible for supervising the clean up of contaminated sites through the superfund program, created in 1980. This was a mandatory participation approach for the parties responsible for the pollution. In the 1980s, states began creating their own superfund programs in an effort to clean up sites that the federal EPA would not reach because it focuses only on sites with the most debilitating contamination. The states, in turn, established guidelines for mandatory cleanup, hoping to restore these sites to productive use. In general, contaminated sites, (including brownfields) were subject to the strict liability standards of the Comprehensive Environmental Response, Compensation and Liability Act (CERCLA or Superfund) of 1980 and its state equivalents (Environmental Law Institute [1]).

These mandatory programs, however, failed to provide much incentive for participation. Liability was a concern for both developers and lenders. In recent years, changes in legislation at both the federal and state level have reduced that liability and focused on the remediation and productive re-use of these sites. These changes culminated with the state voluntary clean-up programs (VCPs).

One of the major components of the VCPs is to facilitate cleanup of sites that are viable for redevelopment. However, the state environmental authorities rarely track the sites once cleanup is complete. This study reviews the success of those sites that have entered the State of Illinois Site Remediation Program (SRP). Sites that successfully completed the program as well as languishing sites are addressed. The feature of this research is that it provides preliminary evidence regarding the success of the program by combining the SRP database with publicly recorded data that includes ownership, structural, and financing characteristics of a site.

2 Literature and Background

2.1 The brownfield problem

The predominance of brownfields and the problems they cause make redevelopment one of the hottest topics in US state capitals.   Recent research shows the number of brownfields as being between 450,000 and 600,000 nationwide (Simons [2 and 3]). Not only are these sites an aesthetic problem, they are often blighted and contribute to the overall economic decline of neighborhoods. Legislators and local communities are looking for alternatives to re-use these properties while at the same time improving tax bases, expanding employment opportunities and restoring environmental quality. As of mid 2001, there were 49 different state voluntary cleanup programs (VCPs) in the US, in 46 states. All total, 11,500 sites had successfully completed cleanups through these VCPs, and another 5,500 sites were in the pipeline (Simons, Pendergrass and Winson-Geideman [4]).

Although brownfield remediation costs can be considerable, for most lightly-to-moderately contaminated sites liability is the main barrier to volunteer capital stepping forward with reinvestment funds. Many US states have tried to protect innocent purchasers of contaminated property. In 1983, a New Jersey law required specific industries with a history of contaminating property to "adequately clean up any contamination of their property before closure, sale, or transfer of their operations." The intent of this law was to reduce the potential liability of future owners, but like CERCLA, affected redevelopment prospects negatively (Page and Rabinowitz [5]).

Changes in environmental laws have hindered redevelopment for two reasons. First, CERCLA liability affects the value of property more than the actual contamination itself (Page and Rabinowitz [6]). Boyd, Harrington and Mccauley [7] studied the behavior of buyers and sellers of contaminated land in an industrial market. They concluded that buyers were at a disadvantage in the flow of information and therefore tended to overestimate the cost of site remediation and risk. In a study on the impact of contamination on the industrial land market in the Canton/Southeast Baltimore, Maryland area, Howland [8] found brownfield sites sold an average of 55% below the price for clean sites.

She also found that the brownfield sites that did sell sold for 25% less than the asking price of brownfield sites that did not sell.

Second, because environmental cleanup programs are still administratively immature and dynamic, changes may result in delays, higher cost and extended cleanup periods (Arrandale [9]). Furthermore, depending on the extent of the contamination, CERCLA mandated cleanups could extend a project's completion by months or even years, rendering it completely uneconomical (Bartsch, Andress, Seitzman & Cooney [10]).

Even though fears are that environmental policy will become more restrictive, the policy process may also result in relaxed environmental laws. An example of this occurred when asbestos was allowed to be managed in place rather than removed [5]. Speculating on the status of laws may cause developers to think it more lucrative to wait to develop.

The inauguration of the VCPs reflects state efforts to make cleanups less expensive and easier to perform. In doing so, sites are made more attractive to developers as costs and cleanup time are controlled. In general, liability concerns remain with the party responsible for the contamination, and new owners and developers are relieved from it [1]. The economic benefits of these programs include increased employment opportunities for residents, tax revenue for communities and states and the reduction of urban blight.

2.2 The State of Illinois program

The Illinois VCP is at the forefront of US remediation programs. Prior to the creation of the voluntary program, the State of Illinois Environmental Protection Agency (IEPA) offered a Section 4(y) Letter pursuant to Section 4(y) of the Environmental Protection Act. Successful participants in the Pre-Notice Site Cleanup Program (1989 through 1995) received Section 4(y) Letters rather than No Further Remediation Letters (NFR), which were not available until 1996 (Environmental Law Institute [11]). The 4(y) letter indicated that the Remediation Applicant ("RA;" i.e., any persons seeking to perform or performing investigative or remedial activities) had demonstrated successful completion of limited remedial actions (e.g., removal of drums, tanks or other containers). The letter did not necessarily represent that all or certain environmental conditions at a site did not constitute a threat to human health and the environment, only that specific actions had been successfully completed.

The Illinois Site Remediation Program (SRP) was created in 1995 to provide RAs the opportunity to receive review and evaluation services, technical assistance and no further remediation determinations from the IEPA. The IEPA intended this program to be flexible and responsive to the requirements of RAs, to project constraints and to variable remediation site conditions. The goal and scope of program actions at remediation sites are normally defined by the RA and subject to the state regulations. The Illinois program has achieved substantial results: over 600 sites have received the NFR, with another 500 sites active in the program. The Chicago area, in Cook County, has been the focal point of a large share of this environmental cleanup and redevelopment.

Successful participation in the SRP program results in the issuance of a No Further Remediation (NFR) letter by the IEPA. A NFR signifies a release from further responsibility for performing approved remedial actions at a site and is considered prima facie evidence that environmental conditions at the site do not constitute a threat to human health and the environment, so long as the site is utilized in accordance with the terms of the NFR. A NFR may contain conditions such as implementation of institutional controls (e.g., groundwater use restrictions), maintenance of engineered barriers (e.g., asphalt or concrete paving), and/or requirements for worker cautions.

Each NFR is categorized both by the type of site characterization performed (comprehensive or focused) and by land use limitations (residential or industrial/commercial). A comprehensive NFR signifies a release from further responsibilities for all recognized environmental conditions and related contaminants of concern at a remediation site. A focused NFR signifies a release from further responsibilities for only specific recognized environmental conditions and related contaminants of concern at a remediation site. Cleaning a site to residential standards indicates the site has been cleaned to the highest level required, with industrial/commercial indicating a lower standard.

In addition, the USEPA and the Illinois EPA have entered into a State Memorandum of Agreement (SMOA) through which the USEPA concurs that further response actions will not be required by the USEPA at sites that have received an NFR letter. The USEPA agrees not to plan or anticipate federal action under CERCLA at an IEPA enrolled site, except in emergency situations. This level of assurance is generally accepted to be among the best available in the US at this time.

Typically, the clean-up process is designed to provide for the protection of public health, welfare, and the environment and to consider the planned use of property. Clean-ups can be accomplished in several ways including the removal and disposal of contaminated soil. On larger sites this is often cost-prohibitive, so other measures can be implemented including the use of a physical barrier such as a building or parking lot pavement that provides a protective barrier between the contamination and the public. When a barrier is used, the regulations concerning it are incorporated into the site deed (deed restrictions) as a partial fulfillment of the requirements of the NFR.

3 Data Collection and Methodology

The research design for this study is predicated upon the merging of two separate data sources. First, the SRP data was acquired from the IEPA to establish the level of the program's activity. This database includes all relevant information regarding site cleanup from the RA's name and address and the type of contamination to the date the NFR was issued. The variables of interest to this study include type of site, 4(y) letter, NFR letter, date of 4(y) letter and NFR, active/non-active, type of institutional barrier, type of physical barrier, worker caution and focused/comprehensive clean-up.

The second set of data that was acquired for this project includes current structural, financial, and transactional information in Cook County, Illinois that is publicly recorded and then distributed by First American Real Estate Services. The variables used for this study included sale date, first mortgage amount, second mortgage amount, land and building value, year built, and lot size. Cook County was chosen for this study due to its dynamic market. Tens of thousands of transactions occur annually in the Chicago area alone.

The two data sets were merged by matching street addresses. Using this method, we were able to correctly match 319 of 783 Cook County sites in the SRP database. In doing so, we were able to combine site remediation information with development data for the sites that had entered the SRP. These 319 correctly matched sites comprise 672 parcels in the Cook County database. The difference is attributable to site subdivision and re-sale.

The methodology for this study includes non-parametric measurements and descriptive statistics of the combined SRP and Cook County databases. These statistics will be used to describe the basic features of the data, and to provide summaries regarding the clean up, financing, and development of these sites. We believe that the bias in using only those sites that we were able to match is minimal. Of the 783 Cook County sites that have been active in the state programs, 65 received a 4(y) Letter (8%), 273 received a NFR (35%), and 216 sites are currently active in the program (28%). These numbers are generally consistent with our sample. Note that our research was limited to what was available in the two data sets, each of which contained some unpopulated fields.

Using the 319 correctly matched sites, the outcomes this study seeks to measure includes program information such as the number of sites that have completed the program and their relevant letter, the number of sites that are currently active, and the level of clean-up that has been required. It attempts to look at the influence of prior land use to the type of remediation required and explore to the extent possible, post-letter land use changes. It will also address development issues including the number redeveloped and financed.

4 Results

4.1 Program results

The results of this study provide a preliminary look into the activity of the State of Illinois Site Remediation Program. As was previously discussed, the level of liability relief is relative to the letter that was issued by the state. Of the 319 sites that were studied, 35 (11%) received a 4(y) letter and 140 (44%) received a NFR. Of those sites receiving an NFR, 31 sites were comprehensively cleaned and 86 sites were cleaned in focused, specific areas. There are 96 sites currently active in the program and 47 that are no longer in the program. Of those 47 out of the SRP, at least 2 were denied letters. Details are listed in Table 1.

Table 1. Cook County sites obtaining either a Section 4(y) Letter or No Further Remediation letter.

Type of Letter	Number Complete	% of Total	Compre-hensive Clean-up	Focused Clean-up
Sites with 4(y) Letter	35	10.97%	N/A	N/A
Sites with 4(y) Letter and NFR	1	0.31%	0	1
Sites with NFR	140	43.88%	31	86
Total Sites with Letters	**176**	**55.17%**	**31**	**87**
Total Sites in Database (with/without letters)	**319**	**100.00%**		

Not surprisingly, sites are being cleaned to industrial/commercial standards at a 2 to 1 ratio over residential standards. Even some sites with prior history as single-family residential are only being cleaned to industrial/commercial standards, most likely indicating a future land-use change or focused clean up with special restrictions on a portion of the property. Of the 78 sites known to be previously industrial, 42% have been cleaned to residential standards. This too could indicate a change in land use, although it may also indicate minimal contamination with manageable clean-up expenses.   It is possible that developers or landowners may justify residential level clean-ups based on concerns of future policy changes.   Prior site use and clean-up standards are listed in Table 2.

Table 2: Standards to which prior land uses are cleaned.

Land Use Prior to Program Entry	Residential	Industrial/ Commercial	Not Indicated	Total
Residential	5	7	1	19
Commercial	17	37	11	65
Industrial	24	58	22	104
Mixed Use Residential/ Commercial	2	0	1	3
Tax Exempt	1	2	0	3
Vacant Land	4	11	6	21
Unidentified	4	4	0	8
Totals:	**57**	**119**	**47**	**223**

Table 3 describes the types of barriers required for sites that have received 4(y) and NFR letters from the SRP. Asphalt is the most predominate barrier with different combinations comprising 68% of all barriers and typically indicating what is or will eventually be a parking facility. Concrete is also fairly common and probably indicates the same assumption as asphalt. Surprisingly, only 10% require a clean soil barrier.

Table 3: Barrier descriptions for sites with letters.

Description of Barrier	Total	% of Total
Concrete Barriers	11	18%
Compacted Limestone/Concrete Barrier	1	2%
Clean Soil Barrier	4	6%
Building Foundation	4	6%
Asphalt/Concrete Barrier/Bldg Foundation	1	2%
Asphalt Barrier/Concrete Cap	1	2%
Asphalt Barrier/Concrete Barrier	2	3%
Asphalt Barrier/Clean Soil/Bldg Foundation	1	2%
Asphalt Barrier/Clean Soil Barrier	1	2%
Asphalt Barrier/Building Foundation	12	19%
Asphalt Barrier	24	39%
Total Sites Requiring Barriers	**62**	**100%**

4.2 Sale and financing information

Almost half of the sites receiving 4(y) letters sold after receiving their letter. About a quarter of the sites receiving an NFR have sold. Even though an NFR provides greater liability protection, the market appears to be dynamic for both types of sites. Because the NFR has been available for about half the time as the 4(y) letter, it is expected that NFR sales will increase in the future. Owners of 4(y) sites tend to hold their property longer, possibly indicating less speculation in that market or indicating greater market time due to the reduced liability protection. In general, sites must be sold before being redeveloped.

Table 4: Sale and financing information.

	NFR	% of Total	4(y) Letter	% of Total	Total
Average number of days between letter issued and sale	489.38	--	711.82	--	--
Number of sites sold post-letter	34	24%	17	49%	51
Number financed post-letter	20	14%	8	23%	28
Number financed pre-letter (1980 to date letter issued)	45	32%	2	6%	47

Table 5 was constructed using the individual parcel information that was extracted from the 319 site sample. There were 672 individual parcels, of which 296 had populated fields indicating the county assessed property values. The average value of a site completing the SRP had a value of $288,338 when a NFR was obtained. 4(y) properties had a value of $396,854. Building values for sites obtaining a NFR and 4(y) Letter ranged from $121 to $4,672,483, and from $182 to $3,103,116, respectively. Land values ranged from $356 to $5,740,011 and $815 to $1,033,221, respectively. The average lot size was 36.24 acres, with a range from .01 to 2,106.41 for sites with a NFR and .07 to 21.8 for sites with a 4(y) Letter.

The higher building and land values for those sites obtaining a 4(y) Letter is most likely related to the established uses of those sites. Of the 140 sites receiving an NFR, one-third (47 sites) were cleaned to residential standards indicating greater subdivision potential and therefore lower individual values.

Table 5: Construction and development information.

	NFR Letter	4(y) Letter	Total w/letters
Number of parcels w/buildings	213	54	267
Total building value	$41,466,585	$16,574,529	$58,041,114
Average building value	$194,678	$306,935	$217,382
Number of parcels with land value (w/29 vacant)	237	59	296
Total land value	$26,869,653	$6,839,878	$33,709,531
Average land value	$113,375	$115,930	$113,884
Total number of parcels	237	59	296
Total land & building value	$68,336,238	$23,414,407	$91,750,645
Average land & building value	$288,338	$396,854	$309,968
Number of parcels w/land area	197	47	244
Total land (acres)	8775.05	332.26	8842.79
Average lot size (acres)	44.54	7.07	36.24

5 Conclusions and further research

The purpose of this study is to provide an introductory analysis of the outcomes of the environmental clean-up programs in the United States. More specifically, it addresses the number and type of liability waivers that are issued in Cook County, Illinois and attempts to provide some insight into the accomplishments of the Illinois program. This study examines 176 sites that have successfully completed the SRP. There are 96 sites currently active in the program and 47 that are no longer in the program. Of those 47, at least 2 were denied letters.

The availability of the NFR and its SMOA with the USEPA appears to have motivated participation in the state VCP, hence the greater numbers receiving NFRs. Of the 319 sites studied, about 55% have received a closure letter. Most of these obtained NFRs, which offer the greatest liability protection of the two available letters. About one third of sites were sold after cleanup.

Building and property values remain high for both types of letters. The average value of a site completing the SRP was $288,338 when a NFR was obtained and $396,854 when a 4(y) Letter was received. Even though the 4(y) letter offers less protection than the NFR, we suspect that 4(y)s were typically sought by owners of industrial property, therefore dictating higher average property values, prior to any redevelopment.

Additional work is necessary to arrive at more complete conclusions regarding this data. To our knowledge, this is the only data set that combines remediation data with development data. Therefore, we have an opportunity to provide additional research into development probabilities, effects of deed restrictions on property values and numerous other topics.

Acknowledgements

Thanks to the US EPA for funding this project (under assistance agreement CR-826755-01). US EPA will not review the findings of this research, and no official endorsement should be inferred.

References

[1] Environmental Law Institute. An analysis of state Superfund programs: 50-state study, Update, 1998.

[2] Simons, R.A. How many urban brownfields are out there? *Public Works Management and Policy*, **2(3)**, pp. 267-273, 1998.

[3] Simons, R.A. *Turning brownfields into greenbacks*. Urban Land Institute: Washington D.C., 1998.

[4] Simons, R.A., Pendergrass, J. and Winson-Geideman, K. Quantifying long-term environmental regulatory risk of brownfields: Are reopeners really an issue? *Submitted to Journal of Environmental Planning and Management*, 2001.

[5] Page, W.G. and Rabinowitz, H.Z., Potential for redevelopment of contaminated brownfield sites. *Economic Development Quarterly*, **8(4)**, pp. 353-363, 1994.

[6] Page, W.G. and Rabinowitz, H.Z., Groundwater contamination: its effects on property values and cities. *Journal of the American Planning Association*, **59**, pp. 473-82, 1993.

[7] Boyd, J., Harrington, W., and Macauley, M., The effects of environmental liability on industrial real estate development, *Journal of Real Estate Finance and Economics*, **12**, pp. 37-58, 1996.

[8] Howland, M. The impact of contamination on the Canton/Southeast Baltimore land market. *Journal of the American Planning Association*, **66**, pp. 411-420, 2000.

[9] Arrandale, T., Developing the contaminated city. *Governing*, **6(3)**, pp. 44-47, 1992.

[10] Bartsch, C., Andress, C., Seitzman, J. and Cooney, D. *New life for old buildings: confronting environmental and economic issues to industrial reuse.* Washington: Northeast-Midwest Institute, 1991.

[11] Environmental Law Institute. Enhancing state Superfund capabilities: nine-state study, part 2, 1990.

Index of Authors

Waste Management

Editors: **D. ALMORZA**, *Universidad de Cadiz, Spain*, **C.A. BREBBIA**, *Wessex Institute of Technology, UK*, **D. SALES**, *Universidad de Cadiz, Spain and* **V. POPOV**, *Wessex Institute of Technology, UK*

Of interest to environmental engineers, local authority representatives, waste disposal experts, research scientists, and civil and chemical engineers, this title presents contributions from the First International Conference on Waste Management and the Environment.

Topics covered include: Waste Management Strategies and Planning; Methodologies and Practices; Physical, Biological, Chemical and Thermal Treatment Methods; Waste Reduction; Landfills, Design, Construction and Monitoring; Water and Wastewater Treatment; Separation, Transformation and Recycling; Organic Waste for Soil Improvement; Biosolids, Composting and Agricultural Re-Use; Hazardous Waste; Liquid Sludge Treatment; Energy Recovery; Clean Technologies; Environmental Management; Pollution Monitoring and Control; Air Pollution; Rehabilitation of Landfills; Institutional, Legal and Economic Issues; Pollution Reduction; Community Involvement and Education; and Case Studies.

ISBN: 1-85312-919-4   2002
apx 800pp
apx £248.00/US$384.00/€399.00

Risk Analysis III

Editor: **C. A. BREBBIA**, *Wessex Institute of Technology, UK*

The analysis and management of risk and mitigation of hazards is essential in our increasingly complex society.

Containing edited versions of papers presented at the Third International Conference on Computer Simulation in Risk Analysis and Hazard Mitigation (RISK), this volume covers a series of important research topics which are of current interest and which have practical applications. The contributions included are concerned with all aspects of risk analysis and hazard mitigation, ranging from specific assessment of risk to mitigation associated with both natural and anthropogenic hazards.

Over 70 papers are included and these cover specific topics within the following subject headings: Estimation of Risks; Hazard Prevention, Management and Control; Emergency Response; Data Collection and Analysis; Air Quality Studies; Soil and Water Contamination; Floods and Droughts; Seismic Hazards; Landslides and Slope Movements.

Series:
Management Information Systems, Vol 5

ISBN: 1-85312-915-1   2002
784pp
£245.00/US$379.00/€398.74

WITPress
Ashurst Lodge, Ashurst, Southampton, SO40 7AA, UK.
Tel: 44 (0) 238 029 3223
Fax: 44 (0) 238 029 2853
E-Mail: witpress@witpress.com

WITPRESS

Urban Transport VIII
Urban Transport and the Environment in the 21st Century

Editors: **L.J. SUCHAROV** and **C.A. BREBBIA**, *Wessex Institute of Technology, UK and F. BENITEZ, University of Seville, Spain*

The issue of urban transportation, together with its inter-related environmental and social concerns, continues to rise up the agenda of all city authorities and central governments. Sometimes this concern is measured, while on other occasions it is prompted by specific crises such as fatal railway accidents. The ever present need for better urban transport systems in general and for a healthier environment has led to a steadily increasing level of study throughout the world.

Reflecting the wide range of research being carried out in many countries today, and highlighting areas of particular current concern such as accessibility, safety and noise pollution, this volume features papers presented at the Eighth International Conference on Urban Transport and the Environment in the 21st Century.

80 papers are included and these are grouped under the following headings: Urban Accessibility and Mobility; Urban Transport Systems; Traffic Control; Simulation; Information Systems; Finance and Planning; Emissions; Environmental Noise; Economic and Social Impact; Safety; Vehicle Technology.

Series:
Advances in Transport, Vol 12

ISBN: 1-85312-905-4   2002
864pp
£249.00/US$385.00/€404.00

The Sustainable City II
Urban Regeneration and Sustainability

Editors: **C.A. BREBBIA**, *Wessex Institute of Technology, UK,* **J.F. MARTIN-DUQUE**, *Universidad Complutense, Spain,* **L.C. WADHWA**, *James Cook University, Australia and* **A. FERRANTE**, *ISC Group, Italy*

Urban areas produce a series of environmental problems that arise from the consumption of natural resources and the consequent generation of waste and pollution. These problems are continuing to grow and new solutions, without adverse effects, therefore need to be developed in order to maintain the quality of life desired by the community.

This book contains most of the papers presented at the Second International Conference on the Sustainable City, and addresses the many interrelated aspects of the urban environment together with the importance of finding sustainable solutions. Almost 100 contributions from a variety of specialists working in many different countries are included. These are divided under the following headings: Strategy and Development; Planning, Development and Management; Restructuring and Renewal; Cultural Heritage and Architectural Issues; Land Use and Management; Environmental Management; The Community and the City; Public Safety and Security; Traffic and Transportation; Transport Environment and Integration; Agriculture and the City; and Energy Resources.

Series:
Advances in Architecture, Vol 14

ISBN: 1-85312-917-8   2002
1072pp
£298.00/US$457.00/€485.00

WITPRESS

The Sustainable Street
The Environmental, Human and Economic Aspects of Street Design and Management

Editors: **C. JEFFERSON** *and* **J. ROWE**, *University of the West of England, UK and* **C.A. BREBBIA**, *Wessex Institute of Technology, UK*

Cities have long been recognised as the engines of cultural and social change, as well as the drivers of the international economy, yet the many, usually inter-related, problems of urban living persist. Among these are air and noise pollution, problems of equality of access for opportunities in work, education and leisure, and the maintenance and enhancement of cultural cohesion and the control of crime. The many problems of the urban environment, as well as the roles and advantages, may be seen in microcosm in the urban street.

The concept of the street as the unit of the urban environment, and as a useful scale for the delivery of urban development towards sustainability, lies at the heart of this book. Bringing together individual experiences from different countries, it informs the debate not only as to what is desirable, but also how, practically, we should move forward. The chapters included will be of interest to many professionals and practitioners including architects, economists, engineers, environmentalists and planners.

Series:
Advances in Architecture, Vol 12

ISBN: 1-85312-832-5   2001
288pp
£115.00/US$178.00/€184.00

Environmental Health in the Aral Sea Basin
The Present and Future

Editor: **D. FAYZIEVA**, *Academy of Sciences, Uzbekistan*

During the last few years new directions in environmental health research have been successfully developed in the Central Asian republics.

This volume contains a broad scientific review of possible influences of environmental conditions such as ambient air quality, water supply, sanitation, nutrition, and industrial and agricultural activity on the health of those living in the Aral Sea Basin.

Promoting a better understanding of current environmental health problems in this region, the book includes the following chapters: Ambient Air Quality and Dust Exposure; Water Quality, Water Supply and Sanitation; Industrial Activity and Health; Epidemiological Studies of Non-Infectious Diseases; Infectious Diseases, Microbiology and Immunology; Nutritional Aspects; Children's Health; Use of Pesticides in Agriculture and Health Effects; Strategies for the Future.

ISBN: 1-85312-945-3   2002
apx 300pp
apx £99.00/US$153.00/€160.00

WITPRESS

Natural Wetlands for Wastewater Treatment in Cold Climates

Editors: Ü. MANDER, University of Tartu, Estonia and P.D. JENSSEN, Agricultural University of Norway, Ås, Norway

This volume presents 13 high-quality, up-to-date contributions about the potential, and use of, natural wetland ecosystems for wastewater treatment in cold climate areas. In this instance the term - natural wetlands - includes slightly man-influenced/changed (i.e. seminatural) wetland ecosystems, and free water surface treatment wetlands which normally have less constructive elements than subsurface flow treatment wetlands.
Series:
Advances in Ecological Sciences, Vol 12

ISBN: 1-85312-859-7   2002   264pp
£89.00/US$139.00/€144.85

Constructed Wetlands for Wastewater Treatment in Cold Climates

Editors: Ü. MANDER, University of Tartu, Estonia and P.D. JENSSEN, Agricultural University of Norway, Ås, Norway

Focusing on the potential and use of constructed wetland ecosystems for wastewater treatment in cold climate areas, this book includes further contributions developed from papers presented at recent international conferences/workshops.
Series:
Advances in Ecological Sciences, Vol 11

ISBN: 1-85312-651-9 2002 apx 362pp
apx £119.00/US$179.00/€187.00

Environmental Health Risk

Editors: C.A. BREBBIA, Wessex Institute of Technology, UK and D. FAYZIEVA, Academy of Sciences, Uzbekistan

As environmental problems increase, related health issues are also becoming a source of major worldwide concern.
Covering the latest developments in this field, this book features edited versions of papers presented at the First International Conference on the Impact of Environmental Factors on Health.
The contributions are divided under the following headings: Air Pollution; Water Supply and Quality; Waste Water and Surface Water Problems; Radiation and Noise; Microbial Risk; Population Health; and Solid Waste.

ISBN: 1-85312-875-9   2001
328pp
£129.00/US$199.00/€208.00